Royal Statistical Society Lecture Note Series

SERIES EDITORS

RAYMOND J. CARROLL   JOHN B. COPAS
DAVID J. HAND   RICHARD L. SMITH

Royal Statistical Society Lecture Note Series

Stochastic Networks
Theory and Applications

Edited by

F. P. KELLY

Statistical Laboratory
University of Cambridge

S. ZACHARY

Department of Actuarial Mathematics and Statistics
Heriot–Watt University, Edinburgh

I. ZIEDINS

Department of Statistics
University of Auckland, New Zealand

CLARENDON PRESS · OXFORD
1996

Oxford University Press, Walton Street, Oxford OX2 6DP

Oxford New York
Athens Auckland Bangkok Bombay
Calcutta Cape Town Dar es Salaam
Delhi Florence Hong Kong Istanbul Karachi
Kuala Lumpur Madras Madrid Melbourne
Mexico City Nairobi Paris Singapore
Taipei Tokyo Toronto
and associated companies in
Berlin Ibadan

Oxford is a trade mark of Oxford University Press

Published in the United States by
Oxford University Press Inc., New York

A catalogue record for this book is available from the British Library

Library of Congress Cataloging in Publication Data
(Data available)
ISBN 0 19 852399 8

Typeset using LaTeX
Printed in Great Britain by
Bookcraft (Bath) Ltd
Midsomer Norton, Avon

Preface

The theory of stochastic networks is presently undergoing a period of intensive research, motivated in part by the need to understand and control the behaviour of modern communications and manufacturing networks, and thus to improve their design and performance.

This volume is a collection of invited papers written after the event by some of the participants at the Royal Statistical Society research workshop on stochastic networks, held at Heriot-Watt University, Edinburgh, from 1 to 11 August 1995. This workshop gathered together most of the leading researchers in this field, including many from industry. All the major themes of current interest were considered, and these are well represented in this volume.

One such theme is the study of approximations to network queueing models via various natural time and space scalings, and the use of these approximations in determining optimal controls. Two scalings are of special interest. The first gives an (essentially deterministic) fluid model in the limit, and fluid models have become an important tool in the investigation of open multiclass queueing networks. Maury Bramson discusses the asymptotic behaviour of fluid models of important classes of FIFO and processor sharing queueing networks. He considers recent results of Dai and others on the connection between the stability of the fluid limit and that of the network itself, and uses powerful ideas of entropy to show that such networks are stable whenever the traffic intensity is strictly less than one. Gideon Weiss considers directly the problem of optimal control for fluid models of transient queueing networks appropriate to manufacturing systems with re-entrant lines (in which components may be required to be processed by the same machine several times, as is typically the case in modern semi-conductor wafer fabrication plants). He shows how continuous linear programming techniques may be applied to the solution of such problems, and solves fully a range of instructive examples.

The second natural scaling typically gives a Brownian model in the limit and is particularly appropriate to studying the more detailed behaviour of stable but heavily loaded networks. Ruth Williams describes the current state of research in this area. She concludes her paper with an important example (due to Harrison and Williams) of an unconventional heavy traffic limit theorem, where it is necessary to combine different scalings to obtain a full characterisation of the limiting behaviour. Mike Harrison uses the Brownian model as the basis for a new, very general and effective approach to dynamic flow control in stochastic networks. See also the papers of Majewski and Kurtz, mentioned below.

A second major theme deals with the use of large deviations theory in analysing queueing and loss systems. In modern telecommunications networks it is necessary to ensure that data are transmitted with a high degree of reliability (for example, such that packets have a maximum loss probability of the order of 10^{-9}). The theory of large deviations is central to the study of such rare events, and large deviations bounds are used in many of the papers in this volume (see, for instance, those of Hajek, Harrison, and Kelly). Neil O'Connell discusses the general problem of using large deviations theory to establish probability approximations for aspects of a system (such as queue lengths) under very general ergodicity and mixing assumptions about the network inputs. He then applies this theory to the characterisation of queue length and sojourn time distributions for single-server resources with multiple arrival streams. Kurt Majewski studies similar questions for sequences of "feedforward" queueing networks, whose traffic intensities approach one, by developing the large deviations theory of the corresponding stationary reflected Brownian motions, thus combining both of the major themes already mentioned.

A third theme deals with the statistical analysis, modelling and control of network arrival processes. Recent work has shown that traffic in modern packet-based high-speed networks frequently exhibits a self-similar, essentially fractal, behaviour over a wide range of time scales. Such behaviour arises in models with long-range time dependence. It has major implications for the statistical analysis of such traffic, and can call into question the validity of traditional modelling techniques based on the assumption of arrival processes with short-range dependence. The paper by Walter Willinger, Murad Taqqu and Ashok Erramilli gives a very comprehensive bibliography of recent work on these issues, including references to closely related work in other branches of science and engineering such as hydrology, financial economics, and biophysics.

Of course it is highly desirable to understand how such arrival process behaviour arises, and to develop realistic analytic models for it. Much recent work (referenced in the above bibliography of Willinger *et al.*) suggests that it may sometimes arise as the result of the superposition of inputs from many on–off sources. Tom Kurtz considers general models for such workload input processes, and gives limit theorems for the formulation of both fluid and heavy traffic approximations for systems with these inputs. He shows that under appropriate assumptions fractional Brownian motion, which possesses long-range dependence and self-similarity properties, can be obtained as the limiting workload input process.

Frank Kelly provides an overview of the use of effective bandwidths as a summary of the statistical characteristics of sources over different time and space scales, as well as in various models of statistical sharing. His framework assumes only stationarity of sources, and illustrative examples include Brownian bridge and fractional Brownian models, as well as various large deviations models. Richard Gibbens uses the notion of effective bandwidth as a graphical descriptor in several examples, including an MPEG-1 video encoding of the *Star Wars* movie and traces of ethernet traffic taken at Bellcore, and compares the latter with fractional Brownian motion. There remains much work to be done on the analysis of data collected in networks, and on the development of more formal methodologies. In a contribution to this area, Susan Pitts describes results on nonparametric estimation for queues, and on the parallels with methods developed in risk theory.

The mathematics of loss networks continues to play a very important role in the analysis and control of modern telecommunications and computer networks—for example, in ATM networks where effective bandwidth takes the role of a capacity requirement, and in cellular radio networks (where there remain severe capacity constraints). Of particular interest here is the study of control processes which, while themselves operating on a very fast time scale, both determine and are determined by the much longer time-scale behaviour of the network. Stan Zachary reviews some recent work in the application of these ideas to the study of the dynamic and equilibrium behaviour of loss networks with high arrival rates and capacities, while Iain MacPhee and Ilze Ziedins use similar ideas to consider design and optimal control in networks with diverse routing. Murat Alanyali and Bruce Hajek give important new results for dynamic load balancing in loss networks, focusing on the simple and robust control strategy of *least load routing*. They use fluid approximations to study typical behaviour under this strategy, and large deviations theory to analyse significant departures from such behaviour. Suzanne Evans takes a different approach to the analysis of system behaviour on different time scales, potentially applicable to both loss and queueing networks.

Problems of job selection and scheduling are of great importance in manufacturing. The papers by Weiss and Harrison, previously mentioned, consider these problems. Additionally, Renate Garbe and Kevin Glazebrook give a careful review of the very recent and powerful theory of Bertsimas and Niño-Mora on the optimal control of stochastic systems satisfying generalized conservation laws. They further develop significant extensions to this theory.

Analysis of the effects of unreliability in queueing networks is notoriously difficult. In general only approximate analyses are possible, and even then such approximations are often unsatisfactory. Ram Chakka and Isi Mitrani introduce new and improved approximation techniques.

François Baccelli, Serguei Foss and Jean Mairesse develop a comprehensive theory of generalized Jackson networks in which classical Markovian assumptions are replaced by those of stationarity and ergodicity. (One motivation for this is the need to model long-range dependence in arrival processes, as discussed above.) They consider in particular the stability and transience of such networks, and give also a range of important counter-examples. One classical queueing model that still generates much interest is that of a tandem queue (a number of queues in series). Tom Mountford and Balaji Prabhakar present new results on the convergence of the successive departure processes from an infinite series of $\cdot/GI/I$ queues, again subject to stationary and ergodic inputs. Yurii Suhov and David Rose study fully connected queueing networks in which the number of nodes is allowed to grow. They show that under certain conditions the so-called Poisson-independence hypothesis holds in the limit, permitting the deduction of the limiting distribution of the end-to-end delay.

Financial support for the workshop was provided primarily by a Visiting

Fellowship Grant from the Engineering and Physical Sciences Research Council. Further very generous assistance was provided by British Telecom Laboratories and by Hewlett-Packard's Basic Research Institute in the Mathematical Sciences. The Royal Statistical Society contributed to the expenses of a number of the research students. The editors wish to thank all of these bodies for their support.

We are also very grateful to the many colleagues (some participants at the workshop) who carefully reviewed the papers in the present volume, and made many helpful suggestions for their improvement. We are grateful to Elizabeth Johnston, Julia Tompson, Anna Drage and the staff of Oxford University Press for their patient assistance in the production of the volume. We are especially grateful to Richard Gibbens for being our LaTeX/PostScript expert and providing simple and elegant solutions to many of the problems involved in putting together a multi-authored work.

Further information relating to this volume is available at the WWW location `http://www.ma.ac.uk/~stan/stochnet/`.

Cambridge	F. K.
Edinburgh	S. Z.
Auckland	I. Z.
March 1996	

Contents

Contributors

Murat Alanyali
University of Illinois, Urbana
Champaign
alanyali@manti.csl.uiuc.edu

François Baccelli
INRIA Sophia-Antipolis
francois.baccelli@sophia.inria.fr

Maury Bramson
University of Wisconsin, Madison
bramson@math.wisc.edu

Ram Chakka
Imperial College, London
rsc@doc.ic.ac.uk

Ashok Erramilli
Bellcore
ashok@bellcore.com

Suzanne P. Evans
Birkbeck College, London
s.evans@statistics.bbk.ac.uk

Serguei Foss
Novosibirsk State University
foss@math.nsk.su

Renate Garbe
University of Newcastle
renate.garbe@newcastle.ac.uk

Richard Gibbens
University of Cambridge
r.j.gibbens@statslab.cam.ac.uk

Kevin Glazebrook
University of Newcastle
kevin.glazebrook@newcastle.ac.uk

Bruce Hajek
University of Illinois, Urbana
Champaign
hajek@shannon.csl.uiuc.edu

J. Michael Harrison
Stanford University
fharrison@gsb-lira.stanford.edu

Frank Kelly
University of Cambridge
f.p.kelly@statslab.cam.ac.uk

Thomas G. Kurtz
University of Wisconsin, Madison
kurtz@math.wisc.edu

Iain MacPhee
University of Durham
i.m.macphee@durham.ac.uk

Jean Mairesse
BRIMS, HP Laboratories, Bristol
jem@hplb.hpl.hp.com

Kurt Majewski
Ludwig-Maximilians-Universität,
München
majewski@borel.zfe.siemens.de

Isi Mitrani
University of Newcastle
isi.mitrani@newcastle.ac.uk

Tom Mountford
University of California, Los Angeles
malloy@math.ucla.edu

Neil O'Connell
BRIMS, HP Laboratories, Bristol
noc@hplb.hpl.hp.com

Susan M. Pitts
University of Cambridge
s.pitts@statslab.cam.ac.uk

Balaji Prabhakar
BRIMS, HP Laboratories, Bristol
balaji@hplb.hpl.hp.com

David Rose
University of Cambridge
d.rose@statslab.cam.ac.uk

Yurii Suhov
University of Cambridge
y.m.suhov@statslab.cam.ac.uk

Murad S. Taqqu
Boston University
murad@math.bu.edu

Gideon Weiss
Haifa University
gweiss@stat.haifa.ac.il

Ruth J. Williams
University of California, San Diego
williams@russel.ucsd.edu

Walter Willinger
Bellcore
walter@bellcore.com

Stan Zachary
Heriot-Watt University, Edinburgh
s.zachary@ma.hw.ac.uk

Ilze Ziedins
University of Auckland
ziedins@stat.auckland.ac.nz

1

Convergence to equilibria for fluid models of certain FIFO and processor sharing queueing networks

Maury Bramson

University of Wisconsin–Madison

Abstract

The qualitative behavior of open multiclass queueing networks is currently a topic of considerable activity. An important goal is to formulate general criteria for when such networks possess equilibria, and to characterize these equilibria when possible. Fluid models have recently become an important tool for such purposes. We are interested here in two families of such models. The first consists of FIFO fluid models of Kelly type, that is, the discipline is first-in, first-out, and the service rate depends only on the station. The second consists of a family of processor sharing fluid models, where the fraction of time spent serving a class present at a station is proportional to the quantity of the class there, and all of the service goes into the "first customer" of each class. We refer to members of this second family as head-of-the-line proportional processor sharing (HLPPS) fluid models. To study both families of models, we introduce entropy functions associated with the states of the systems. Appropriate estimates show that for both families, if the traffic intensity function is at most 1, then the fluid models converge exponentially fast to equilibria with fixed concentrations of customer types at each queue. When the traffic intensity function is strictly less than 1, the limit is always the empty state and occurs after a finite time. A consequence is that generalized Kelly and HLPPS networks with traffic intensity strictly less than 1 are positive Harris recurrent, and hence possess unique equilibria. The results for FIFO models presented here are demonstrated in Bramson (1996a); the results for HLPPS models will appear in Bramson (1996b).

1 Introduction

In these notes, we analyze the asymptotic behavior of two families of fluid models. Although these models are of interest in their own right, motivation is provided by their connection with certain families of queueing networks. We therefore first briefly discuss such networks.

We consider here queueing networks which consist of single-server stations j, $j = 1, 2, \ldots, d$. The networks will be *open*; that is, after entering the network,

1

customers move about from station to station until exiting from the system at some future time. The customers are members of classes k, $k = 1, 2, \ldots, c$, with a customer of class k being served at a unique station j, written $k \in \mathcal{C}(j)$. Denote by m_k the mean service time for the class k. Associated with each class k, there is a stationary external arrival process; let α_k be the rate of these arrivals. The queueing networks are also assumed to be *multiclass queueing networks* in the sense of Harrison (1988): upon being served at j, a customer of class k immediately becomes a customer of class ℓ with probability $P_{k\ell}$ independently of its past history, where $P = (P_{k\ell})$ is a given transient subprobability matrix. The probability of a customer of class k leaving the network upon completion of service is thus $1 - \Sigma_\ell P_{k\ell}$. The service and interarrival times are typically assumed to be independent of one another. To make the network Markovian, these times are often assumed to be exponentially distributed as well. The waiting buffer at each station is typically stipulated as having infinite capacity, with the server remaining busy as long as there are customers present. We are interested here in two disciplines in this setting. We will investigate *first-in, first-out* (FIFO) networks, where customers are served in the order of their arrival at each station without regard to their class. We will also consider *head-of-the-line proportional processor sharing* (HLPPS) networks, where the fraction of time spent serving a class present at a station is proportional to the quantity of the class there, and all of the service goes into the first customer of each class.

To investigate open multiclass queueing networks, one employs the solutions λ_ℓ, $\ell = 1, 2, \ldots, c$, of the *traffic equations*

$$\lambda_\ell = \alpha_\ell + \sum_{k=1}^{c} \lambda_k P_{k\ell}, \tag{1.1}$$

or equivalently, in vector form, of

$$\lambda = \alpha + P^t \lambda. \tag{1.2}$$

(All vectors are to be envisioned as column vectors; "t" denotes the transpose here.) Since P is transient, one can set $Q = (I - P^t)^{-1} = I + P^t + (P^t)^2 + \cdots$; the unique solution λ of (1.2) is then given by $\lambda = Q\alpha$. The term λ_k is referred to as the *total arrival rate* for class k. Employing m and λ, one defines the *traffic intensity* ρ_j for the jth server as

$$\rho_j = \sum_{k \in \mathcal{C}(j)} m_k \lambda_k. \tag{1.3}$$

When $\rho_j \leq 1$ for all j, we refer to the queueing network as *subcritical*, when $\rho_j < 1$ for all j, as *strictly subcritical*, and when $\rho_j = 1$ for all j, as *critical*.

A fundamental question is under what conditions a given network possesses a unique equilibrium (i.e., invariant probability measure). A sufficient condition for this to hold is that the underlying Markov process of the network be *positive Harris recurrent*; see Dai (1995) or Dai–Meyn (1995) for more detail and references. We refer here to networks satisfying the latter condition as *stable*. For a queueing network to be stable, it must be strictly subcritical. One can ask what can be said in the opposite direction. For irreducible networks, it is tempting to guess that strict subcriticality is also sufficient for stability. A spate of recent counterexamples has shown that this is not correct. (See Lu–Kumar (1991) and Rybko–Stolyar (1992) for the first counterexamples, and Bramson (1994a, 1994b) for those in the FIFO context.) The feedback within these networks tends to create "cycles", over which there is a forced idleness at certain servers while customers are being served elsewhere, thus reducing the effectiveness of these servers. This behavior causes the number of customers in the network to grow as $t \to \infty$. The exact conditions for stability of queueing networks are not known and are most likely difficult to obtain.

The above difficulties do not arise for *Kelly networks* and HLPPS networks. Kelly networks are those FIFO queueing networks where all service and interarrival times are independent and exponentially distributed, with each customer class served at a given station being assumed to possess the same mean service time. If all traffic intensities are strictly less than 1, then an equilibrium distribution always exists, and factors into a product of geometric-like distributions corresponding to the different stations. (See Baskett *et al.* (1975) and Kelly (1975, 1979).) For HLPPS networks, we assume that the service and interarrival times are independent and exponentially distributed, but we do not make any assumptions about the mean service times. HLPPS networks are a simpler variant of the usual processor sharing networks, where service at a station is equally divided between all customers present there. (HLPPS networks with one station were studied in Johnson (1983).) Here, we will investigate the fluid model analogs of Kelly and HLPPS networks.

Fluid models are continuous, deterministic analogs of queueing networks. The concept of customer is replaced here by that of "customer mass". This customer mass may be of any of k types, or classes, $k = 1, 2, \ldots, c$, which changes as the mass moves through the system. In accordance with its analog for queueing networks, customer mass enters the system, moves from station to station, and at some point leaves the system. It does this according to "arrival rates" α_k, "mean service times" m_k and "transition probabilities" $P_{k\ell}$, which are to be interpreted as before. Fluid models are rigorously defined in terms of accompanying *fluid model equations*; versions appropriate to our setting are given by (2.3)–(2.7) and (2.12)–(2.15). Fluid models, in the context of *fluid limits*, have recently served as an important tool in the study of the stability of queueing networks. Fluid limits are the limits of sample paths along subsequences obtained from a given queueing network under scaling which is linear in both time and the number of customers. Employed in Rybko–Stolyar (1992) and systematized and generalized in Dai (1995) (see also Stolyar (1995)), these limits, when well-behaved, imply the stability of the associated queueing network. Presumably, fluid models also preserve much of the behavior of the corresponding queueing networks.

Fluid models also occur in the context of heavy traffic limits for open multiclass queueing networks. Without going into details, we recall that a heavy traffic limit is obtained from a sequence of queueing networks by scaling time and the number of customers in the system so as to produce a reflected Brownian motion in the positive orthant. Substantial technical problems are present in the rigorous justification of this procedure. Generalized Jackson networks (networks with one class per station) were analyzed in Reiman (1984), multiclass networks with a single station were analyzed in Reiman (1988) and Dai–Kurtz (1995), and reentrant lines with first-buffer, first-served priority were analyzed in Chen–Zhang (1995b). General approximation schemes were given in Harrison–Nguyen (1990, 1993). As is the case with the stability of queueing networks, the anticipated convergence results do not hold in the most general setting; in Dai–Wang (1993), it was shown that the approximations in Harrison–Nguyen (1990) are not always valid.

As in Harrison (1995), one can view critical fluid models as corresponding to the reflected Brownian motion limits of queueing networks when observed over an infinitesimal stretch of time. On this shorter time scale, the process is still deterministic, and tends (hopefully) to a limit as time increases. The possible states for the Brownian motion should then consist of these instantaneous limits, with the process moving amongst them on its own time scale. Fluid models, from this perspective, are thus a step in understanding their reflecting Brownian motion analogs. In Harrison (1995), the equivalence of the convergence of a fluid model starting from any initial state as $t \to \infty$, and the existence of the heavy traffic limit for the corresponding queueing network, was conjectured.

The goal here is to examine the convergence for FIFO and HLPPS fluid models. The first class of models is examined under the restriction that they be of *Kelly type*. That is, we assume that the service rate depends on the station of the customer mass, but not otherwise on the mass' class. In view of the preceding discussion on the instability of certain queueing networks and the stability of Kelly networks, such a restriction is most natural. (Unstable fluid models along the lines of the networks in Bramson (1994a) are given in Dumas (1994).) For HLPPS fluid models, this restriction is not needed. In Bramson (1996a), it is shown for FIFO fluid models of Kelly type that if the traffic intensity function is at most 1, then the states of the system converge exponentially fast to equilibria with fixed concentrations of customer types throughout each queue. When the traffic intensity function is strictly less than 1, the limit is always the empty state, and is attained after a finite time. In Bramson (1996b), analogous results are shown for HLPPS fluid models. In these notes, we summarize the results and arguments for both papers. We observe in passing that it is not difficult to show the existence of solutions corresponding to the fluid model equations of each class of models (equations (2.3)–(2.7) for FIFO fluid models and (2.12)–(2.15) for HLPPS fluid models; see the appendix in Bramson (1996a) for more detail). The uniqueness of such solutions appears, however, to be a much more delicate matter, and remains unknown.

The main technique in Bramson (1996a, 1996b) consists of employing an appropriate *entropy function* in each case. The entropy measures the amount by which the concentrations of customer types vary from the proportion in equilibrium in the different queues. When the traffic intensity function is at most 1, the entropy decreases exponentially fast in t to 0. This estimate is obtained from bounds employing the convexity of the associated functions $h_k(x)$ together with the specific form of the entropy. The exponential bounds on the entropy then quickly yield corresponding bounds on the rate of convergence of the system. When the traffic intensity is strictly less than 1, quicker estimates involving the entropy give the limit of the system. Our main results are Theorems 1 and 2 for FIFO fluid models and Theorem 3 for HLPPS fluid models.

Although we are primarily interested here in the limiting behavior of these fluid models, we note their connection with queueing networks in the two contexts mentioned earlier. Recall that fluid models can be interpreted in terms of rescaled limits of queueing networks. Their convergence to the empty state thus provides information about these networks. In this context, one can consider *generalized Kelly networks*. By this, we mean those open multiclass queueing networks where the assumption that the service and interarrival times are all exponential is dropped in the definition of Kelly networks. One can define *generalized HLPPS networks* similarly. Employing Theorem 4.2 in Dai (1995) in conjunction with our Theorems 1 and 3, it is routine to show that strictly subcritical generalized Kelly and HLPPS networks are stable under mild conditions on their service and interarrival laws.

We also mentioned fluid models in the context of heavy traffic limits, where they can be thought of as giving the instantaneous evolution for the corresponding reflected Brownian motions. The evolution of such a Brownian motion is governed by the random input on its time scale. This randomness is absent from the corresponding fluid model. The fluid model nonetheless exhibits a restoring force which is also present in the random model, and should determine the submanifold on which the Brownian motion can live. Hopefully, then, our results will provide an important ingredient for the resolution of the heavy traffic problem for generalized Kelly and HLPPS networks.

These notes are structured as follows. In Section 2, we rigorously define the fluid models we will be working with. We introduce some further basic terminology in Section 3, and then state Theorems 1–3, and the accompanying corollary on generalized Kelly and HLPPS networks. Basic bounds for the entropy function of FIFO fluid models of Kelly type are discussed in Section 4. A brief sketch of the ideas behind the proofs of Theorems 1 and 2 is then given. In Section 5, the basic bounds and ideas leading to Theorem 3 are discussed.

2 The models

We first introduce the systems we will be dealing with, namely the FIFO fluid models of Kelly type and HLPPS fluid models. For the FIFO fluid models, we borrow wholesale from the terminology in Harrison (1995), making only minor changes when convenient. The notation is in both cases chosen so as to resemble that for queueing networks, and the reader should think in terms of these analogs to get a feel for the quantities involved. In our setting, these quantities are continuous and deterministic, and correspond to concepts such as "customer mass" rather than the number of customers, although we will use a hybrid of this vocabulary when convenient.

FIFO fluid models of Kelly type, denoted by $\mathfrak{X}$, will consist of 5-tuples of column vectors $(A(t), D(t), W(t), Y(t), Z(t))$ with $t \geq 0$. The vectors $A(t)$, $D(t)$ and $Z(t)$ are indexed by subscripts $k = 1, 2, \ldots, c$, and $W(t)$ and $Y(t)$ are indexed by $j = 1, 2, \ldots, d$, where $c, d \in \mathbb{Z}^+$ with $c \geq d$. These vectors are required to solve the *fluid model equations* (2.3)–(2.7) and to possess certain basic regularity properties. The terms α, m and P employed in (2.3)–(2.7) are given in advance. The above quantities should be conceptualized as follows. The system $\mathfrak{X}$ is the continuous deterministic analog of the process associated with a queueing network. The index k denotes the *class* of a customer, and j the corresponding *station*, with $j = s(k)$. Denote by $\mathcal{C}(j)$ the *constituency* of a station j, which is the set of classes k satisfying $s(k) = j$, and by C the accompanying *constituency matrix*,

$$\begin{aligned} C_{jk} &= 1 \qquad \text{if } k \in \mathcal{C}(j), \\ &= 0 \qquad \text{otherwise.} \end{aligned} \tag{2.1}$$

(It is assumed that $\mathcal{C}(j)$ is nonempty for each j.) The terms $\alpha_k \geq 0$ and μ_j, with $\mu_j = 1/m_j$, $m_j > 0$, are the rates at which customers enter the system at the kth class and are served at the jth station. The transition matrix $P = (P_{k\ell})$, $k, \ell = 1, 2, \ldots, c$, gives the probability of a class k customer entering the ℓth class upon being served at $s(k)$. The analogs of these terms for queueing networks were introduced in Section 1. (Here, m depends on j rather than on k.) As before, P is assumed to be transient, so that one can set $Q = (I - P^t)^{-1} = I + P^t + (P^t)^2 + \cdots$. The equations

$$\lambda_\ell = \alpha_\ell + \sum_{k=1}^{c} \lambda_k P_{k\ell}, \qquad \ell = 1, \ldots, c, \tag{2.2}$$

or equivalently, $\lambda = \alpha + P^t \lambda$, correspond to (1.1)–(1.2), and possess a unique solution λ given by $\lambda = Q\alpha$. We assume here, without loss of generality, that $\lambda_k > 0$ for each k. The term λ_k is referred to as the *total arrival rate* for class k. The vectors defining $\mathfrak{X}$ have the same interpretation as their queueing network analogs from the literature (see, e.g., Harrison–Nguyen (1993)). The quantities $A_k(t)$ and $D_k(t)$ are to be interpreted as the amount of fluid arriving at and departing from class k by time t, and $Z_k(t)$ is the length of the queue. $W_j(t)$ is the *immediate workload* at station j, that is, the amount of time required for server j to process all of the fluid already present at j at time t. The quantity $Y_j(t)$ is the *cumulative idleness* or unused capacity at station j up to time t.

The fluid model equations given below specify the evolution of the system. They are the analogs of (2.9)–(2.15) in Harrison (1995), with minor changes. We assume here that the vectors $A(t)$, $D(t)$, $W(t)$, $Y(t)$ and $Z(t)$ are all continuous and nonnegative in each component, with $A(t)$, $D(t)$ and $Y(t)$ being nondecreasing. (Most of these regularity conditions can be derived from the others.) The term M below denotes the $d \times d$ matrix $M = \text{diag}(m_1, \ldots, m_d)$ and e denotes the d-vector of all ones.

$$A(t) = \alpha t + P^t D(t), \tag{2.3}$$

$$Z(t) = Z(0) + A(t) - D(t), \tag{2.4}$$

$$W(t) = W(0) + MCA(t) - et + Y(t), \tag{2.5}$$

$$Y_j \text{ can only increase when } W_j = 0, \quad j = 1, \ldots, d, \tag{2.6}$$

$$D_k(t + W_j(t)) = Z_k(0) + A_k(t), \quad k = 1, \ldots, c. \tag{2.7}$$

In (2.6), we mean that $Y_j(t_2) > Y_j(t_1)$ implies $W_j(t) = 0$ for some $t \in [t_1, t_2]$. That is, a server is idle only when there is no immediate work at the station. The equality (2.7) defines the FIFO property of the system. The dependence of the service times m_j on j rather than on k specifies that the system is of Kelly type. Here and later on, when the indices j and k appear together, we implicitly set $j = s(k)$.

The systems $\mathfrak{X} = (A(t), D(t), W(t), Y(t), Z(t), t \geq 0)$ satisfying (2.3)–(2.7) will be referred to as *FIFO fluid models of Kelly type*. To study them, we need to specify the initial conditions. Since (2.7) specifies $D_k(t')$ only for $t' \geq W_j(0)$, we employ the following consistency condition for smaller times:

$$m_j \sum_{k \in \mathcal{C}(j)} D_k(t) = t \quad \text{for } t \in [0, W_j(0)], \quad j = 1, \ldots, d. \tag{2.8}$$

Terms $W_j(0)$ and $D_k(t)$, $t \in [0, W_j(0)]$, satisfying (2.8) will be said to constitute the *initial data* for $\mathfrak{X}$. It follows quickly from (2.3), (2.5), (2.7) and (2.8) that

$$A(0) = D(0) = 0, \quad Y(0) = 0, \quad W(0) = MCZ(0). \tag{2.9}$$

Employing (2.3)–(2.9), it is not difficult to show that (2.8) can in fact be extended for all time:

$$MCD(t) + Y(t) = et \quad \text{for } t \geq 0. \tag{2.10}$$

We note that, together with (2.4), (2.5) and (2.9), (2.10) implies that

$$W(t) = MCZ(t) \quad \text{for all } t.$$

HLPPS fluid models, denoted by $\mathcal{Y}$, will consist of 3-tuples of column vectors $(A(t), D(t), Z(t))$ with $t \geq 0$. The vectors $A(t)$, $D(t)$ and $Z(t)$ are indexed by $k = 1, 2, \ldots, c$, with the quantities $A_k(t)$ and $D_k(t)$ again being interpreted as the amount of fluid arriving at and departing from class k by time t, and with $Z_k(t)$ being the length of the queue. The notation through (2.2) will again be employed, although now m will depend on k instead of on j. The equations (2.2) are still assumed. Here, we set

$$Z_j^{\Sigma}(t) = \sum_{k \in \mathcal{C}(j)} Z_k(t). \tag{2.11}$$

The *HLPPS fluid model equations* are specified as follows:

$$A(t) = \alpha t + P^t D(t), \tag{2.12}$$

$$Z(t) = Z(0) + A(t) - D(t), \tag{2.13}$$

$$\sum_{k \in \mathcal{C}(j)} m_k(D_k(t_2) - D_k(t_1)) \leq t_2 - t_1 \quad \text{for } 0 \leq t_1 \leq t_2, \tag{2.14}$$

$$D_k'(t) = \frac{Z_k(t)}{m_k Z_j^{\Sigma}(t)} \quad \text{when } Z_j^{\Sigma}(t) > 0. \tag{2.15}$$

Systems $\mathcal{Y} = (A(t), D(t), Z(t), t \geq 0)$ satisfying (2.12)–(2.15) will be referred to as *HLPPS fluid models*. Note that the pair (2.12)–(2.13) corresponds to (2.3)–(2.4) for the FIFO fluid model. The equality in (2.15) defines the HLPPS property of the present system. Since customers are unordered here, it is enough to specify $Z(0)$ and set $D(0) = 0$ for the initial data of $\mathcal{Y}$.

When dealing with vectors, we will always employ the max norm, although this is a matter of convenience. It is not difficult to check that for the FIFO and HLPPS fluid models, the functions $A(t)$, $D(t)$, $W(t)$, $Y(t)$ and $Z(t)$ are all Lipschitz continuous: that is, for some $N > 0$,

$$|f(t_2) - f(t_1)| \leq N|t_2 - t_1| \quad \text{for all } t_1, t_2 \geq 0, \tag{2.16}$$

if $f(t)$ is any of the above functions. Since $A(t)$, $D(t)$, $W(t)$, $Y(t)$ and $Z(t)$ are Lipschitz, they are also absolutely continuous. Derivatives therefore exist almost everywhere (a.e.), and the fundamental theorem of calculus holds. Using this, one can employ the usual manipulations involving the first derivatives of the components of $\mathfrak{X}$ and $\mathcal{Y}$ when dealing with the entropies $\mathcal{H}(t)$ and $\hat{\mathcal{H}}(t)$ defined in (3.8) and (3.9).

3   Main results

Our goal is to understand the asymptotic behavior of fluid models satisfying properties (2.3)–(2.7) or properties (2.12)–(2.15). As one might expect, this behavior is dependent on the traffic intensity. In analogy with queueing networks, the *traffic intensity* for the jth server is

$$\rho_j = \sum_{k \in \mathcal{C}(j)} m_k \lambda_k. \tag{3.1}$$

(Under (2.3)–(2.7), m_k depends just on j.) We say that the fluid model is *subcritical* if $\rho_j \leq 1$ for all j, and *strictly subcritical* if $\rho_j < 1$ for all j. The fluid model is *critical* (or *balanced* in the terminology of Harrison (1995)) if $\rho_j \equiv 1$. If $\rho_j > 1$ for some j, then the fluid model is *supercritical*.

We are interested in examining the convergence of subcritical fluid models. Before doing this, we briefly consider supercritical fluid models. It is not surprising that these must diverge, as is shown in the following simple result (Bramson (1996a, 1996b)). The notation "$(\cdot)_k$" means the kth coordinate. Observe that $(Q(Z(t)))_k$ is the total mass of customers in the system at time t which eventually visit class k (weighted for repeated visits).

Proposition 3.1 *Assume that the system is a FIFO fluid model of Kelly type or an HLPPS fluid model. Then for all t and j,*

$$\sum_{k \in \mathcal{C}(j)} m_k (Q(Z(t) - Z(0)))_k \geq (\rho_j - 1)t. \tag{3.2}$$

In particular, if the fluid model is supercritical, then

$$\liminf_{t \to \infty} |Z(t)|/t > 0. \tag{3.3}$$

To analyze the asymptotic behavior of subcritical FIFO fluid models, we need to introduce notation describing the state of the system at a given time. (The state of an HLPPS fluid model at time t is given by $Z(t)$.) As in Harrison (1995), we define the vector-valued function $X(t, \cdot)$, $t \geq 0$, which has coordinates $X_k(t, \cdot)$, so that $X_k(t, \cdot) : [0, \infty) \to [0, \infty)$ satisfies

$$\begin{aligned} X_k(t, s) &= D_k(t + s) - D_k(t) && \text{for } 0 \leq s \leq W_j(t), \\ &= Z_k(t) && \text{for } s \geq W_j(t). \end{aligned} \tag{3.4}$$

The function $X_k(t, s)$ is to be interpreted as the quantity of type k customers already at the jth station at time t who will be served by time $t + s$. By time $t + W_j(t)$, all of these customers will have been served. Note that by (2.10),

$$\begin{aligned} m_j \sum_{k \in \mathcal{C}(j)} X_k(t, s) &= s && \text{for } 0 \leq s \leq W_j(t), \\ &= W_j(t) && \text{for } s \geq W_j(t). \end{aligned} \tag{3.5}$$

The functions $x(\cdot) = X(t, \cdot)$ are Lipschitz and nondecreasing in each coordinate, with

$$|x(s_2) - x(s_1)| \leq N|s_2 - s_1|, \tag{3.6}$$

where N is as in (2.16). We equip the set of all such functions (N fixed) with the uniform metric $\| \cdot \|_\infty$.

Our basic tool for analyzing the asymptotic behavior of $X(t, \cdot)$ will be the *entropy function* $\mathcal{H}(t)$. To define this, we set

$$h(x) = x \log x, \quad x \geq 0, \tag{3.7}$$

and

$$h_k(x) = \lambda_k h(x/\lambda_k) = x \log(x/\lambda_k), \quad x \geq 0,$$

for $k = 1, 2, \ldots, c$, where λ_k is given in (2.2). Note that $h_k(0) = h_k(\lambda_k) = 0$, $h_k(x) < 0$ for $0 < x < \lambda_k$ and $h_k(x) > 0$ for $x > \lambda_k$, $h'_k(\lambda_k) = 1$, and that $h_k(x)$ is convex with $h''_k(x) = 1/x$. We define

$$\mathcal{H}(t) = \sum_k \int_t^{t+W_j(t)} h_k(D'_k(r)) \, dr, \quad t \geq 0. \tag{3.8}$$

$\mathcal{H}(t)$ measures the "distance" at time t from the equilibrium $D'_k(r) = \lambda_k$,

$r \in [t, t + W_j(t)]$, for subcritical $\mathfrak{X}$. In Section 4, we will discuss the main properties of $\mathcal{H}(t)$, such as $\mathcal{H}(t) \geq 0$ and $\mathcal{H}'(t) \leq 0$, and the rate at which $\mathcal{H}(t) \to 0$ as $t \to \infty$.

In order to analyze the asymptotic behavior of HLPPS fluid models, we introduce the entropy function

$$\hat{\mathcal{H}}(t) = \sum_k Z_k(t) \log(D'_k(t)/\lambda_k) \qquad (3.9)$$

or, equivalently,

$$\hat{\mathcal{H}}(t) = \sum_k m_k Z_j^{\Sigma}(t) h_k(D'_k(t)). \qquad (3.10)$$

$\hat{\mathcal{H}}(t)$ has properties similar to $\mathcal{H}(t)$, although the computations that are involved are different. In Section 5, we discuss the behavior of $\hat{\mathcal{H}}(t)$.

We are now ready to state our main results. Theorem 1 says that for subcritical FIFO fluid models of Kelly type, the immediate workload $W(t)$ for the system converges exponentially fast at a rate proportional to $|W(0)|$. For strictly subcritical models, the limit is the empty state, and occurs after a finite time. Here and throughout the paper, B_i, $i = 1, 2, \ldots$, will denote nonnegative constants whose exact values do not concern us.

Theorem 1 *Assume that $\mathfrak{X}$ is a subcritical FIFO fluid model of Kelly type. For appropriate B_1 and $B_2 > 0$,*

$$|W(t) - W(\infty)| \leq B_1 \sqrt{\mathcal{H}(0)|W(0)|} \exp\{-B_2 t/|W(0)|\} \qquad (3.11)$$

for some $W(\infty)$ and all $t \geq 0$. If for a given j, $\rho_j < 1$, then $W_j(\infty) = 0$. If $\mathfrak{X}$ is strictly subcritical, then $W(t) = 0$ for $t \geq B_3 |W(0)|$ and appropriate B_3.

The constants B_1, B_2 and B_3 in Theorem 1 are allowed to depend on the choice of α, m and P used in (2.3)–(2.7). They are not uniformly bounded for ρ_j less than but close to 1, even for α, m and P well-behaved, since although $W(\infty) = 0$ holds, the rate of convergence can be made arbitrarily slow. One can check by using (3.8) that $\mathcal{H}(0)$ is always at most a fixed multiple of $|W(0)|$. Also note that if $\rho_j = 1$ for some choice of j, then by (3.2),

$$\sum_{k \in \mathcal{C}(j)} (Q(Z(t) - Z(0)))_k \geq 0 \quad \text{for all } t. \qquad (3.12)$$

So with respect to the weighting given by Q here, the customer mass in the system cannot decrease. In particular, if $Z_k(0) > 0$ for some class k which eventually leads to j under P, then by (3.12), $W(\infty) \neq 0$.

We set $Z_k(\infty) = \lambda_k W_j(\infty)$, where $W(\infty)$ is chosen as in (3.11). We define the vector-valued function $X(\infty, \cdot)$ so that its coordinates satisfy

$$\begin{aligned} X_k(\infty, s) &= \lambda_k s && \text{for } 0 \leq s \leq W_j(\infty), \\ &= Z_k(\infty) && \text{for } s \geq W_j(\infty). \end{aligned} \qquad (3.13)$$

Theorem 2 says that $X(t, \cdot)$ converges to $X(\infty, \cdot)$ at the rates given in Theorem 1.

Theorem 2 *Assume that $\mathfrak{X}$ is a subcritical FIFO fluid model of Kelly type. Choose B_2 and B_3 as in Theorem 1. For appropriate $B_4 > 0$,*

$$\|X(t, \cdot) - X(\infty, \cdot)\|_\infty \le B_4 \sqrt{\mathcal{H}(0)|W(0)|} \exp\{-B_2 t/|W(0)|\} \qquad (3.14)$$

for all $t \ge 0$. If for a given j, $\rho_j < 1$, then $X_k(\infty, \cdot) = 0$ for $k \in C(j)$. If $\mathfrak{X}$ is strictly subcritical, then $X(t, \cdot) = 0$ for $t \ge B_3 |W(0)|$.

Some thought shows that the factors $|W(0)|$ in the exponent and $\sqrt{\mathcal{H}(0)|W(0)|}$ in front are appropriate choices for the bounds given in (3.11) and (3.14). In particular, they are consistent with the following changes in scale. Suppose the initial data satisfying (2.8) is scaled by $\gamma > 0$ with

$$^\gamma D_k(t) = \gamma D_k(t/\gamma) \quad \text{for } k = 1, \dots, c \quad \text{and} \quad t \ge 0. \qquad (3.15)$$

One can check that, by scaling $A(t)$, $D(t)$, $W(t)$, $Y(t)$ and $Z(t)$ correspondingly, the new system $^\gamma \mathfrak{X}$ thus defined also satisfies (2.3)–(2.7). That is, by dilating both the space and time scales by the factor γ, one obtains another solution. Applying this scaling to (3.11) and (3.14), with $\gamma = 1/|W(0)|$, gets rid of the factor $|W(0)|$ in the exponent. Since $^\gamma \mathcal{H}(t) = \gamma \mathcal{H}(t/\gamma)$, the factor $\sqrt{\mathcal{H}(0)|W(0)|}$ in front is reduced by $|W(0)|$. Observe that these bounds are just the same as what one obtains for $^\gamma \mathfrak{X}$ by employing (3.11) and (3.14) directly.

For HLPPS fluid models, Theorem 3 is the analog of Theorems 1 and 2. Here, $|Z(0)|$ plays the role of $|W(0)|$.

Theorem 3 *Assume that $\mathcal{Y}$ is a subcritical HLPPS fluid model. For appropriate B_5 and $B_6 > 0$,*

$$|Z(t) - Z(\infty)| \le B_5 \sqrt{\mathcal{H}(0)|Z(0)|} \exp\{-B_6 t/|Z(0)|\} \qquad (3.16)$$

for some $Z(\infty)$ and all $t \ge 0$. If for a given j, $\rho_j < 1$, then $Z_j^\Sigma(\infty) = 0$. If $\mathcal{Y}$ is strictly subcritical, then $Z(t) = 0$ for $t \ge B_7 |Z(0)|$ and appropriate B_7.

As an application of Theorems 1 and 3 in the strictly subcritical case, one can conclude that a wide class of generalized Kelly and HLPPS networks is stable. Using the formulation given in Dai (1995), we assume that the exogenous interarrival times $\xi_1, \dots, \xi_c$ and the service times $\eta_1, \dots, \eta_c$ of the network are mutually independent and i.i.d. sequences which satisfy the following properties:

$$
\begin{aligned}
E[\xi_k(1)] &< \infty \quad && \text{for classes } k \text{ with nonnull exogenous arrivals,} \\
E[\eta_k(1)] &< \infty \quad && \text{for all } k,
\end{aligned}
\qquad (3.17)
$$

$$P(\xi_k(1) \ge x) > 0 \quad \text{for all } x, \qquad (3.18)$$

and, for some $\ell_k > 0$ and some nonnegative $q_k(\cdot)$, with $\int_0^\infty q_k(x)dx > 0$,

$$P(\xi_k(1) + \cdots + \xi_k(\ell_k) \in dx) \ge q_k(x)dx. \qquad (3.19)$$

One employs fluid limits, which were given in Dai (1995) as the sample path limits along subsequences obtained from a given queueing network under scaling which is linear in both time and the number of customers. Such limits are almost surely

solutions of the accompanying *delayed fluid model equations.* By Theorem 5.3 of
Chen (1995), it is enough to consider solutions of the accompanying (undelayed)
fluid model equations, such as (2.3)–(2.7) and (2.12)–(2.15) in our context.

Assume now that (3.17)–(3.19) are satisfied for a given generalized Kelly or
HLPPS queueing network. It follows from Theorem 4.2 of Dai (1995) that if

$$Z(t) = 0 \quad \text{for } t \geq B_8|Z(0)|,$$

appropriate B_8, for all solutions of the corresponding fluid model equations,
then the queueing model is stable. One therefore has the following consequence
of Theorems 1 and 3.

Corollary 1 Any strictly subcritical generalized Kelly network or HLPPS network whose exogenous interarrival times and service times satisfy (3.17)–(3.19)
is stable.

One can also obtain related ergodicity results for these queueing networks
along the lines of Dai–Meyn (1995) if one replaces (3.17) by analogous conditions
on the pth moments. (The conditions (3.17)–(3.19) and their analogs can be
weakened further.) The recent papers Chen–Zhang (1995a) and Foss–Rybko
(1995) present results related to Corollary 1 for FIFO queueing networks. The
assumption of equal service times for classes at the same station is omitted in
both papers, and replaced by spectral radius bounds associated with the queueing
network in the first paper, and by restrictions on the transition matrix P in the
second paper.

4 Basic estimates for Theorems 1 and 2

In this section, we discuss our basic bounds for the entropy of FIFO fluid models
of Kelly type, and briefly say how these bounds lead to Theorems 1 and 2; for
more detail, the reader is referred to Bramson (1996a). We begin our study of
the entropy function

$$\mathcal{H}(t) = \sum_k \int_t^{t+W_j(t)} h_k(D'_k(r)) \, dr, \quad t \geq 0, \tag{4.1}$$

by observing that since $D(r)$ and $W(t)$ are Lipschitz, it follows that $\mathcal{H}(t)$ is also
Lipschitz. One can check, by using Jensen's inequality, that $\mathcal{H}(t) \geq 0$ if the
system $\mathfrak{X}$ is subcritical. It is therefore natural to examine $\mathcal{H}'(t)$, which we do by
first checking that $\mathcal{H}'(t) \leq 0$ a.e.

We first note that differentiation of (4.1) gives

$$\mathcal{H}'(t) = \sum_k [(1 + W'_j(t))h_k(D'_k(t + W_j(t))) - h_k(D'_k(t))] \tag{4.2}$$

a.e. in t. By (2.7), this equals

$$\sum_k [(1 + W'_j(t))h_k(A'_k(t)/(1 + W'_j(t))) - h_k(D'_k(t))]. \tag{4.3}$$

One can check that this equals

$$\sum_k [h_k(A_k'(t)) - h_k(D_k'(t))] - \sum_j \left(\sum_{k \in \mathcal{C}(j)} A_k'(t) \right) \log(1 + W_j'(t)). \qquad (4.4)$$

If $Y_j'(t) \neq 0$ and $W_j'(t)$ exists, it follows from (2.6) that $W_j'(t) = 0$. The summands in the second half of (4.4) are therefore a.e. equal to

$$\left(\sum_{k \in \mathcal{C}(j)} A_k'(t) + \frac{1}{m_j} Y_j'(t) \right) \log(1 + W_j'(t)) = \frac{1}{m_j} h(1 + W_j'(t)), \qquad (4.5)$$

where the equality follows from (2.5). Together, (4.2)–(4.5) imply that a.e.,

$$\mathcal{H}'(t) = \sum_k [h_k(A_k'(t)) - h_k(D_k'(t))] - \sum_j \frac{1}{m_j} h(1 + W_j'(t)). \qquad (4.6)$$

To analyze $\mathcal{H}'(t)$, it thus suffices to examine the two sums on the right side of (4.6). One can show that, a.e. in t,

$$\sum_j \frac{1}{m_j} h(1 + W_j'(t)) \geq \sum_k Z_k'(t) \qquad (4.7)$$

and

$$\sum_k [h_k(A_k'(t)) - h_k(D_k'(t))] \leq \sum_k Z_k'(t), \qquad (4.8)$$

from which $\mathcal{H}'(t) \leq 0$ a.e. follows. The first inequality follows almost immediately upon linearizing (4.7); the second inequality requires some manipulation involving Jensen's inequality. (See Lemmas 4.1 and 4.2 in Bramson (1996a) for more detail.)

Some intuition for the above inequalities is provided by the following observations. The sum $\sum_k Z_k'(t)$ on the right side of (4.7) and (4.8) is due to the system being open; for the closed analog, the sum would vanish. In (4.7), the inequality is due to the averaging of mass within a station which occurs because of the imbalance between the arrival and service rates, which is reflected in the term $W'(t)$. In particular, one can envision arrivals at j over time $[t, t + dt]$ as corresponding to an infinitesimal rectangle $\left(\sum_{k \in \mathcal{C}(j)} A_k'(t) \right) \times dt$, which immediately undergoes a measure preserving transformation to the rectangle

$$\left(\sum_{k \in \mathcal{C}(j)} D_k'(t + W_j(t)) \right) \times (1 + W_j'(t)) dt$$
$$= \left(\left(\sum_{k \in \mathcal{C}(j)} A_k'(t) \right) / (1 + W_j'(t)) \right) \times (1 + W_j'(t)) dt.$$

The latter rectangle then remains at j until time $t + W_j(t)$. In (4.8), the averaging is due to the randomness inherent in the transition matrix P. For closed re-entrant lines, for example, the terms on the left telescope, and both sides are equal to 0.

In order to analyze the asymptotic behavior of $\mathfrak{X}$, one in fact needs bounds on the rate at which $\mathcal{H}(t) \to 0$ as $t \to \infty$, which means one needs a more careful analysis of the terms in (4.6). The inequalities (4.7) and (4.8) followed from the convexity of $h(x)$; for sharper results, one needs to estimate this convexity. Using such estimates, it is not difficult to improve (4.7) to the inequality

$$\sum_j \frac{1}{m_j} h(1 + W_j'(t)) \geq \sum_k Z_k'(t) + B_9 \sum_j (W_j'(t))^2, \tag{4.9}$$

where $B_9 > 0$. Together with (4.8), this implies that

$$\mathcal{H}'(t) \leq -B_9 \sum_j (W_j'(t))^2 \quad \text{a.e. in } t. \tag{4.10}$$

When $\mathfrak{X}$ is strictly subcritical, one has elementary upper bounds on the time it takes for a given station to empty based on the lengths of all of the queues. Together with (4.10) and some manipulation, these bounds allow one to demonstrate Theorems 1 and 2 in the strictly subcritical state without too much trouble.

The derivation of Theorems 1 and 2 in the general subcritical case is considerably longer. Besides estimates which also strengthen (4.8), one needs upper and lower bounds on $\mathcal{H}(t)$ of the form

$$B_{10} \sum_k \int_t^{t+W_j(t)} (D_k'(r) - \lambda_k)^2 dr, \tag{4.11}$$

with $B_{10} > 0$. One then obtains lower bounds on the decrease of $\mathcal{H}(t)$ over "cycles" $[T_i, T_{i+1}]$ which are long enough to allow a certain proportion of the mass to escape from anywhere in the system. These bounds on $\mathcal{H}(t)$ induce upper bounds on T_i and $|W(t)|$, which in turn imply the exponential convergence of $\mathcal{H}(t)$. The bounds on $W(t)$ and $X(t, \cdot)$ in (3.11) and (3.14) of Theorems 1 and 2 then follow without difficulty. This work on the general subcritical case is done in Sections 5–7 of Bramson (1996a).

5 Basic estimates for Theorem 3

Here, we discuss our basic bounds for the entropy of HLPPS fluid models, and briefly explain how they lead to Theorem 3; for more detail, the reader is referred to Bramson (1996b). Formal differentiation of the entropy function

$$\hat{\mathcal{H}}(t) = \sum_k Z_k(t) \log(D_k'(t)/\lambda_k), \tag{5.1}$$

together with (2.15), gives

$$\hat{\mathcal{H}}'(t) = \sum_k Z_k'(t) \log(D_k'(t)/\lambda_k). \tag{5.2}$$

(The other sum drops out.) So, as before, the derivative of the entropy function has a simple form. One can show, with some work, that $D_k'(t)$ is bounded away from 0 on intervals $[t_1, \infty)$ with $t_1 > 0$. Using this, one can verify that (5.2) in fact holds a.e., and that $\hat{\mathcal{H}}(t)$ is Lipschitz on $[t_1, \infty)$.

Substitution of (2.12) and (2.13) into (5.2) gives

$$\hat{\mathcal{H}}'(t) = \sum_k \left[\alpha_k + \sum_\ell P_{\ell k} D_\ell'(t) - D_k'(t) \right] \log(D_k'(t)/\lambda_k). \tag{5.3}$$

Along the lines of (4.7)–(4.8) for $\mathcal{H}(t)$, we first wish to show $\hat{\mathcal{H}}'(t) \le 0$ before investigating $\hat{\mathcal{H}}(t)$ further. The idea is to note that the right side of (5.3) is of the form

$$\sum_k \left[\alpha_k + \sum_\ell P_{\ell k} \lambda_\ell d_\ell - \lambda_k d_k \right] \log d_k, \tag{5.4}$$

where $d_k > 0$, and $\log x$ is, of course, monotone, with $\log 1 = 0$. One then applies Abel partial summation ("summation by parts") to (5.4). Together with a little work, this shows that (5.4) ≤ 0, and hence that $\hat{\mathcal{H}}'(t) \le 0$ a.e.

By redoing the above calculation more carefully, one can, in fact, show that

$$\hat{\mathcal{H}}'(t) \le -B_{11} \sum_k (D_k'(t) - \lambda_k)^2 \tag{5.5}$$

a.e., where $B_{11} > 0$. Note that under $Z_j^\Sigma(t) > 0$,

$$\sum_{k \in \mathcal{C}(j)} m_k D_k'(t) = 1. \tag{5.6}$$

It follows from (3.1) and (5.6) that there is a $k \in \mathcal{C}(j)$ with $D_k'(t) \ge \lambda_k/\rho_j$. Employing (5.5), it is therefore easy to see that for a strictly subcritical system with $Z_j^\Sigma(t) > 0$ for some j,

$$\hat{\mathcal{H}}'(t) < -B_{12} \tag{5.7}$$

for some $B_{12} > 0$. From this, it follows without difficulty that for a strictly subcritical system,

$$\hat{\mathcal{H}}(t) = 0 \quad \text{for } t \ge B_7 |Z(0)|,$$

for some B_7. This implies that $Z(t) = 0$ for $t \ge B_7|Z(0)|$, which implies Theorem 3 in the strictly subcritical case. Related reasoning using (5.7) also shows that $Z_j^\Sigma(\infty) = 0$ when $\rho_j < 1$.

In order to show exponential convergence as in (3.16) for the general subcritical case, one can derive from (3.10) that

$$\hat{\mathcal{H}}(t) \le B_{13} \sum_k Z_j^\Sigma(t)(D_k'(t) - \lambda_k)^2, \tag{5.8}$$

for appropriate B_{13}. With some work, one can show that

$$Z_j^\Sigma(t) \le B_{14}|Z(0)| \tag{5.9}$$

for all t, j and some B_{14}. It follows from (5.5), (5.8) and (5.9) that

$$\hat{\mathcal{H}}'(t) \leq -B_6 \hat{\mathcal{H}}(t)/|Z(0)|$$

a.e. for some $B_6 > 0$, and hence that

$$\hat{\mathcal{H}}(t) \leq \hat{\mathcal{H}}(0)\exp\{-B_6 t/|Z(0)|\}. \tag{5.10}$$

The bound (3.16) on $|Z(t) - Z(\infty)|$ then follows from (5.10) after a little more estimation.

Acknowledgments The author thanks Jim Dai for useful conversations on queueing networks. The author thanks Mike Harrison for making his manuscript Harrison (1995) available, but in particular for many helpful conversations on fluid models.

Bibliography

1. Baskett, F., Chandy, K. M., Muntz, R. R. and Palacios, F. G. (1975). Open, closed and mixed networks of queues with different classes of customers. *J. ACM*, **22**, 248–260.

2. Bramson, M. (1994a). Instability of FIFO queueing networks. *Ann. Appl. Probab.*, **4**, 414–431.

3. Bramson, M. (1994b). Instability of FIFO queueing networks with quick service times. *Ann. Appl. Probab.*, **4**, 693–718.

4. Bramson, M. (1995). Two badly behaved queueing networks. *Stochastic Networks, IMA Volumes in Mathematics and its Applications*, **71**, Springer Verlag, New York, pp. 105–116.

5. Bramson, M. (1996a). Convergence to equilibria for fluid models of FIFO queueing networks. *Queueing Syst.*, to appear.

6. Bramson, M. (1996b). Convergence to equilibria for fluid models of certain processor sharing queueing networks. *Queueing Syst.*, to appear.

7. Chen, H. (1995). Fluid approximations and stability of multiclass queueing networks: work-conserving disciplines. *Ann. Appl. Probab.*, to appear.

8. Chen, H. and Zhang, H. (1995a). Stability of multiclass queueing networks under FIFO service discipline, submitted to *Math. Oper. Res.*

9. Chen, H. and Zhang, H. (1995b). Diffusion approximations for re-entrant lines with a first-buffer-first-served priority discipline, submitted to *Queueing Syst.*

10. Dai, J. (1995). On the positive Harris recurrence for multiclass queueing networks. *Ann. Appl. Probab.*, **5**, 49–77.

11. Dai, J. and Kurtz, T. (1995). A multiclass station with Markovian feedback in heavy traffic. *Math. Oper. Res.*, **20**, 721–742.

12. Dai, J. and Meyn, S. (1995). Stability and convergence of moments for multiclass queueing networks via fluid models. *IEEE Trans. Autom. Control*, **40**, 1889–1904.

13. Dai, J. and Wang, Y. (1993). Nonexistence of Brownian models for certain

multiclass queueing networks. *Queueing Syst.*, **13**, 41–46.

14. Dumas, V. (1994). Unstable cycles in fluid Bramson networks. *Rapport de Recherche*, **2318**, *INRIA*, August, 1994.

15. Foss, S. and Rybko, A. (1995). Stability of multiclass Jackson-type networks. Preprint.

16. Harrison, J. M. (1988). Brownian models of queueing networks with heterogeneous customer populations. *Stochastic Differential Systems, Stochastic Control Theory and their Applications, IMA Volumes in Mathematics and its Applications*, **10**, Springer Verlag, New York, pp. 147–186.

17. Harrison, J. M. (1995). Balanced fluid models of multiclass queueing networks: a heavy traffic conjecture. *Stochastic Networks, IMA Volumes in Mathematics and its Applications*, **71**, Springer Verlag, New York, pp. 1–20.

18. Harrison, J. M. and Nguyen, V. (1990). The QNET method for two-moment analysis of open queueing networks. *Queueing Syst.*, **6**, 1–32.

19. Harrison, J. M. and Nguyen, V. (1993). Brownian models of multiclass queueing networks: current status and open problems. *Queueing Syst.*, **13**, 5–40.

20. Johnson, D. P. (1983). Diffusion approximations for optimal filtering of jump processes and for queueing networks. Ph.D. thesis, University of Wisconsin.

21. Kelly, F. P. (1975). Networks of queues with customers of different types. *J. Appl. Probab.*, **12**, 542–554.

22. Kelly, F. P. (1979). *Reversibility and Stochastic Networks*. Wiley, New York.

23. Lu, S. H. and Kumar, P. R. (1991). Distributed scheduling based on due dates and buffer priorities. *IEEE Trans. Autom. Control*, **36**, 1406–1416.

24. Reiman, M. I. (1984). Open queueing networks in heavy traffic. *Math. Oper. Res.*, **9**, 441–458.

25. Reiman, M. I. (1988). A multiclass feedback queue in heavy traffic. *Adv. Appl. Probab.*, **20**, 179–207.

26. Rybko, S. and Stolyar, A. L. (1992). Ergodicity of stochastic processes that describe the functioning of open queueing networks. *Problems Inf. Trans.*, **28**, 3–26 (in Russian).

27. Stolyar, A. (1995). On the stability of multiclass queueing networks. *Proceedings of the 2nd Conference on Telecommunication Systems – Modeling and Analysis*, Nashville, March 24–27, 1994, pp. 1020–1028.

2

Optimal draining of fluid re-entrant lines: some solved examples

Gideon Weiss

Haifa University

Abstract

In a fluid re-entrant line fluid moves in sequence through buffers $1, \ldots, K$ which are partitioned between I machines. Each machine can divide its effort continuously between the buffers it serves and flow rates out of the buffers are proportional to the effort. It is desired to find flow rates over period $(0, T)$ that empty the line with minimum average inventory, equivalently cycle time. This formulation is a first approximation to a transient re-entrant line queueing network, which in turn models many industrial processes, e.g. semiconductor wafer fabrication. The problem can be formulated as a continuous linear program, with a very special structure. In this paper we discuss some properties of the optimal solutions, and these properties enable us to solve several examples. The solutions themselves are quite instructive. We also outline some ideas for an efficient algorithm to solve the problem.

1 Introduction

A manufacturing system in which all the parts undergo the same sequence of operations, but these operations may require each part to revisit some of the machines several times, is called a re-entrant line. Re-entrant lines are typical of the ultra expensive and sophisticated semiconductor wafer fabrication plants that form the major step in computer chip manufacturing. Parts that have completed all the operations preceding k, and are waiting for (machine, work-) station $i = \sigma(k)$ to complete their kth operation, are said to be in buffer (or queue) k. The dynamics of re-entrant line manufacturing systems can be modeled by a re-entrant queueing network, with the stations as (single or multiserver) nodes, and the buffers as the queues—this is in fact a so called multiclass queueing network, because each node serves several distinct classes of parts (in different stages of completion). The theory of multiclass queueing networks is far from complete. The question of stability of such networks turned out to be hard, because stability is no longer a property of the network alone but also of the

Research supported by US NSF grant DDM-9215233 and US-Israel BSF grant 9400196.

policy exercised at the multiclass nodes. Recent results have shown that the stability of multiclass queueing networks is intimately related to the stability of their deterministic, transient, fluid approximations.

In the present paper we consider fluid approximations to re-entrant queueing networks (and their originating manufacturing system). We assume that the content of the buffers is continuous fluid rather than discrete parts. We assume that if a station devotes a constant fraction of its capacity to a buffer, its contents decrease linearly. Finally, we assume that the station capacity is arbitrarily divisible between the buffers it serves. We investigate the problem of draining (emptying) the contents of the entire line, with no outside input, in an optimal fashion. In fact our objective is minimal inventory, equivalently minimal flow time. References on re-entrant lines and fluid models are listed in the bibliography.

Consider a fluid re-entrant line with fluid moving in sequence through buffers $1, \ldots, K$ and out. Buffers are partitioned into sets C_i where $k \in C_i$ is served by station $i = \sigma(k)$, $i = 1, \ldots, I$. Service times per unit of fluid are m_k and initial fluid levels are a_k. We wish to find control flow rates, bounded measurable functions $u_1(t), \ldots, u_K(t)$, for time $0 \leq t \leq T$, so as to:

$$\text{maximize} \int_0^T (T - t) u_K(t) dt \tag{1.1}$$

subject, for all $t \in [0, T]$, to:

$$\int_0^t u_1(s)\, ds \;\leq\; a_1, \tag{1.2}$$

$$\int_0^t (u_k(s) - u_{k-1}(s))\, ds \;\leq\; a_k, \qquad k = 2, \ldots, K \tag{1.3}$$

$$\sum_{k \in C_i} m_k u_k(t) \;\leq\; 1, \qquad i = 1, \ldots, I \tag{1.4}$$

$$u_k(t) \;\geq\; 0, \qquad k = 1, \ldots, K. \tag{1.5}$$

Denote by $x_k(t)$ the slacks of the flow balance constraints (1.2, 1.3); these are the buffer fluid levels. Denote by $z_i(t)$ the slacks of the station capacity constraints (1.4); these are unused resources. In the objective, $u_K(t) dt$ is the amount of fluid departing at time $(t, t + dt)$, and it will bring a reward at the rate $T - t$, the length of the useful period, from completion of the part at t to the time horizon T. This objective is easily seen to be equivalent to minimizing the average total inventory over $[0, T]$, which is the same as minimizing the average cycle time. Without loss of generality we assume no exogenous inflow; inflow over $[0, T]$ can be replaced by an initial buffer. Let $l_k(t) = \sum_{j=1}^k x_k(t)$; in particular $l_K(t)$ is the total fluid inventory at time t. A useful graphic description of the dynamics of the re-entrant line is to plot the lines $l_k(t)$.

The problem (1.1–1.5) was introduced in Weiss (1995). It can be analyzed through Pontryagin's maximum principle (Avram, Bertsimas and Ricard 1995);

it is also a special case of an SCLP (Separated Continuous Linear Program) studied extensively by Anderson and Nash (1987), Pullan (1993, 1995a, 1995b, 1996a, 1996b) and Anderson and Pullan (1995). Pullan has formulated a dual, proved strong duality, characterized the form of the optimal solution, and developed an algorithm for the solution of general SCLP. However, the solution of SCLP of even modest size requires a large amount of computation. We hope that the special structure of (1.1–1.5) will allow a more efficient and informative solution.

In this paper we solve several examples of the problem. Our purpose here is twofold. Firstly, the solutions are extremely informative, in that they provide qualitative ideas on how to control re-entrant lines: To minimize inventory one should myopically give priority to fluid close to the end of the line—in particular to the fluid in the last nonempty buffer in the system, as well as to the last nonempty buffer at each station, a policy called LBFS (Last Buffer First Served). However, the system may contain bottleneck buffers, which, once all the fluid above them has been drained, will delay the outflow. This introduces a trade-off between LBFS and the necessity of pumping fluid through bottleneck buffers. The resulting optimal schedule, transient in nature, will locate all the bottlenecks. This is important in itself, since bottlenecks of multiclass queueing networks are difficult to define, leading some researchers to coin the term hidden bottlenecks. The optimal schedule will then pinpoint the exact times at which to shift priority from LBFS to flow through these bottlenecks, lower down the line. Secondly, the study of the examples indicates how to construct a general purpose efficient algorithm for the solution of (1.1–1.5).

The paper is structured as follows. In Section 2 we present, without proof, some results on the structure of the optimal solution, a formulation of a dual problem, and complementary slackness conditions. These provide a framework to prove optimality. In Section 3 we present our examples. They include a comprehensive study of 1-, 2-, and 3-station re-entrant lines. Finally, in Section 4 we give a brief outline of a possible algorithm to solve the general problem.

2 Preliminary results

In this section we summarize some of the properties of optimal solutions for the problem (1.1–1.5). These properties essentially follow from the results of Pullan, though some of the proofs are quite lengthy. For this reason we omit the proofs. For our purpose, the results on duality and complementary slackness provide the tool to show that the solutions presented in Section 3 are indeed optimal. We list some further results which highlight some special features of fluid re-entrant lines.

2.1 Basic solutions

Consider the matrix $\mathcal{K}$ of coefficients of the left hand side of (1.2–1.4), of $u_k(t)$ or of the integrand $u_k(s)$. The matrix $\mathcal{K}$ is of dimension $(K + I) \times (2K + I)$, and of the form:

$$\mathcal{K} = \begin{bmatrix} 1 & 0 & \cdots & 0 & 0 & | & 1 & 0 & & & & | & \\ -1 & 1 & & & & | & 0 & 1 & & & & | & \\ & & \ddots & & & | & & & \ddots & & & | & \mathbf{0} \\ & & & 1 & 0 & | & & & & 1 & 0 & | & \\ & & & -1 & 1 & | & & & & 0 & 1 & | & \\ - & - & - & - & - & | & - & - & - & - & - & | & - & - & - \\ & & & & & | & & & & & & | & 1 & \cdots & 0 \\ & & \mathbf{m} & & & | & & & \mathbf{0} & & & | & & \ddots & \\ & & & & & | & & & & & & | & 0 & & 1 \end{bmatrix} \qquad (2.1)$$

where $\mathbf{m}$ is the resource requirement matrix, containing m_k in the $(\sigma(k), k)$ position. We partitioned the matrix horizontally into K rows for the integrated constraints and I rows for the instantaneous constraints. Vertically the partition is into K, K, and I columns for the variables u_k, x_k, and z_i, respectively.

Theorem 2.1 *There exists a basic (i.e. extreme point) optimal solution to (1.1–1.5) with the following properties:*

(i) *The solution is piecewise constant, i.e. there exist breakpoints $0 = t_0 < t_1 < \cdots < t_N = T$, such that $u(t)$ is constant in $t_{n-1} < t < t_n$.*

(ii) *In each interval the support of the solution (positive components) corresponds to a set of linearly independent columns of $\mathcal{K}$.*

(iii) *For each interval there exists L, $L \leq I$, buffers $k_1 < k_1' \leq \cdots \leq k_L < k_L' \leq K + 1$, and a partition of the stations into subsets I_1, I_2, of sizes L and $I - L$, such that for $t_{n-1} < t < t_n$:*

- *For $l = 1, \ldots, L$, $x_{k_l}(t) > 0$, $x_{k_l'}(t) > 0$, and $x_j(t) = 0$ for $k_l < j < k_l'$.*
- *For $l = 1, \ldots, L$, $u_j(t) = U_l > 0$ for $k_l \leq j < k_l'$, while for all other j, $u_j(t) = 0$.*
- *For all $i \in I_1$ (L in number), $z_i(t) = 0$.*

(iv) *Let $I_2 = \{i_1, \ldots, i_{I-L}\}$, let e_i denote the ith unit vector, e the vector of all ones, and δ the Kronecker delta. Let M be the $I \times L$ matrix defined by:*

$$M_{i,l} = \sum_{j=k_l}^{k_l'-1} m_j \delta(i, \sigma(j)) \qquad l = 1, \ldots, L \quad i = 1, \ldots, I. \qquad (2.2)$$

Then the values of $U_1, \ldots, U_L$ and $z_{i_1}, \ldots, z_{i_{I-L}}$ are determined by:

$$MU + \sum_{l=1}^{I-L} e_{i_l} z_{i_l} = e. \qquad (2.3)$$

Proof Parts (i) and (ii) are due to Pullan (1995a), parts (iii) and (iv) follow by straightforward though lengthy arguments. Note that (iv) uses notation defined in (iii). □

We call the successive buffers in $[k_l, k_l')$, $l = 1, \ldots, L$, the L flow sections. In each time interval (t_{n-1}, t_n) the solution consists of L flow sections $[k_l, k_l')$, through which there is a constant flow U_l, moving fluid from nonempty buffer k_l to nonempty buffer k'_l, through empty buffers $k_l < j < k'_l$. All other buffers have no flows through them, and their contents are unchanged. The L flow sections fully utilize the stations $i \in I_1$, while some of the remaining stations, $i \in I_2$, may have slack capacities $z_i > 0$. We can also show that:

Proposition 2.2 *In an optimal basic solution:*

(i) *So long as $l_K(t) > 0$ the flow out of the last nonempty buffer is positive.*

(ii) *The flow out of the system $U_L(t) = u_K(t)$ is non-increasing in t.*

2.2   The dual problem and complementary slackness

The dual problem to (1.1–1.5) is formulated in reversed time (i.e. t of the dual problem corresponds to $T - t$ of the primal problem) as:

$$\text{minimize} \sum_{k=1}^{K} a_k P_k(T) + \int_0^T \sum_{i=1}^{I} r_i(t)\, dt \qquad (2.4)$$

subject, for all $t \in [0, T]$, to:

$$m_k r_{\sigma(k)}(t) \geq P_{k+1}(t) - P_k(t), \qquad k = 1, \ldots, K-1 \qquad (2.5)$$
$$m_K r_{\sigma(K)}(t) \geq t - P_K(t), \qquad\qquad\qquad\qquad (2.6)$$
$$P_k(0) = 0, \qquad P_k(t) \text{ nondecreasing}, \quad k = 1, \ldots, K \qquad (2.7)$$
$$r_i(t) \geq 0 \text{ a.e.} \qquad i = 1, \ldots, I. \qquad\qquad\qquad (2.8)$$

Let $q_k(t)$ denote the slack variables for the dual constraints (2.5, 2.6). Complementary slackness is defined to hold if almost everywhere $0 < t < T$:

$$x_k(t) > 0 \quad \Rightarrow \quad P_k(T - t) \text{ is nonincreasing at } T - t, \qquad (2.9)$$
$$z_i(t) > 0 \quad \Rightarrow \quad r_i(T - t) = 0, \qquad\qquad\qquad\qquad (2.10)$$
$$u_k(t) > 0 \quad \Rightarrow \quad q_k(T - t) = 0. \qquad\qquad\qquad\qquad (2.11)$$

We denote the complete primal solution by $w = (u, x, z)$, and the complete dual solution by $w^* = (r, P, q)$. Based on the results of Pullan (1996a) we can state:

Theorem 2.3 **(i)** *Strong duality holds between (1.1–1.5) and (2.4–2.8); that is, there exist optimal solutions to the two problems which achieve the same objective values.*

(ii) *There exists a pair of complementary slack optimal solutions. Conversely: if w is a feasible solution of the primal, w^* is a feasible solution of the dual, and if they are complementary slack, then w, w^* are optimal.*

At optimality $r_i(t)$ is the value rate of additional capacity at station i at time $T - t$, while $P_k(t)$ is the cost of additional fluid in buffer k over the time interval $(T - t, T)$.

2.3   Optimal solutions

We now consider some of the implications of complementary slackness. Let w, w^* be a pair of optimal, basic, complementary slack solutions. Let $t \in (t_{n-1}, t_n)$, and let $[k_l, k_l']$, $l = 1, \ldots, L$, be the flow sections where, by Proposition 2.2, $k_L' = K + 1$. Let M be the reduced matrix defined by (2.2) and let I_1, I_2 be the partition of I. Make the nondegeneracy assumption that $z_i(t) > 0$, $i \in I_2$. (This can always be assured by perturbing the m_k infinitesimally, as usual in linear programming contexts. It is of no practical consequence.) By complementary slackness $r_i(T - t) = 0$ for $i \in I_2$. Let $\bar{M}$ be the matrix obtained from the $L = |I_1|$ rows $i \in I_1$ of M. The values of the flows are determined (see 2.3) by:

$$\bar{M} \begin{bmatrix} U_1 \\ \vdots \\ U_L \end{bmatrix} = \begin{bmatrix} 1 \\ \vdots \\ 1 \end{bmatrix} \qquad (2.12)$$

Proposition 2.4 *The values of $r_i(t)$, $i \in I_1$, are linear functions of t in the time interval $(T - t_n, T - t_{n-1})$, and their slopes $\dot{r}_i$ are determined by:*

$$\bar{M}^T \begin{bmatrix} \dot{r}_{i_1} \\ \vdots \\ \dot{r}_{i_{L-1}} \\ \dot{r}_{i_L} \end{bmatrix} = \begin{bmatrix} 0 \\ \vdots \\ 0 \\ 1 \end{bmatrix}. \qquad (2.13)$$

We now list some further properties of optimal solutions. These properties are not used directly in the solution of the examples in Section 3.

Theorem 2.5 *There exists a pair of basic, optimal, complementary slack solutions to the primal and dual problems such that:*

(i) *The dual variables $r_i(t)$, $i = 1, \ldots, I$, and $P_k(t)$, $k = 1, \ldots, K$, are continuous.*

(ii)

$$t \geq P_K(t), \qquad P_{k+1}(t) \geq P_k(t), \qquad k = 1, \ldots, K - 1. \qquad (2.14)$$

(iii) *If $l_K(t) > 0, 0 < t < T$, then in addition to (2.14), $P_1(t) = 0$, $0 < t < T$.*

(iv) *If $T = \inf\{t : \sum_{k=1}^{K} x_k(t) = 0\}$, then in addition to (i) and (ii), the following holds for almost all $0 < t < T$:*

$$\sum_{k=1}^{K} x_k(T - t) = \sum_{k=1}^{K-1} u_k(T - t)(P_{k+1}(t) - P_k(t)) + u_K(T - t)(t - P_K(t))$$

$$= \sum_{i=1}^{I} r_i(t). \qquad (2.15)$$

The result (i) is quite recent and is due to Pullan (1996b).

3 Some solved examples

Each of the following examples illustrates an additional feature of the optimal solutions for this problem, which emerges as the examples increase in complexity. For each example we describe the solution by listing the time intervals and describing the flow sections in each of the intervals. The actual flow values, $U_1, \ldots, U_L$, and the slopes of the dual variables $\dot{r}_i$, $\dot{P}_k$, are then easily calculated. Also, equations to obtain the interval endpoints can then be given. This is a full description of the pair of primal and dual solutions. It is easy in each case to verify that the solutions are primal and dual feasible and complementary slack, hence optimal. We give the complete solution only for some of the examples. The interested reader should have no difficulty in completing the other examples.

3.1 Single station and 2 stations with no bottlenecks

It was shown in Weiss (1995) that (trivially):

Proposition 3.1 *The LBFS buffer priority policy will minimize the inventory* $l_K(t)$ *of a single-station re-entrant line at every t (such a solution which minimizes the integrand of the objective function for every point in the range is called pathwise optimal).*

We can also show:

Proposition 3.2 *For any k, let $m_k^{(+)} = \sum_{j \in C_{\sigma(k)}, j \geq k} m_j$. If $m_k^{(+)}$ is non-increasing in k, and if $I = 2$, then LBFS is pathwise optimal.*

Remark If $m_k^{(+)} > m_{k-1}^{(+)}$ then k is a potential bottleneck. However, it is not clear in general what constitutes a bottleneck in a re-entrant line. In particular we note that Proposition 3.2 is false if $I > 2$.

3.2 The 3-buffer 2-station re-entrant line

The simplest nontrivial re-entrant line has 2 stations, 3 buffers, with $C_1 = \{1, 3\}$, $C_2 = \{2\}$, and with $m_2 > m_1 + m_3$. It is our first example in which the optimal solution does not minimize the inventory at every time point. Flow out of buffer 3 needs to slow down so that fluid can move from buffer 1 through buffer 2, to start utilizing buffer 2 early, because eventually the flow out of the system is limited by buffer 2, with machine 1 only partly utilized.

The exact optimal policy for this line is, as we shall presently see, to use LBFS while $x_2(t) > 0$. Once $x_2(t) = 0$, continue to use LBFS for as long as $\frac{x_3(t)}{x_1(t)} > \frac{m_2 - m_1 - m_3}{m_1 + m_3}$. When the ratio $\frac{x_3(t)}{x_1(t)}$ falls lower start a flow from buffer 1 through 2 to fully utilize station 2, and use the remaining capacity of station 1 for buffer 3. Once $x_2(t) = x_3(t) = 0$ use flow rate $u_1 = u_2 = u_3 = \frac{1}{m_2}$, until the system is empty.

For the rest of this section we take $a_2 = 0$ for simplicity. The optimal solution can be constructed geometrically as follows: draw horizontal and vertical lines meeting at the origin O. Put points A, B on the vertical line, with lengths $OB = a_1$, $BA = a_3$. Draw line AC with slope $\frac{1}{m_3}$. Draw horizontal line from B to meet AC at D. Draw line DE with slope $\frac{1}{m_1 + m_3}$. Draw a vertical line from

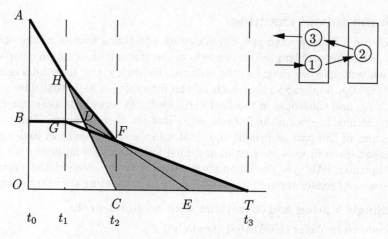

FIG. 1. Details of the geometric construction

C to meet DE at F. Draw a line of slope $\frac{1}{m_2}$ through F, to meet BD at G and the continuation of OE at T. Draw a vertical line from G to meet AC at H. Connect HF. Figure 1 describes the construction.

In the solution constructed in Fig. 1, the flows are piecewise constant over 3 time intervals, with flow sections:

$$
\begin{array}{lll}
(t_0, t_1) & [3, \text{out}] & z_2 > 0 \\
(t_1, t_2) & [1, 3] \text{ and } [3, \text{out}] & z_1 = z_2 = 0 \\
(t_2, t_3) & [1, \text{out}] & z_1 > 0
\end{array}
$$

To derive the actual values of the flows, consider for example the time interval (t_1, t_2). We construct first the matrix M for this time interval, from the matrix $\mathbf{m}$ (see 2.2):

$$
\mathbf{m} = \begin{array}{c|ccc}
 & 1 & 2 & 3 \\
\hline
1 & m_1 & 0 & m_3 \\
2 & 0 & m_2 & 0
\end{array}.
$$

We need to collect the first 2 columns for the flow section $[1, 3)$, and column 3 for the flow section $[3, \text{out})$:

$$
M = \begin{bmatrix} m_1 & m_3 \\ m_2 & 0 \end{bmatrix}.
$$

The equations for the values of U_1, U_2 and of $\dot{r}_1, \dot{r}_2$ are (see 2.3, 2.12, 2.13):

$$
\begin{bmatrix} m_1 & m_3 \\ m_2 & 0 \end{bmatrix} \begin{bmatrix} U_1 \\ U_2 \end{bmatrix} = \begin{bmatrix} 1 \\ 1 \end{bmatrix} \qquad \begin{bmatrix} m_1 & m_2 \\ m_3 & 0 \end{bmatrix} \begin{bmatrix} \dot{r}_1 \\ \dot{r}_2 \end{bmatrix} = \begin{bmatrix} 0 \\ 1 \end{bmatrix}.
$$

Similar (easier) equations define the values for the other time intervals. In summary one obtains, for the 3 time intervals:

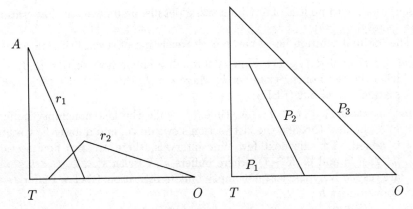

FIG. 2. The dual solution for the 3-buffer 2-stations line

	(t_0, t_1)	(t_1, t_2)	(t_2, t_3)
u	$(0, 0, \frac{1}{m_3})$	$(\frac{1}{m_2}, \frac{1}{m_2}, \frac{1}{m_3}(1 - \frac{m_1}{m_2}))$	$(\frac{1}{m_2}, \frac{1}{m_2}, \frac{1}{m_2})$
z	$(0, 1)$	$(0, 0)$	$(1 - \frac{m_1 + m_3}{m_2}, 0)$
$\dot{r}$	$(\frac{1}{m_3}, 0)$	$(\frac{1}{m_3}, -\frac{m_1}{m_2 m_3})$	$(0, \frac{1}{m_2})$
$\dot{P}$	$(0, 0, 0)$	$(0, \frac{m_1}{m_3}, 0)$	$(0, 0, 1)$

The dual solution is described in Fig. 2. Once we have all the flow rates of the primal problem and all the slopes for the dual variables, it remains to determine the time breakpoints t_1, t_2, t_3. The events that happen at these points are: at t_1 the value of $r_2(T - t)$ reaches 0; at t_2 buffer 3 empties; at t_3 the whole system empties. Equations for the time intervals $\tau_n = t_n - t_{n-1}, n = 1, 2, 3$, are:

$$\frac{1}{m_3}\tau_1 + \left(\frac{m_2 - m_1}{m_2 m_3} - \frac{1}{m_2}\right)\tau_2 = a_3, \quad \text{empty buffer 3,}$$
$$\frac{1}{m_2}(\tau_2 + \tau_3) = a_1, \quad \text{empty buffer 1,}$$
$$-\frac{m_1}{m_2 m_3}\tau_2 + \frac{1}{m_2}\tau_3 = 0, \quad \text{reach } r_2(T - t_1) = 0,$$

from which we obtain:

$$\tau_1 = m_3\left(a_3 - \frac{m_2 - m_1 - m_3}{m_1 + m_3}a_1\right), \quad \tau_2 = m_2\frac{m_3}{m_1 + m_3}a_1, \quad \tau_3 = m_2\frac{m_1}{m_1 + m_3}a_1.$$

This completes the construction of a pair of complementary, slack feasible solutions and proves the optimality.

We remark that the shaded area in Fig. 1 has height r_2, thus confirming the result of Theorem 2.5.

3.3 A re-entrant line with a multibuffer and a single-buffer station

An easy generalization of the previous example has $I = 2$ stations and K buffers, $C_1 = \{1, \ldots, L-1, L+1, \ldots, K\}$, $C_2 = \{L\}$. In the nontrivial case buffer L is a 'bottleneck', i.e., $m_L > \sum_{k=L+1}^{K} m_k + m_{L-1}$. On the other hand, to illustrate some new features, assume $m_L < \sum_{k=L+1}^{K} m_k + \sum_{k=1}^{L-1} m_k$. To simplify the presentation, with no loss of any interesting details, assume again that station 2 starts empty, $a_L = 0$.

The optimal solution has 4 parts, each consisting of several time intervals.

Time intervals $0 < t < T_1$: Flow is through a single flow section $[k, K + 1)$, from the last nonempty buffer k, where $k > L$, through buffers $k, \ldots, K$ of station 1, and out (LBFS).

Time intervals $T_1 < t < T_2$: At time T_1, while the last nonempty buffer is $k^{(1)}$, the flow through the last section slows down, and a lower flow section is added. For the next few time intervals, there are two flow sections, $[k', L + 1)$ and $[k, K + 1)$, where buffers k', k (with $k' < L$, $L < k \le k^{(1)}$) are respectively the last nonempty buffers of station 1 below buffer L and above buffer L.

Time intervals $T_2 < t < T_3$: At time T_2 buffer $L + 1$ empties—at this point the last nonempty buffer is buffer $k^{(2)} < L$, and we still have: $m_L > \sum_{j=L+1}^{K} m_j + \sum_{j=k^{(2)}}^{L-1} m_j$. In the following time intervals, buffer L is nonempty—it starts to fill up at time T_2. There are again two flow sections, $[k', L)$ and $[L, K + 1)$, where $k' \le k^{(2)}$ is the last nonempty buffer of station 1.

Time intervals $T_3 < t < T$: At time T_3 buffer L is again empty, and at that time the last nonempty buffer is $k^{(3)} < k^{(2)}$, where by now $m_L < \sum_{j=L+1}^{K} m_j + \sum_{j=k^{(3)}}^{L-1} m_j$. From time T_3 until the system empties at time T there is one flow section, $[k', K + 1)$, out of the last nonempty buffer $k' \le k^{(3)}$.

This example illustrated in a little more detail the role of a bottleneck. Here station 2 was a bottleneck station, and buffer L a single bottleneck buffer. Some vagueness remains as to whether it is the station or the buffer which constitutes the bottleneck—we have no answer to that. Note that buffer L fills up during the time intervals (T_2, T_3).

3.4   Re-entrant lines with 2 stations and 5 buffers

The next generalization is a 2-station 5-buffer line with $C_1 = \{1, 3, 5\}$, $C_2 = \{2, 4\}$, where buffer 2 is a bottleneck buffer, in the sense that:

$$m_5 < m_4 < m_5 + m_3 < m_5 + m_3 + m_1 < m_4 + m_2,$$

and initial state:

$$a_1, a_3, a_5 > 0, \qquad a_2 = a_4 = 0.$$

There are two cases to consider. The first case, when $\frac{m_2}{m_1} > \frac{m_4}{m_3}$, is a straightforward generalization of the previous examples, LBFS is used for a while, and then flow from buffer 5 slows down and a flow from buffer 1 through buffer 2, to fully utilize station 2, is added.

The case $\frac{m_2}{m_1} < \frac{m_4}{m_3}$ has a new feature: when the flow out of buffer 5 is slowed down, the new flow section is from buffer 3 through buffer 4, at rate $1/m_4$, to fully utilize station 2. Flow through the bottleneck buffer 2 starts only when buffer 3 is empty. This again illustrates the difficulty of defining a bottleneck.

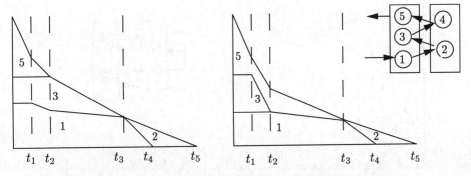

FIG. 3. Draining a 2-station 5-buffer line

Figure 3 illustrates these two cases.

3.5 Re-entrant lines with 2 stations and 6 buffers

In our last example of a 2-station re-entrant line we encounter 2 interfering bottlenecks. We have $C_1 = \{1,3,4,6\}$, $C_2 = \{2,5\}$, where both buffers 2 and 5 are bottlenecks, in the sense that:

$$m_6 < m_6 + m_4 < m_5 < m_6 + m_4 + m_3 < m_6 + m_4 + m_3 + m_1 < m_5 + m_2,$$

and $a_2 = a_4 = 0$ (station 2 initially empty). The interesting feature in this example is how the two bottlenecks interfere.

We will consider mainly the case $m_2/m_1 > m_5/m_4$, illustrated in Fig. 4. Here flow starts out of buffer 6, with station 2 not utilized. Later, in time interval (t_2, t_3) flow out of buffer 4 through buffer 5 is at rate $1/m_5$, while flow out of buffer 6 is slowed down; flow out of buffer 5 continues in the time intervals (t_3, t_5) at the same rate until buffer 5 is empty at time t_5. The new feature is the flow out of buffer 1 through buffer 2 in the time interval (t_1, t_2): Together with a similar flow in (t_5, t_6) it moderates the starvation period (t_6, t_8). The flow sections and the flow values are described in the following table, where $D = m_2(m_3 + m_4 + m_6) - m_1 m_5$:

Sections	(t_0, t_1)	(t_1, t_2)	(t_2, t_3)	(t_3, t_4)
	$[6, \text{out})$	$[1,3), [6, \text{out})$	$[4,6), [6, \text{out})$	$[4,5), [5, \text{out})$
U_2	$\frac{1}{m_6}$	$\frac{m_2 - m_1}{m_2 m_6}$	$\frac{m_5 - m_4}{m_5 m_6}$	$\frac{1}{m_5}$
U_1		$\frac{1}{m_2}$	$\frac{1}{m_5}$	$\frac{m_5 - m_6}{m_4 m_5}$
z	$0,1$	$0,0$	$0,0$	$0,0$
$\dot r_1$	$\frac{1}{m_6}$	$\frac{1}{m_6}$	$\frac{1}{m_6}$	0
$\dot r_2$	$r_2 = 0$	$-\frac{m_1}{m_2 m_6}$	$-\frac{m_4}{m_5 m_6}$	$\frac{1}{m_5}$

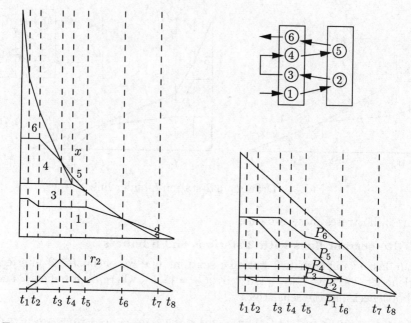

FIG. 4. A 2-station 6-buffer line with interference of the two bottlenecks

	(t_4, t_5)	(t_5, t_6)	(t_6, t_7)	(t_7, t_8)
Sections	$[3, 5), [5, \text{out})$	$[1, 3), [3, \text{out})$	$[1, 2), [2, \text{out})$	$[2, \text{out})$
U_2	$\frac{1}{m_5}$	$\frac{m_2 - m_1}{D}$	$\frac{1}{m_2 + m_5}$	$\frac{1}{m_2 + m_5}$
U_1	$\frac{m_5 - m_6}{(m_3 + m_4)m_5}$	$\frac{m_3 + m_4 + m_6 - m_5}{D}$	$\frac{m_2 + m_5 - m_3 - m_4 - m_6}{m_1(m_2 + m_5)}$	
z	$0, 0$	$0, 0$	$0, 0$	$> 0, 0$
$\dot{r}_1$	0	$\frac{m_2}{D}$	0	$r_1 = 0$
$\dot{r}_2$	$\frac{1}{m_5}$	$-\frac{m_1}{D}$	$\frac{1}{m_2 + m_5}$	$\frac{1}{m_2 + m_5}$

The behavior of r is of interest here. $r_1(t)$ (not shown in Fig. 4) increases for $T - t_6 < t < T$. r_2 is decreasing when the flow out of the system slows down and is not LBFS. Figure 4 also shows the dual variables P. It can be checked that the P satisfy the dual constraints and complementary slackness.

The time intervals $\tau_n = t_n - t_{n-1}$ are determined from the equations:

$$\frac{1}{m_6}\tau_1 + \frac{m_2 - m_1}{m_6}\tau_2 + \left(\frac{m_5 - m_4}{m_5 m_6} - \frac{1}{m_5}\right)\tau_3 = a_6, \qquad \text{empty buffer 6,}$$

$$\frac{1}{m_5}\tau_3 + \left(\frac{m_5 - m_6}{m_4 m_6} - \frac{1}{m_5}\right)\tau_4 = a_4, \qquad \text{empty buffer 4,}$$

$$\left(\frac{1}{m_5} - \frac{m_5 - m_6}{m_4 m_6}\right)\tau_4 + \left(\frac{1}{m_5} - \frac{m_5 - m_6}{(m_3 + m_4)m_5}\right)\tau_5 = 0, \qquad \text{empty buffer 5,}$$

$$-\frac{1}{m_2}\tau_2 + \frac{m_5 - m_6}{(m_3 + m_4)m_5}\tau_5 + \frac{m_2 + m_5 - m_1 - m_3 - m_4 - m_6}{D}\tau_6 = a_3, \qquad \text{empty buffer 3,}$$

$$\frac{1}{m_2}\tau_2 + \frac{m_3 + m_4 + m_6 - m_5}{D}\tau_6 + \frac{1}{m_2 + m_5}(\tau_7 + \tau_8) = a_1, \qquad \text{empty buffers 1,2,}$$

$$-\frac{m_4}{m_5 m_6}\tau_3 + \frac{1}{m_5}(\tau_4 + \tau_5) = 0, \qquad r_2(T - t_2) = r_2(T - t_5),$$

$$-\frac{m_1}{m_2 m_6}\tau_2 - \frac{m_1}{D}\tau_6 + \frac{1}{m_2 + m_5}(\tau_7 + \tau_8) = 0, \qquad r_2(T - t_1) = 0.$$

For the picture to be as described here, one needs to assume that a_1 is large enough (compared to a_3, a_4), so that the 2 buffers interfere, and that a_6 is large enough so that there will be an initial time interval with flow out of buffer 6 only.

The cases $m_5/m_4 > m_2/m_1 > m_5/(m_3 + m_4)$ and $m_5/(m_3 + m_4) > m_2/m_1$ are different: instead of time interval (t_1, t_2) with flow from buffer 1 through buffer 2, flow from buffer 4 through buffer 5 starts earlier, so the interval (t_2, t_3) lasts longer. We skip the remaining details.

3.6 3 stations, 4 buffers, 1 hidden bottleneck

Here $C_1 = \{4\}, C_2 = \{1, 3\}, C_3 = \{2\}$, with buffer 2 a bottleneck in the sense that $m_4 < m_1 + m_3 < m_2$; initially $a_2 = 0$, $a_1, a_3, a_4 > 0$. Buffer 2 is a hidden bottleneck; whether it becomes a bottleneck depends on the fluid level in buffer 1. Figure 5 shows the optimal solution for 3 different levels of buffer 1. The solution for the case when a_1 is large enough and 2 becomes a bottleneck is given in the following table.

Sections	(t_0, t_1) $[3,4), [4, \text{out})$	(t_1, t_2) $[1,3), [3,4), [4, \text{out})$	(t_2, t_3) $[1,3), [3, \text{out})$	(t_3, t_4) $[1, \text{out})$
U_3	$\frac{1}{m_4}$	$\frac{1}{m_4}$	$\frac{1}{m_3}(1 - \frac{m_1}{m_2})$	$\frac{1}{m_2}$
U_2	$\frac{1}{m_3}$	$\frac{1}{m_3}(1 - \frac{m_1}{m_2})$	$\frac{1}{m_2}$	0
U_1	0	$\frac{1}{m_2}$	0	0
z	$0, 0, 1$	$0, 0, 0$	$> 0, 0, 0$	$> 0, > 0, 0$
$\dot{r}_1$	$\frac{1}{m_4}$	$\frac{1}{m_4}$	$r_1 = 0$	$r_1 = 0$
$\dot{r}_2$	0	0	$\frac{1}{m_3}$	$r_2 = 0$
$\dot{r}_3$	$r_3 = 0$	0	$-\frac{m_1}{m_2 m_3}$	$\frac{1}{m_2}$

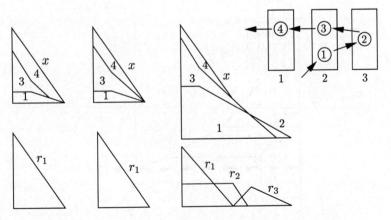

FIG. 5. A 3-station line with a hidden bottleneck

The equations to determine the interval lengths are:

$$(\tfrac{1}{m_4} - \tfrac{1}{m_3})\tau_1 + (\tfrac{1}{m_4} - \tfrac{1}{m_3}(1 - \tfrac{m_1}{m_2}))\tau_2 = a_4, \quad \text{empty buffer 4,}$$

$$\tfrac{1}{m_3}\tau_1 + (\tfrac{1}{m_3}(1 - \tfrac{m_1}{m_2}) - \tfrac{1}{m_2})(\tau_2 + \tau_3) = a_3, \quad \text{empty buffer 3,}$$

$$\tfrac{1}{m_2}(\tau_2 + \tau_3 + \tau_4) = a_1, \quad \text{empty buffer 2,}$$

$$-\tfrac{m_1}{m_2 m_3}\tau_3 + \tfrac{1}{m_2}\tau_4 = 0, \quad r_3(T - t_2) = 0.$$

4   Proposal for a general algorithm

The complete problem calls for optimal draining of $A = \sum_{j=1}^{K} a_j$ fluid. Assume for some k^* and some value $a < a_{k^*}$ that we know how to solve the problem with:

$$\tilde{a}_j = \begin{cases} 0 & j < k^*, \\ a & j = k^*, \\ a_j & j > k^*. \end{cases}$$

Let $Y = a + \sum_{j=k^*+1}^{K} a_j = \sum_{j=1}^{K} \tilde{a}_j$. The idea of the algorithm is to allow Y to go from 0 to A, in a number of steps, keeping an optimal solution at each step. Equivalently, let $\tilde{T}$ be the emptying time for a system with initial state $\tilde{a}$; allow $\tilde{T}$ to grow from 0 to T. The procedure is described schematically in Fig. 6.

We outline this in a little more detail. Consider the solution for a given $Y, \tilde{T}$. Let the time breakpoints be $0 = t_0 < \cdots < t_N = T$, with interval lengths $\tau_n = t_n - t_{n-1}$. In an optimal solution there are some events which define the breakpoints, which are of two forms—either a buffer empties, or the flow from a buffer slows down and a new flow is initiated from a lower buffer of the same station. These events define a set of linear equations for the interval lengths, of the form:

$$A\tau = b$$

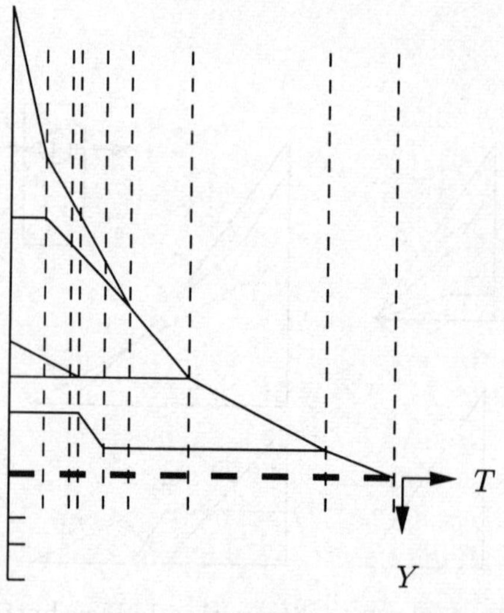

FIG. 6. Schematic description of an algorithm

where the elements of $\mathcal{A}$ are rational functions of the m_k, and the elements of b are either the a_k or 0.

Assume all $\tau_n > 0$. Then increasing a will change the τ linearly, while the solution remains optimal. We thus can increase a to extend the solution to a larger Y, until one of the following happens:

a reaches the value of a_k^*: If $k^* = 1$, the solution is complete. Else, decrease k^* by 1, and set $a = 0$.

A certain τ_n reaches 0: Eliminate the interval (t_{n-1}, t_n).

In either case, new intervals need to be introduced and the equations for τ modified.

Acknowledgements I have discussed this problem with many colleagues in three continents, including Florin Avram, Eddie Anderson, Malcolm Pullan, Jim Dai, Frank Kelly, Bruce Hajek, Dimitri Bertsimas, Celia Glass, Michal Penn, Shmuel Gal, Ellis Johnson, Nimrod Megiddo and Rolf Möhring. Much of the modest progress I made on this fascinating problem is due to their generous comments and suggestions.

Bibliography

1. Anderson, E. J. (1981). A new continuous model for job-shop scheduling. *International Journal of Systems Science* **12**, 1469–1475.

2. Anderson, E. J. and Nash, P. (1987). *Linear Programming in Infinite Dimensional Spaces*. Wiley-Interscience, Chichester.

3. Anderson, E. J. and Pullan, M. C. (1995). Purification for separated continuous linear programs. *Mathematical Methods of Operations Research, formerly ZOR* **43**, 9–33.

4. Avram, F., Bertsimas, D. and Ricard, M. (1995). Fluid models of sequencing problems in open queueing networks; an optimal control approach. In *Stochastic Networks—IMA Volumes in Mathematics and its Applications*, Editors: F.P. Kelly and R. J. Williams, pp. 199-234. Springer, New York.

5. Bramson, M. (1994). Instability of FIFO queueing networks. *Annals of Applied Probability* **4**, 414–431.

6. Chen, H. and Yao, D. (1993). Dynamic scheduling of a multi class fluid network. *Operations Research* **41**, 1104–1115.

7. Dai, J. G. (1995). On positive Harris recurrence for multiclass queueing networks: a unified approach via fluid limit models. *Annals of Applied Probability* **5**, 49–77.

8. Dai, J. G. and Vande-Vate, J. (1995). The stability regions for two station queueing networks. Preprint.

9. Dai, J. G. and Weiss, G. (1996). Stability and instability of fluid models for re-entrant lines. *Mathematics of Operations Research* **21**, 115–134.

10. Kumar, P. R. (1993). Re-entrant lines. *Queueing Systems: Theory and Applications: Special issue on queueing networks* **13**, 87–110.

11. Kumar, P. R. (1994). Scheduling manufacturing systems of re-entrant lines. Chapter 8 in *Stochastic Modeling and Analysis of Manufacturing Systems*, Editor: David D. Yao, pp. 325–360. Springer-Verlag, New York.

12. Kumar, P. R. (1995). Scheduling queueing networks: stability, performance analysis and design. In *Stochastic Networks—IMA Volumes in Mathematics and its Applications*, Editors: F.P. Kelly and Ruth Williams, pp. 21–70. Springer-Verlag, New York.

13. Lu, S. H. and Kumar, P. R. (1991). Distributed scheduling based on due dates and buffer priorities. *IEEE Transactions on Automatic Control* **36**, 1406–1416.

14. Pullan, M. C. (1993). An algorithm for a class of continuous linear programs. *SIAM Journal of Control and Optimization* **31**, 1558–1577.

15. Pullan, M. C. (1995a). Forms of optimal solutions for separated continuous linear programs. *SIAM Journal of Control and Optimization* **33**, 1952–1977.

16. Pullan, M. C. (1995b). Convergence of a general class of algorithms for separated continuous linear programs. Cambridge University Report CUED/E-MS/TR.

17. Pullan, M. C. (1996a). A duality theory for separated continuous linear programs. *SIAM Journal of Control and Optimization* **34**.

18. Pullan, M. C. (1996b). Existence and duality theory for separated continuous linear programs. Preprint, submitted to *Mathematical Modelling of Systems*.

19. Rybko, A. N. and Stolyar, A. L. (1992). Ergodicity of stochastic processes describing the operation of open queueing networks. *Problemy Peredachi Informatsii* **28**, 3–26.

20. Wein, L. M. (1988). Scheduling semiconductor wafer fabrication. *IEEE Transactions on Semiconductor Manufacturing* **1**, 115–130.

21. Weiss, G. (1995). Optimal draining of a fluid re-entrant line. In *Stochastic Networks—IMA Volumes in Mathematics and its Applications*, Editors: F.P. Kelly and Ruth Williams, pp. 93–105. Springer-Verlag, New York.

3

On the approximation of queueing networks in heavy traffic

R. J. Williams

University of California, San Diego

Abstract

This paper reports on the current state of research concerning approximations to the queue length or workload processes in stable but heavily loaded queueing networks. Most of the approximations that have been considered to date are justified by heavy traffic limit theorems that have a certain *conventional* form, namely (i) the queue length or workload processes in a sequence of heavily loaded networks are normalized by a central limit theorem type of scaling, and (ii) the weak limit of these normalized processes is a reflecting Brownian motion. Such conventional limit theorems for both open and closed networks are surveyed first, with a view to outlining their general form. Whilst there is a reasonable collection of limit theorems for single class networks and some multiclass networks (especially those with feedforward structure), recent developments have indicated that not all multiclass networks with feedback have conventional heavy traffic behavior. These developments are surveyed here and in particular, the Harrison–Williams example of an unconventional heavy traffic limit theorem for a closed analog of the Lu–Kumar/Rybko–Stolyar priority network is described.

1 Introduction

Apart from product form networks (see [41, 42, 1, 45]), most queueing networks have not proven amenable to exact analysis. One natural method for analyzing such networks is to *approximate* them by more tractable entities. Heavily loaded networks, where congestion is a compelling problem, are of particular interest. Under certain conditions [39, 40, 26, 66, 76, 43, 62, 67, 17, 13, 10, 59, 60] it has been shown that one can approximate a normalized version of the queue length or workload process in a heavily loaded queueing network by a diffusion process called a reflecting Brownian motion which lives in a convex polyhedron. Over the last fifteen to twenty years, a substantial theory has been developed for these diffusion processes (see [80] for a recent survey), and in general they are more tractable than the original queueing networks. A further attractive feature of

Research supported in part by NSF Grant GER 9023335.

this approximation is that the diffusion characteristics are determined from just first and second moment information of the arrival and service processes for the network, together with a Markovian routing matrix and the service discipline at each station. Thus the approximation does not depend on satisfying stringent distributional assumptions such as those required for product form networks, and even though the approximating diffusion process is Markovian, the original queue length or workload process need not satisfy this assumption. An aim of analysis of these approximations is to better understand the relationship between network parameters (e.g., interarrival and service time variances, service disciplines, routing matrices) and measures of congestion and delay (e.g., long run average queue lengths and long run average idleness at each station). Such an analysis may not only provide approximate measures of performance for a heavily loaded queueing network, but play an integral role in solving problems of optimal design or control for queueing networks (see, e.g., [33, 74, 75, 46, 53] and references therein).

The limit theorems in [39, 40, 26, 66, 76, 43, 62, 67, 17, 13, 10, 59, 60] justify a diffusion approximation to the queue length or workload process in a heavily loaded network under various conditions. These limit theorems cover different kinds of networks including those that are open [39, 40, 26, 66, 76, 43, 62, 67, 17, 13, 59] (customers arrive from outside the system, receive a finite number of services at various stations and then exit the system), closed [10] (a fixed number of customers circulate perpetually in the system), or mixed [60] (a hybrid of open and closed systems). These networks may be further classified according to whether they are single class [39, 40, 26, 66, 43, 10, 59] (customers at a given station are essentially indistinguishable from one another), or multiclass [76, 43, 62, 67, 17, 13, 10, 60] (different classes of customers, perhaps having different service or routing characteristics, may be served at a station). For multiclass networks the mapping from classes to stations is many-to-one. Assumptions common to these limit theorems are that the arrival and service processes satisfy a functional central limit theorem (this holds for example if the interarrival and service times are given by independent sequences of i.i.d. random variables with finite first and second moments), and the routing is Markovian. Note that in the above, the term *customers* was used generically to denote those receiving service in a queueing network. One may equally well think of these as *jobs* in a manufacturing system or *packets* of data in a computer communication network, etc.

As a precursor to a discussion of conventional versus unconventional heavy traffic limit theorems, a synopsis of existing limit theory is given in the next section. The main aim of this description is to outline the *general form* that conventional limit theorems have taken. Accordingly, an exhaustive bibliography will not be given, but papers which illustrate the progression of the theory will be mentioned. For more on the history and other references, see the surveys of Whitt [77], Lemoine [54], Coffman and Reiman [14], Flores [24] and Glynn [25]. Recent developments on heavy traffic approximations for state- and/or time-dependent queueing networks will not be surveyed—for references see the introduction to Pats [61] or see Krichagina *et al.* [50] and Mandelbaum and Massey [56]. Furthermore, the limit theorems surveyed below are *weak* convergence theorems, i.e., normalized processes taking values in a certain path space converge *in distribution* to a limit process. Under some conditions one can prove strong approximation results which when applicable have certain theoretical and practical advantages. For more on the existence and application of strong approximations, see Horváth [38] and Chen and Mandelbaum [11]. Finally, interchangeability of the heavy traffic limit (as $n \to \infty$ here) and the limit as time tends to infinity will not be discussed, though such a result is needed if one wants to approximate the equilibrium distribution of a queueing network by the stationary distribution of its heavy traffic limit. For some work on this question in the context of closed networks, see Kaspi and Mandelbaum [44].

For simplicity, the following network assumptions will be implicitly assumed, unless stated otherwise. There is a single server at each station, the arrival process for each customer class is a renewal process for which the interarrival times have positive finite mean and variance, the service times for each class are given by a sequence of i.i.d. random variables having positive finite mean and variance, the routing is Markovian, and the arrival processes, service times and customer routing are mutually independent. For single class networks there is just one customer class per station, whereas for multiclass networks there may be several different classes of customers served at a single station (recall that in multiclass networks the mapping from classes to stations is many-to-one). The service discipline at a station specifies the order in which customers are served. This will be assumed to be work conserving, i.e., a server is never idle when there are customers waiting to be served at its station. For single class networks, in which all of the customers at a given station are indistinguishable, the egalitarian first-in-first-out (FIFO) service discipline is commonly used. In multiclass networks, all customers of the same class are treated similarly, but the service discipline may distinguish between classes, e.g., one class may be given priority over another. For the specific structure of the queueing networks considered below and details of the limit theorems, the reader is referred to the papers cited. Emphasis is placed below on the approximation of networks that are stable (mean queue lengths are bounded for all time), but in which each station is heavily loaded. For open networks this means that the limit will be taken through a sequence of stable networks in which the network parameters approach the boundary between stability and instability. (Networks in which some stations are lightly loaded whilst others are heavily loaded have also been treated in the literature, see, e.g., Reiman [65], Johnson [43] and Chen and Mandelbaum [10]. A major feature of these works is that under heavy traffic scaling the workloads and queue lengths for the lightly loaded stations vanish in the heavy traffic limit, and consequently the analysis reduces to that of a (smaller) network with all stations heavily loaded.)

2   Conventional heavy traffic limit theorems

Heavy traffic theory has been concerned largely with open networks and the
theory is most well developed for single class networks with FIFO service discipline. Heavy traffic analysis of single class FIFO networks began with the case
of a single station. Early work of Kingman [47, 48, 49], Prohorov [64], Borovkov
[4, 5], and others, emphasized approximation of steady state distributions or
finite dimensional distributions of waiting time and queue length processes for
such a station. A process level heavy traffic approximation was first clearly
elucidated in the work of Iglehart and Whitt [39, 40], who considered a single
station with multiple servers. Their work is generally regarded as the prototype
for conventional heavy traffic limit theorems. In particular, in [40], a sequence
of suitably normalized queue length processes is shown to converge weakly to
a one-dimensional reflecting Brownian motion as the traffic intensity converges
upwards to one. (The station is stable whenever the traffic intensity parameter is
less than one.) Iglehart and Whitt [40] note that their analysis of a single station
could be viewed as part of a network analysis, provided there is no feedback (i.e.,
the output from a station can never become part of its input). However, their
network limit process is given as a complicated function of multidimensional
Brownian motion. Following this work, Harrison [26] proved a heavy traffic limit
theorem for two stations in tandem and first identified a sample path representation for the two-dimensional limit process, which is a reflecting Brownian motion
living in the positive quadrant. His limit theorem was generalized by Reiman
[66] to single class FIFO networks with feedback. These networks are sometimes
referred to as generalized Jackson networks or non-parametric Jackson networks
because they are generalizations of the (product form) networks considered by
Jackson [41, 42] in the sense that the exponential distributions of the interarrival
and service times used by Jackson [41, 42] are replaced by general distributions
satisfying the finite first and second moment conditions described in the introduction above. In [66], the limit process is a reflecting Brownian motion living
in the positive orthant. A pathwise construction of this process, in particular,
the continuity of a reflection mapping defining the reflecting Brownian motion
from a driving Brownian motion, due to Harrison and Reiman [32], played a key
role in the proof of the weak convergence in [66].

For single class open queueing networks as considered by Reiman [66], heavy
traffic is quantified by requiring that the traffic intensity at each station be
(uniformly) close to one. More precisely, for such a network with d stations,
if α_i denotes the mean rate of arrival to station i of customers from outside
the system, $i = 1, \ldots, d$, and P_{ij}, $i, j = 1, \ldots, d$, denotes the probability that a
customer completing service at station i goes next to station j, then since P is
substochastic for an open queueing network, there is a unique (vector) solution
$\lambda = (I - P')^{-1}\alpha$ of the traffic or flow-balance equation:

$$\lambda = \alpha + P'\lambda, \tag{2.1}$$

where $'$ denotes transpose. (Strictly speaking, the element $P'\lambda$ in (2.1) should be replaced by $P'(\lambda \wedge \mu)$, where the ith component μ_i of μ is the mean service rate at station i and $(\lambda \wedge \mu)_i = \min(\lambda_i, \mu_i)$. This incorporates the fact that μ_i is the maximal mean rate at which customers can flow out of station i [57, 10]. However, only stable systems will be considered here in which for each i, $\lambda_i \leq \mu_i$ is automatically satisfied, and so it will not be necessary to consider a more complex form than (2.1).) The traffic intensity parameter ρ_i at station i is defined by

$$\rho_i = \lambda_i m_i, \quad i = 1, \ldots, d, \tag{2.2}$$

where $m_i = 1/\mu_i$ is the mean service time at station i. The open single class network is stable if $\rho_i < 1$ for $i = 1, \ldots, d$ (see, e.g., [71, 58]). Then, for each i, λ_i can be interpreted as the long run average number of customer visits to station i per unit time, and ρ_i can be interpreted as the long run average rate at which work (measured in units of required service time) arrives at station i. A simple version of the heavy traffic limit theorem proved by Reiman [66] is the following.

Consider a sequence of stable networks indexed by n, all with the same common structure except that the exogenous arrival rate vector $\alpha^{(n)}$ for the nth system tends to a value α in such a way that the traffic intensity vector $\rho^{(n)}$ for the nth system tends upwards to unity in the following manner:

$$\sqrt{n}(\rho_i^{(n)} - 1) \to c_i \quad \text{as} \quad n \to \infty, \tag{2.3}$$

where c_i is a finite (negative) constant and $i = 1, \ldots, d$. (The negativity of all of the c_i guarantees that the limiting diffusion process will be positive recurrent [34]. One may also consider some $c_i \geq 0$, to incorporate null recurrent and transient cases, but this will not be done here.) Normalize the d-dimensional queue length process $Q^{(n)}$ for the nth system using a central limit theorem (or diffusion) type of scaling:

$$\hat{Q}^{(n)}(\ \cdot \ ) = \frac{Q^{(n)}(n \ \cdot \ )}{\sqrt{n}}. \tag{2.4}$$

Now, $\hat{Q}^{(n)}$ takes values in the space of d-dimensional r.c.l.l. (right continuous with finite left limits) paths defined on $[0, \infty)$. When this space is endowed with the usual Skorokhod $\mathbf{J_1}$-topology [72, 23], then $\hat{Q}^{(n)}$ converges weakly as $n \to \infty$ to a reflecting Brownian motion Z that lives in the d-dimensional positive orthant and has a semimartingale decomposition of the form

$$Z = X + RY, \tag{2.5}$$

where X is a d-dimensional Brownian motion with constant drift and non-degenerate covariance matrix, $R = I - P'$ is called the reflection matrix, and Y is a d-dimensional continuous non-decreasing process such that Y_i can increase only when Z_i is zero (in fact, $m_i Y_i$ is the weak limit of the normalized cumulative idle time process for station i, where the normalization is the same central limit theorem type of scaling (2.4) used for the queue length processes). As pointed out in [62, 30], one is not restricted to convergence of the traffic intensity parameter at the rate indicated in (2.3). The essential aspect is that the queue length process is multiplied by a factor $(1/\sqrt{n}$ above) that measures the (uniform) proximity of the traffic intensity parameters to one, and time is divided by the square of this (small) parameter (i.e., multiplied by n in the above). One can also allow suitable variations in the initial queue lengths, arrival and service time variances, service rates, and Markovian switching matrices, as one passes through the sequence of networks. Such variations have not been described here to keep the illustration as simple as possible (see [66] for such refinements).

For multiclass networks it is natural to consider other service disciplines besides FIFO. The extant heavy traffic limit theorems for such networks have largely focussed on those with (static) priority service across classes and FIFO service within a priority class (by allowing all classes to be of the same priority, this includes the case of straight FIFO service across classes). Let us consider open multiclass networks. In such networks, the dimension of the queue length process is equal to the number of classes served in the system (recall that the mapping from classes to stations is many-to-one). Another process of interest, especially for multiclass networks, is the (immediate) workload process W whose dimension is equal to the number of stations and is such that $W_i(t)$ represents the amount of work (measured in units of service time) embodied in the customers at station i at time t. In a system with straight FIFO service discipline at each station, W also represents the virtual waiting time process. Whitt [76] proved heavy traffic limit theorems for a single multiclass station with preemptive resume priority service. Although he did not consider a sequence of stable systems with traffic intensities approaching one, he did consider a single station with traffic intensity equal to one and two priority classes. He showed for such a "saturated station" that the two-dimensional queue length process, normalized with a central limit theorem type of scaling, converges weakly to a process in which the component corresponding to the high priority class is identically zero and that corresponding to the low priority class is a one-dimensional driftless reflecting Brownian motion. Knowing the results of Reiman [66] for single class networks and Whitt [76] for a single station with priorities, it is natural to try to combine the features of these two situations in a single limit theorem for multiclass networks with priority service at each station. Johnson [43] was able to do this for multiclass networks having two types of customers, those of high priority and those of low priority, where a customer retains the same priority designation during its entire sojourn through the network, i.e., once a high priority customer, always a high priority customer etc. A network of this kind will be said to have *separated priorities*. (Johnson [43] also verified the results of Reiman [66] and developed a methodology which has proven useful in subsequent work [62, 10].)

Peterson [62] proved a heavy traffic limit theorem for multiclass networks with priority service (two levels, not separated), but he restricted to the case of feedforward routing. In this case, the traffic equations (2.1) still hold, but now α is the vector of *class* level arrival rates, P is the class level Markovian switching matrix, and for each class k, λ_k is interpreted as the long run average number of customer visits to class k per unit time. The feedforward structure is quantified by the fact that P is upper triangular. A traffic intensity parameter for each station $i = 1, \ldots, d$ is defined by

$$\rho_i = \sum_{k \in \mathcal{C}_i} \lambda_k m_k, \qquad (2.6)$$

where m_k denotes the mean service time for class k customers and $\mathcal{C}_i$ denotes the *constituency* of station i, i.e., it consists of those classes that are served at station i (note that $\{\mathcal{C}_i, i = 1, \ldots, d\}$ forms a partition of the customer classes for the network). The feedforward network is stable provided $\rho_i < 1$ for $i = 1, \ldots, d$. The Peterson analog of the version of Reiman's single class limit theorem sketched above can be described as follows. Assume (2.3) holds with the traffic intensity parameters as defined in (2.6) above, and suppose the low priority traffic intensities at all stations do not vanish in the limit as $n \to \infty$. (The low priority traffic intensity for station i is defined by (2.6), but with summation only over those indices k corresponding to the low priority classes at station i.) Then the d-dimensional workload process $W^{(n)}$, when normalized with the same central limit theorem type of scaling as in (2.4), i.e.,

$$\hat{W}^{(n)}(\cdot) = \frac{W^{(n)}(n \cdot)}{\sqrt{n}}, \qquad (2.7)$$

converges weakly to a reflecting Brownian motion Z that lives in the d-dimensional positive orthant and has a semimartingale decomposition of the form (2.5), with $R = (I + \Gamma)^{-1}$ where Γ is non-negative and lower triangular. The class level queue length processes, with the same normalization as for the workload processes, also converge weakly. As in the single station case [76], for each station, the limit for the high priority queue length processes is identically zero and for the low priority queue length processes is proportional to the limit of the normalized workload processes for the station. This phenomenon, that the limit of the workload processes determines the limit for the higher dimensional queue length processes, has been called *state space collapse*. Other examples of this phenomenon can be found in Reiman [65] and a generalized version of it can be seen in Coffman *et al.* [13]. In the latter paper, the authors prove a heavy traffic limit theorem for a single station with an *exhaustive polling service discipline*. In this system with a single station and two customer classes, the server alternates between the two classes and serves all customers of one class in FIFO order before switching instantaneously to serve the other class. The service times for the two classes are assumed to have the same distribution, but the distributions of the interarrival times may vary with the class. As in the Peterson limit theorem, the one-dimensional workload process normalized with the central limit theorem type of scaling (cf. (2.7)) converges weakly to a one-dimensional reflected Brownian motion as the traffic intensity parameter converges to one. However, limits of the *individual* normalized queue length processes do not exist in the usual sense, although they do exist in a time-averaged sense and these limits are recoverable from the limiting workload process. One may view this as a generalized form of state space collapse.

Although heavy traffic limit theorems have been proved for a single multiclass *station with feedback* and certain service disciplines such as round-robin [67] and FIFO [17], there is currently no general heavy traffic limit theorem for *multiclass networks with feedback*. (Chen and Zhang [12] have recently claimed a limit theorem for multiclass networks with feedback and FIFO service discipline, but they still require a restrictive spectral radius condition on one of the data matrices.)

Closed queueing networks are natural models for manufacturing systems with a dependable supply of raw materials. For instance, one may envisage a fixed number of palettes circulating perpetually in a production facility. One places raw materials on a palette at the beginning of processing, the palette carries the job through the facility during the entire production process, and the finished product is removed from the palette at the end of processing. The palette then returns to the beginning of processing for a new load of raw materials. Analogs of the open network heavy traffic limit theorems of Reiman [66] and Johnson [43] have been proved by Chen and Mandelbaum [10] for closed networks. In particular, they considered closed networks that are single class or have separated priorities. A version of their results is described below.

Closed networks are automatically stable and so in contrast to the open network case, it is not necessary to consider a sequence of network parameters approaching the boundary between stability and instability. Instead one may consider each station to be in "bottleneck" status, as described in the next paragraph. (One may also allow stations to approach bottleneck status through a sequence of networks, or even allow some stations to be near bottleneck status and others to only be lightly loaded. However, such refinements will not be considered here in order to keep the illustration simple—see Chen and Mandelbaum [10] for an extensive treatment.) Furthermore, for a closed network the heavy traffic parameter n has a natural interpretation as the fixed number of customers in the system. Indeed, a natural spatial scaling for the queue length process $Q^{(n)}$ is to divide it by n so that its normalized component $\frac{1}{n}Q_i^{(n)}$ represents the *fraction* of the total customer population that is in queue at station i (or in class i for a multiclass network). Given this natural spatial scaling, to achieve a central limit theorem type of scaling for $Q^{(n)}$ one must accelerate time by n^2 to obtain scaled queue length processes:

$$\hat{Q}^{(n)}(\cdot) = \frac{Q^{(n)}(n^2 \cdot)}{n}. \tag{2.8}$$

For a closed network there is no exogenous arrival process and no departure process. Hence α in the flow-balance equation (2.1) is zero and P is a stochastic matrix. Assuming P is irreducible, there is a one-parameter family of solutions λ for equation (2.1). Each station is said to be in bottleneck status if there is a positive solution λ of (2.1) such that for ρ_i as defined by (2.6), $\rho_i = 1$ for

$i = 1, \ldots, d$. The results of Chen and Mandelbaum [10] show that under this bottleneck condition, for a *single class* closed network with n customers in the system and a limiting distribution for the fraction of the total customer population initially at each station, the d-dimensional queue length process normalized as in (2.8) converges weakly as $n \to \infty$ to a reflecting Brownian motion that lives in the d-dimensional simplex (see Harrison *et al.* [37] for some analysis of this limit process). For a closed network with separated priorities of the two types high and low, P is not irreducible, but provided each station is in bottleneck status and the low priority traffic intensities at all stations do not vanish in the limit as $n \to \infty$, the low priority queue length process obeys a limit theorem similar to that for the single class case, and the normalized high priority queue length process converges to the (identically) zero process (here convergence may need to be taken in the space of r.c.l.l. paths on $(0, \infty)$ unless the initial fraction of high priority customers is assumed to converge to zero as $n \to \infty$). For further details, the reader is referred to Chen and Mandelbaum [10]. Despite the positive results outlined above, as in the open network case, there is currently no general heavy traffic limit theorem for multiclass closed queueing networks.

In addition to the work of Chen and Mandelbaum [10], Nguyen [59, 60] has proved heavy traffic limit theorems for some networks motivated by manufacturing considerations. In particular, fork-and-join networks [59] with feedforward structure have a heavy traffic limit which is a reflecting Brownian motion living in a non-simple polyhedral cone, where more than d-faces meet at the vertex of the cone in d-dimensions. She has also considered hybrid networks [60] that have some open and some closed aspects. These arise in modeling manufacturing systems with both make-to-order and make-to-stock requirements.

All of the above-mentioned heavy traffic limit theorems are of a so-called *conventional type*. This notion of conventionality, described by Harrison in [28], has two salient characteristics, namely (i) the queue length or workload processes in a sequence of heavily loaded networks are normalized by a central limit theorem type of scaling, and (ii) the weak limit of these normalized processes is a reflecting Brownian motion. In an effort to generalize the pattern perceived for open networks in [39, 40, 26, 66, 43, 62, 67], Harrison [27] and Harrison and Nguyen [29, 30] proposed a reflecting Brownian motion approximation (or an approximate "Brownian model") for open multiclass queueing networks with FIFO service discipline. An analogous approximation is proposed in the Appendix to Harrison and Williams [35], where, extrapolating from the work of Whitt [76] and Johnson [43], static priority and processor sharing service disciplines are included in addition to FIFO. One could similarly formulate approximate Brownian models for closed queueing networks.

The approximation procedures described in [27, 29, 30, 35] are formal in the sense that although conventional heavy traffic limit theorems that would justify these approximations are outlined there, except for the situations covered by [39, 40, 26, 66, 76, 43, 62, 67, 17], no heavy traffic limit theorem has been *proved* to justify these approximations. In contrast, there is a rigorous existence and uniqueness theory for the reflecting Brownian motions proposed as approximations. These have a semimartingale form as in (2.5) but with a more general reflection matrix R than the $I - P'$ associated with single class networks. Reiman and Williams [68] established a necessary condition for the existence of such a semimartingale reflecting Brownian motion (SRBM), namely that the reflection matrix R is completely-$\mathcal{S}$, i.e., for each principal submatrix $\tilde{R}$ of R there is a positive vector $\tilde{u}$ such that $\tilde{R}\tilde{u} > 0$. Taylor and Williams [73] established the sufficiency of this condition for weak existence and uniqueness, provided that X minus its drift process is a martingale relative to the filtration generated by X, Y, Z. For an extension of these results to convex polyhedral state spaces, which is relevant to closed network approximations, see Dai and Williams [22].

3 The Dai–Wang counterexample and stability of open multiclass queueing networks

In [20], Dai and Wang produced a surprising example which showed that not all open multiclass queueing networks can have an approximation as proposed in [27, 29, 30, 35]. Indeed the proposed approximation for the two-station FIFO example of Dai and Wang would have a reflection matrix with negative values on the diagonal, which is clearly not completely-$\mathcal{S}$. Further investigation by Dai and Nguyen [18] of a similar example showed that the normalized workload processes could not converge weakly to a continuous limit. Contemporaneous work of Whitt [79] for a deterministic network suggests that the non-convergence may be caused by large fluctuations in the workload processes.

Around the time that the Dai–Wang counterexample was produced, there was a separate growing interest in the stability of open multiclass networks. For single class queueing networks the condition for stability is that $\rho_i < 1$ for each station i, where the traffic intensity ρ_i is defined by (2.1)–(2.2), see Meyn and Down [58]. It is natural to conjecture that this is also the condition for stability of *multiclass* networks, provided ρ_i is now defined by (2.1), (2.6). Despite some cases [43, 62] where this is true, Kumar and coworkers [52, 55] have given simple two-station deterministic examples which show that the conjecture is false in general for multiclass networks with feedback, i.e., they gave examples in which $\rho_i < 1$ for each i, but the network is unstable. Rybko and Stolyar [69] gave the first stochastic counterexample, for a two-station network with priority service discipline. (This network has a similar structure to the Lu–Kumar example.) Whilst it might be argued that one can see (with hindsight) that the priorities in the Lu–Kumar [55] and Rybko–Stolyar [69] examples are "bad" for stability, it was a further surprise when Bramson [6, 7] gave a two-station stochastic counterexample having FIFO service discipline. (Also, Seidman [70] gave a deterministic FIFO counterexample.)

The appearance of the counterexamples described above was followed by an explosion of interest in finding conditions for the stability of open multiclass networks. A frequent tool in these investigations has been the use of mathematical programming to determine appropriate Lyapunov functions which prove sufficiency of certain conditions for stability. For example, Kumar and Meyn

[51] and Bertsimas *et al.* [2] used this approach for networks directly. A significant advance was made when Dai [15] showed that the stability of associated fluid limits (obtained as a limit of a Markovian state descriptor for the network under a law of large numbers type of scaling) was sufficient for stability of the original queueing network. This result has been exploited by a number of authors to prove sufficient conditions for the stability of open multiclass networks. For a recent survey of this work, in particular to see the use of piecewise linear Lyapunov functions in proving stability of fluid limits, see Dai [16]. Since the writing of [16], Bramson [8, 9] has proved the stability of open *Kelly-type* networks with FIFO service discipline and of open networks with a head-of-the-line proportional processor sharing service discipline (HLPPS), provided $\rho_i < 1$ for each i. Here *Kelly-type* means that the service rate is the same for all customers at a given station, i.e., m_k is the same for all $k \in \mathcal{C}_i$, and HLPPS means that service at each station is divided amongst the customers at the head-of-the-line of each class in proportion to the number of customers of that class present at the station.

In light of this work on stability, one might be tempted to conjecture that the Dai–Wang example fails to have a Brownian approximation because it is not stable, i.e., one has the wrong notion of heavy traffic ($\rho_i \approx 1$ for $i = 1, 2$) for this example. Simulations of Dai (private communication) suggest that the example is unstable. However, the example of Whitt [79], in which large oscillations of the queue length processes are exhibited, suggests that one might also entertain other possible explanations for the lack of a Brownian approximation, such as the wrong scaling, the wrong topology on path space, or the wrong limit process. For a similar example, Dai and Nguyen [18] showed that under the usual stability condition with conventional scaling the normalized workload processes cannot be tight with the Skorokhod $\mathbf{J_1}$-topology on path space. However, in general one might consider any of the possibilities mentioned above as explanations for why a multiclass network did not follow a conventional heavy traffic limit theorem.

In an effort to investigate these possibilities, Harrison and Williams [36] exhibited a heavy traffic limit theorem for a closed queueing network in which (i) the notion of heavy traffic was different from (or at least a refinement of) that used by Chen and Mandelbaum [10], (ii) the normalization of the queue length processes was different from the usual central limit theorem type of scaling, and (iii) the limit of the normalized processes was not obtained from a reflecting Brownian motion. Moreover, to fully describe the limit process, non-trivial limits of all of the normalized queue length processes were needed and the weak convergence for some of the components was in a weaker topology than the usual Skorokhod $\mathbf{J_1}$-topology. The Harrison–Williams [36] example is a closed network analog of the Lu–Kumar [55] and Rybko–Stolyar [69] open priority networks mentioned above. Accordingly it gives some intuition concerning the behavior of those examples. The *closed priority* network was considered in [36] so that a simple example could be exhibited with the unconventional features mentioned above. An open or FIFO network example would have required a larger and more complicated state space description. Furthermore, closed priority networks are

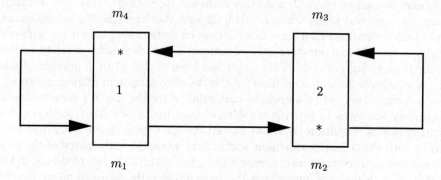

* Classes 2 and 4 have preemptive resume priority
at stations 2 and 1, respectively

FIG. 1. A multiclass closed queueing network with priority service

of interest in their own right.

The example given by Harrison and Williams [36] is described in the next section. A final section concludes with some open problems.

4 A closed queueing network with unconventional heavy traffic behavior

The closed queueing network to be considered is pictured in Fig. 1. This was first studied by Harrison and Nguyen [31]. The network has two single server stations and n customers circulate perpetually with the deterministic routing shown. A customer cycle consists of four services at stations 1, 2, 2 and 1 again, in that order. Customers that are awaiting or undergoing the kth service of their cycle will be called class k customers ($k = 1, 2, 3, 4$). All service times are independent and class k service times are exponentially distributed with mean $m_k > 0$, $k = 1, 2, 3, 4$. Each server follows a preemptive resume priority discipline where classes 2 and 4 have priority over classes 3 and 1, respectively. Consider the symmetric case where the mean service times for the low priority customers are all the same, i.e., $m_1 = m_3$, and similarly for the high priority customers, i.e., $m_2 = m_4$.

Let Q_k denote the queue length process for class k, so that $Q_k(t)$ denotes the number of class k customers existing at time t. To indicate the dependence of the vector queue length process Q on n, a superscript will be added: $Q^{(n)}$. Consider the case where all of the customers start in class 1, so that $Q^{(n)}(0) = (n, 0, 0, 0)$.

For comparison, a (false) conventional heavy traffic limit statement will first be formulated for $Q^{(n)}$. Consider the heavy traffic regime in which $n \to \infty$ and each station is in bottleneck status. For the network pictured in Fig. 1, a solution λ of the flow-balance equation (2.1) with $\alpha = 0$ has all components equal. Denoting this common value by λ_0, by analogy with [10], the bottleneck conditions for the two stations are then

$$\rho_1 = \lambda_0(m_1 + m_4) = 1, \quad \rho_2 = \lambda_0(m_2 + m_3) = 1. \tag{4.1}$$

Since $\lambda_0 > 0$ is a free scale parameter, without loss of generality we may set it to one and then equations (4.1) represent conditions on the mean service times:

$$m_1 + m_4 = m_2 + m_3 = 1. \tag{4.2}$$

Extrapolating from the cases treated by Chen and Mandelbaum [10] and Peterson [62], a conventional heavy traffic limit theorem for this closed network would say that with the central limit theorem type of scaling as in (2.8), the normalized high priority queue length processes $\hat{Q}_2^{(n)}$, $\hat{Q}_4^{(n)}$ converge weakly to the (identically) zero process, and the normalized pair of low priority queue length processes $(\hat{Q}_1^{(n)}, \hat{Q}_3^{(n)})$ converges weakly to a process $(Z, 1 - Z)$ where Z is a reflecting Brownian motion on the unit interval and $Z(0) = 1$. We note for reference below that such a limit theorem would imply that the fluid-scaled process $\bar{Q}^{(n)}$ defined by

$$\bar{Q}^{(n)}(\cdot) \equiv \frac{Q^{(n)}(n \cdot)}{n} = \hat{Q}^{(n)}(n^{-1} \cdot) \tag{4.3}$$

converges weakly as $n \to \infty$ to

$$(Z(0), 0, 1 - Z(0), 0) = (1, 0, 0, 0). \tag{4.4}$$

Harrison and Williams [36] showed that when (4.2) holds and

$$m_2 + m_4 = 1, \tag{4.5}$$

the above conventional limit behavior does not pertain. Indeed, in [36], (i) a different scaling of time and space is used to obtain a stochastic limit of the normalized queue length processes (the scaling is a mix of fluid scaling for the low priority queue lengths and diffusion (or central limit theorem) scaling for the high priority classes), and (ii) the limit process is not obtained from a reflecting Brownian motion, although it is related to Brownian motion.

Before giving a precise statement of the result proved in [36], let us consider the motivation for focussing on the particular combination of parameters in (4.2) and (4.5). Consider the open network shown in Fig. 2. This has the same assumptions as the closed network pictured in Fig. 1, except that customers arrive into class 1 from outside the system according to a Poisson process of rate 1 and a customer after receiving service in class 4 departs the system. This is a stochastic version of the open network studied by Lu and Kumar [55] and has similar characteristics to the priority example given by Rybko and Stolyar [69]. By establishing stability of associated fluid models, Dai and Weiss [21] have shown that the open network pictured in Fig. 2 is stable whenever

$$m_1 + m_4 < 1, \quad m_2 + m_3 < 1 \quad \text{and} \quad m_2 + m_4 < 1. \tag{4.6}$$

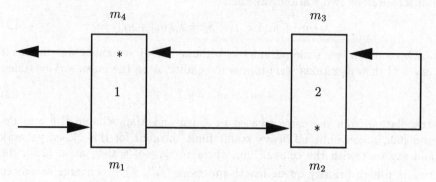

* Classes 2 and 4 have preemptive resume priority
  at stations 2 and 1, respectively

FIG. 2. A multiclass open network with priority service

Furthermore, Dai and VandeVate [19] have recently shown that if any of the inequalities in (4.6) is violated with strict inequality, then the open network will be unstable. Thus, it is natural to think of the conditions expressed in (4.2) and (4.5) as representing the boundary between stability and instability in the open network and hence these are natural bottleneck conditions for the closed network pictured in Fig. 1 (with $\lambda_0 = 1$). These conditions are a refinement of the station level bottleneck conditions seen in conventional heavy traffic limit theorems for closed networks (see Chen and Mandelbaum [10]). Indeed, the conditions in (4.2) represent bottleneck conditions for the two stations, and (4.5) may be thought of as an additional "hidden" bottleneck condition for the high priority classes. Intuitively one might expect this condition because, as shown by Harrison and Nguyen [31], one of the unusual qualities of this closed network is that after one of the queues for class 2 or 4 becomes empty, these high priority classes 2 and 4 can never again be served simultaneously.

A statement of the unconventional heavy traffic limit theorem proved by Harrison and Williams [36] for the closed network pictured in Fig. 1 will now be given. Conditions (4.2) and (4.5), in combination with the symmetry condition $m_1 = m_3$ and $m_2 = m_4$, imply that $m_1 = m_2 = m_3 = m_4 = \frac{1}{2}$. Assuming these conditions and recalling that all customers start in class 1, define scaled queue length processes as follows:

$$\hat{Q}_1^{(n)}(\cdot) = \frac{1}{n} Q_1^{(n)}(n\cdot) \quad \text{and} \quad \hat{Q}_3^{(n)}(\cdot) = \frac{1}{n} Q_3^{(n)}(n\cdot), \qquad (4.7)$$

$$\hat{Q}_2^{(n)}(\cdot) = \frac{1}{\sqrt{n}} Q_2^{(n)}(n\cdot) \quad \text{and} \quad \hat{Q}_4^{(n)}(\cdot) = \frac{1}{\sqrt{n}} Q_4^{(n)}(n\cdot). \qquad (4.8)$$

Note that the scaling of time is the same for all components and the initial

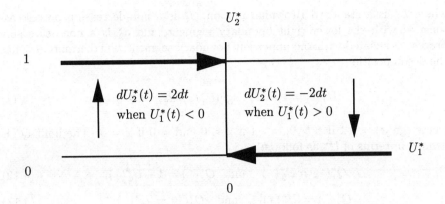

FIG. 3. The Markov process U^*

condition implies that

$$\hat{Q}^{(n)}(0) = (1, 0, 0, 0) \quad \text{for all } n. \tag{4.9}$$

Let D denote the space of real-valued r.c.l.l. (right continuous with finite left limits) paths defined on $[0, \infty)$. Then, as $n \to \infty$, $\hat{Q}^n$ converges weakly to a process Q^*, where Q^* is the stochastic process defined in the next paragraph. For this weak convergence, the path space $D^4 = D \times D \times D \times D$, where $\hat{Q}^n$ and Q^* take their values, is endowed with the product topology where the first and third copies of D are endowed with the usual Skorokhod $\mathbf{J_1}$-topology and the second and fourth copies are endowed with the Skorokhod $\mathbf{M_1}$-topology. (The processes Q_2^* and Q_4^* have jumps that are obtained as the (weak) limits of many coalescing small jumps of the processes $\hat{Q}_2^{(n)}$ and $\hat{Q}_4^{(n)}$. The $\mathbf{M_1}$-topology proved to be the appropriate topology to accommodate this convergence. For details on the different Skorokhod topologies, see Skorokhod [72], Pomarede [63], Whitt [78].)

Actually, the weak limit Q^* is defined in terms of a two-dimensional Markov process U^* whose precise mathematical construction is given in [36]. Informally, however, it is quite easy to explain how U^* behaves. First, its state space is contained in the infinite strip $(-\infty, \infty) \times [0, 1]$, see Fig. 3. Second, when $0 < U_2^* < 1$, the horizontal component U_1^* evolves as a Brownian motion with drift parameter equal to zero and variance parameter equal to 4. Third, U_2^* moves upward at the deterministic rate 2 when U^* is to the left of the vertical axis and moves downward at rate 2 when U^* is to the right of the vertical axis. Finally, when either the upper left or lower right portion of the strip's boundary is hit, there is an immediate horizontal jump to the vertical axis $U_1^* = 0$, and the process U_1^* spends zero time (in the sense of Lebesgue measure) in the state $U_1^* = 0$. Indeed, U_1^* can be decomposed as follows:

$$U_1^*(t) = 2B^*(t) - J_1^*(t) + J_2^*(t), \qquad t \geq 0, \tag{4.10}$$

where B^* is a standard Brownian motion, J_1^* is a non-decreasing process associated with the lower right boundary segment, and J_2^* is a non-decreasing process associated with the upper left boundary segment. Furthermore, U_2^* has the decomposition

$$U_2^*(t) = 1 - 2 \int_0^t \mathrm{sgn}(U_1^*(s))\,ds, \quad t \geq 0, \tag{4.11}$$

where $\mathrm{sgn}(x) = +1$ if $x > 0$, $= -1$ if $x < 0$ and $= 0$ if $x = 0$. The limit Q^* is defined in terms of U^* as follows:

$$Q_1^*(\cdot) = U_2^*(\cdot) \quad \text{and} \quad Q_3^*(\cdot) = 1 - U_2^*(\cdot), \tag{4.12}$$

$$Q_2^*(\cdot) = [U_1^*(\cdot)]^+ \quad \text{and} \quad Q_4^*(\cdot) = [U_1^*(\cdot)]^-, \tag{4.13}$$

where $x^+ = \max(x, 0)$ and $x^- = \max(-x, 0)$. Intuitively, the excursions of U^* from the vertical axis ($U_1^* = 0$) correspond to busy periods for the high priority customers in the closed network pictured in Fig. 1 and whenever U^* hits the upper or lower boundary of the strip, this corresponds to the end of a busy period for one of the servers.

Note that all four of the processes defined by (4.12)–(4.13) are non-deterministic. It is immediate from the limit theorem stated above that the fluid-scaled queue length process $\bar{Q}^{(n)}$, defined in (4.3), converges weakly as $n \to \infty$ to $(Q_1^*, 0, Q_3^*, 0)$, where for the weak convergence the space of four-dimensional r.c.l.l. paths in which $\bar{Q}^{(n)}$ and the limit take their values is endowed with the usual Skorokhod $\mathbf{J_1}$-topology. Thus, the high priority queue length processes are asymptotically zero under fluid scaling, as in conventional heavy traffic theory, but the fluid limit of the low priority queue length processes is a dynamic stochastic process whose full description requires knowledge of the limits of the high priority queue length processes under the finer scaling exhibited in (4.8). By contrast, in conventional heavy traffic theory the fluid limit would be a constant (see (4.4)). This is the basis for the claim that the example considered by Harrison and Williams satisfies an unconventional heavy traffic limit theorem when (4.2) and (4.5) hold.

5   Open problems

The following are natural open problems raised by the discussion above.

(a) Harrison and Williams [36] considered the heavy traffic regime in which the two stations are bottlenecks, i.e., (4.2) holds, and the high priority classes form a hidden bottleneck, i.e., (4.5) holds. This is called the *critical case* in [36]. It is natural to ask what heavy traffic behavior occurs in the subcritical ($m_2 + m_4 < 1$) and supercritical ($m_2 + m_4 > 1$) cases. Harrison and Williams [36] conjecture that a conventional heavy traffic limit theorem holds in the subcritical case, but do not even venture a guess at the heavy traffic behavior for the supercritical case.

(b) It is likely that an analog of the unconventional heavy traffic limit theorem exhibited by Harrison and Williams [36] for a closed network could be proved for an open network. In particular, one could consider a sequence of open networks of the form shown in Fig. 2, where the traffic intensity parameters defined by $\rho_1 = m_1 + m_4$, $\rho_2 = m_2 + m_3$, and $\rho_H = m_2 + m_4$ (H for high priority customers) converge upwards to one through a sequence of networks indexed by n, in a similar manner to that in (2.3). In some sense the limit process of Harrison and Williams [36] should be a conditional version (obtained by conditioning on a sum of the components) of the limit process for such a sequence of open networks.

(c) The example of an unconventional heavy traffic limit theorem given by Harrison and Williams [36] grew out of the Lu–Kumar/Rybko–Stolyar examples of networks with unconventional stability behavior. The connection between these two unconventional notions has only begun to be explored. In particular, the refinement of the notion of heavy traffic used in [36] and embodied in the additional bottleneck condition (4.5) prompts the question: what characterizes the heavy traffic regime in multiclass networks?

(d) In section 3 a number of possible reasons are advanced to explain why a conventional heavy traffic limit theorem might not exist for a given multiclass network. It would be interesting to know what combinations of these can occur in general and whether there are other viable explanations.

(e) One would like to identify a "good" class of multiclass networks for which conventional heavy traffic limit theorems can be proved. Conversely, it would be helpful to know the size of the collection of networks that do not have conventional heavy traffic behavior, e.g., is it a small set in some measure theoretic sense?

Acknowledgements The author thanks J. G. Dai, J. M. Harrison, A. Mandelbaum, V. Nguyen, W. P. Peterson, M. I. Reiman and an anonymous referee for their helpful comments on a preliminary version of this paper.

Bibliography

1. Baskett, F., Chandy, K. M., Muntz, R. R., and Palacios, F. G. (1975). Open, closed and mixed networks of queues with different classes of customers. *Journal of the ACM*, **22**, 248–260.

2. Bertsimas, D., Paschalidis, I. Ch., and Tsitsiklis, J. N. (1994). Optimization of multiclass queueing networks: polyhedral and nonlinear characterizations of achievable performance. *Annals of Applied Probability*, **4**, 43–75.

3. Billingsley, P. (1968). *Convergence of Probability Measures*. Wiley, New York.

4. Borovkov, A. A. (1964). Some limit theorems in the theory of mass service, I. *Theory of Probability and Its Applications*, **9**, 550–565.

5. Borovkov, A. A. (1965). Some limit theorems in the theory of mass service, II. *Theory of Probability and Its Applications*, **10**, 375–400.

6. Bramson, M. (1994). Instability of FIFO queueing networks. *Annals of Applied Probability*, 4, 414–431.

7. Bramson, M. (1994). Instability of FIFO queueing networks with quick service times. *Annals of Applied Probability*, **4**, 693–718.

8. Bramson, M. (1996). Convergence to equilibria for fluid models of FIFO queueing networks. To appear in *Queueing Systems: Theory and Applications*.

9. Bramson, M. (1996). Convergence to equilibria for fluid models of certain processor sharing queueing networks. To appear in *Queueing Systems: Theory and Applications*.

10. Chen, H., and Mandelbaum, A. (1991). Stochastic discrete flow networks: diffusion approximations and bottlenecks. *Annals of Probability*, **4**, 1463–1519.

11. Chen, H., and Mandelbaum, A. (1994). Hierarchical modeling of stochastic networks, Parts I and II. In *Stochastic Modeling and Analysis of Manufacturing Systems*, D. D. Yao (ed.), Springer-Verlag, New York, 47–131.

12. Chen, H., and Zhang, H. (1996). Diffusion approximations for some multiclass queueing networks with FIFO service discipline. Preprint.

13. Coffman, E. G. Jr., Puhalskii, A. A., and Reiman, M. I. (1995). Polling systems with zero switchover times: a heavy traffic averaging principle. *Annals of Applied Probability*, **5**, 681–719.

14. Coffman, E. G. Jr., and Reiman, M. I. (1984). Diffusion approximations for computer communication systems. In *Mathematical Computer Performance and Reliability*, G. Iazeolla, P. J. Courtois and A. Hordijk (eds.), North-Holland, Amsterdam, 33–53.

15. Dai, J. G. (1995). On positive Harris recurrence of multiclass queueing networks: a unified approach via fluid limit models. *Annals of Applied Probability*, **5**, 49–77.

16. Dai, J. G. (1995). Stability of open multiclass queueing networks via fluid models. In *Stochastic Networks*, IMA Volumes in Mathematics and Its Applications, F. P. Kelly and R. J. Williams (eds.), **71**, Springer-Verlag, New York, 71–90.

17. Dai, J. G., and Kurtz, T. G. (1995). A multiclass station with Markovian feedback in heavy traffic. *Mathematics of Operations Research*, **20**, 721–742.

18. Dai, J. G., and Nguyen, V. (1994). On the convergence of multiclass queueing networks in heavy traffic. *Annals of Applied Probability*, **4**, 26–42.

19. Dai, J. G., and VandeVate, J. (1996). Virtual stations and the capacity of two-station queueing networks. Submitted to *Operations Research*.

20. Dai, J. G., and Wang, Y. (1993). Nonexistence of Brownian models of certain multiclass queueing networks. *Queueing Systems: Theory and Applications*, **13**, 41–46.

21. Dai, J. G., and Weiss, G. (1996). Stability and instability of fluid models for certain re-entrant lines. *Mathematics of Operations Research*, **21**, 115–134.

22. Dai, J. G., and Williams, R. J. (1995). Existence and uniqueness of semi-

martingale reflecting Brownian motions in convex polyhedrons. *Theory of Probability and Its Applications*, **40**, 3–53 (in Russian), to appear in the SIAM translation journal of the same name.

23. Ethier, S. N., and Kurtz, T. G. (1986). *Markov Processes: Characterization and Convergence*. Wiley, New York.

24. Flores, C. (1985). Diffusion approximations for computer communications networks. In *Computer Communications, Proc. Symp. Appl. Math., Vol. 31*, B. Gopinath (ed.), American Mathematical Society, Providence, RI, 83–124.

25. Glynn, P. W. (1990). Diffusion approximations. In *Handbooks in Operations Research, Vol. 2*, D. P. Heyman and M. J. Sobel (eds.), North-Holland, Amsterdam, 145–198.

26. Harrison, J. M. (1978). The diffusion approximation for tandem queues in heavy traffic. *Advances in Applied Probability*, **10**, 886–905.

27. Harrison, J. M. (1988). Brownian models of queueing networks with heterogeneous customer populations. In *Stochastic Differential Systems, Stochastic Control Theory and Applications*, IMA Volumes in Mathematics and Its Applications, W. Fleming and P.-L. Lions (eds.), Springer-Verlag, New York, 147–186.

28. Harrison, J. M. (1995). Balanced fluid models of multiclass queueing networks: a heavy traffic conjecture. In *Stochastic Networks*, IMA Volumes in Mathematics and Its Applications, F. P. Kelly and R. J. Williams (eds.), **71**, Springer-Verlag, New York, 1–20.

29. Harrison, J. M., and Nguyen, V. (1990). The QNET method for two-moment analysis of open queueing networks. *Queueing Systems: Theory and Applications*, **6**, 1–32.

30. Harrison, J. M., and Nguyen, V. (1993). Brownian models of multiclass queueing networks: current status and open problems. *Queueing Systems: Theory and Applications*, **13**, 5–40.

31. Harrison, J. M., and Nguyen, V. (1995). Some badly behaved closed queueing networks. In *Stochastic Networks*, IMA Volumes in Mathematics and Its Applications, F. P. Kelly and R. J. Williams (eds.), **71**, Springer-Verlag, New York, 117–124.

32. Harrison, J. M., and Reiman, M. I. (1981). Reflected Brownian motion on an orthant. *Annals of Probability*, **9**, 302–308.

33. Harrison, J. M., and Wein, L. M. (1989). Scheduling networks of queues: heavy traffic analysis of a simple open network. *Queueing Systems: Theory and Applications*, **5**, 265–280.

34. Harrison, J. M., and Williams, R. J. (1987). Brownian models of open queueing networks with homogeneous customer populations. *Stochastics*, **22**, 77–115.

35. Harrison, J. M., and Williams, R. J. (1992). Brownian models of feedforward

queueing networks: quasireversibility and product form solutions. *Annals of Applied Probability*, **2**, 263–293.

36. Harrison, J. W., and Williams, R. J. (1996). A multiclass closed queueing network with unconventional heavy traffic behavior. To appear in *Annals of Applied Probability*.

37. Harrison, J. M., Williams, R. J., and Chen, H. (1990). Brownian models of closed queueing networks with homogeneous customer populations. *Stochastics and Stochastics Reports*, **29**, 37–74.

38. Horváth, L. (1992). Strong approximations of open queueing networks. *Mathematics of Operations Research*, **17**, 487–508.

39. Iglehart, D. L., and Whitt, W. (1970). Multiple channel queues in heavy traffic I. *Advances in Applied Probability*, **2**, 150–177.

40. Iglehart, D. L., and Whitt, W. (1970). Multiple channel queues in heavy traffic II. *Advances in Applied Probability*, **2**, 355–364.

41. Jackson, J. R. (1957). Networks of waiting lines. *Operations Research*, **5**, 518–521.

42. Jackson, J. R. (1963). Jobshop-like queueing systems. *Management Science*, **10**, 131–142.

43. Johnson, D. P. (1983). *Diffusion Approximations for Optimal Filtering of Jump Processes and for Queueing Networks*. Ph.D. dissertation, Department of Mathematics, University of Wisconsin, Madison, WI.

44. Kaspi, H., and Mandelbaum, A. (1992). Regenerative closed queueing networks. *Stochastics and Stochastics Reports*, **39**, 239–258.

45. Kelly, F. P. (1979). *Reversibility and Stochastic Networks*. Wiley, Chichester.

46. Kelly, F. P., and Laws, C. N. (1993). Dynamic routing in open queueing networks: Brownian models, cut constraints and resource pooling. *Queueing Systems: Theory and Applications*, **13**, 47–86.

47. Kingman, J. F. C. (1961). The single server queue in heavy traffic. *Proceedings of the Cambridge Philosophical Society*, **57**, 902–904.

48. Kingman, J. F. C. (1962). On queues in heavy traffic. *Journal of the Royal Statistical Society, Series B*, **24**, 383–392.

49. Kingman, J. F. C. (1964). The heavy traffic approximation in the theory of queues. In *Proceedings of the Symposium on Congestion Theory*, University of North Carolina, Chapel Hill, W. L. Smith and R. I. Wilkinson (eds.), 137–169.

50. Krichagina, E. V., Liptser, R. S., and Puhalsky, A. A. (1988). Diffusion approximation for the system with arrival process depending on queue and arbitrary service distribution. *Theory of Probability and Its Applications*, **33**, 124–135.

51. Kumar, P. R., and Meyn, S. P. (1995). Stability of queueing networks and scheduling policies. *IEEE Transactions on Automatic Control*, **40**, 251–260.

52. Kumar, P. R., and Seidman, T. I. (1990). Dynamic instabilities and stabilization methods in distributed real-time scheduling of manufacturing systems. *IEEE Transactions on Automatic Control*, **35**, 289–298.

53. Kushner, H. J. (1995). A control problem for a new type of public transportation system, via heavy traffic analysis. In *Stochastic Networks*, IMA Volumes in Mathematics and Its Applications, F. P. Kelly and R. J. Williams (eds.), **71**, Springer-Verlag, New York, 139–167.

54. Lemoine, A. J. (1978). Networks of queues—a survey of weak convergence results. *Management Science*, **24**, 1175–1193.

55. Lu, S. H., and Kumar, P. R. (1991). Distributed scheduling based on due dates and buffer priorities. *IEEE Transactions on Automatic Control*, **36**, 1406–1416.

56. Mandelbaum, A., and Massey, W. A. (1995). Strong approximations for time-dependent queues. *Mathematics of Operations Research*, **20**, 33–64.

57. Massey, W. A. (1981). *Nonstationary Queueing Networks*. Ph.D. dissertation, Department of Mathematics, Stanford University, Stanford, CA.

58. Meyn, S. P., and Down, D. (1994). Stability of generalized Jackson networks. *Annals of Applied Probability*, **4**, 124–148.

59. Nguyen, V. (1993). Processing networks with parallel and sequential tasks: heavy traffic analysis and Brownian limits. *Annals of Applied Probability*, **3**, 28–55.

60. Nguyen, V. (1995). Fluid and diffusion approximations of a two-station mixed queueing network. *Mathematics of Operations Research*, **20**, 321–354.

61. Pats, G. (1995). *State Dependent Queueing Networks: Approximations and Applications*. Ph.D. dissertation, Department of Industrial Engineering and Management, Technion, Haifa, Israel.

62. Peterson, W. P. (1991). Diffusion approximations for networks of queues with multiple customer types. *Mathematics of Operations Research*, **9**, 90–118.

63. Pomarede, J. L. (1976). *A Unified Approach via Graphs to Skorohod's Topologies on the Function Space D*. Ph.D. dissertation, Department of Statistics, Yale University, New Haven, CT.

64. Prohorov, Y. V. (1963). Transient phenomena in processes of mass service (in Russian). *Litovskii Matematiceskii Sbornik*, **3**, 199–205.

65. Reiman, M. I. (1984). Some diffusion approximations with state space collapse. In *Proceedings International Seminar on Modeling and Performance Evaluation Methodology*, Lecture Notes in Control and Information Sciences, F. Baccelli and G. Fayolle (eds.), Springer-Verlag, New York, 209–240.

66. Reiman, M. I. (1984). Open queueing networks in heavy traffic. *Mathematics of Operations Research* **9**, 441–458.

67. Reiman, M. I. (1988). A multiclass feedback queue in heavy traffic. *Mathe-*

matics of Operations Research, **20**, 179–207.

68. Reiman, M. I., and Williams, R. J. (1988–89). A boundary property of semimartingale reflecting Brownian motions. *Probability Theory and Related Fields*, **77**, 87–97, and **80**, 633.

69. Rybko, A. N., and Stolyar, A. L. (1991). Ergodicity of stochastic processes describing the operation of an open queueing network. *Problemy Peredachi Informatsii*, **28**, 2–26.

70. Seidman, T. I. (1994). 'First come, first served' can be unstable!. *IEEE Transactions on Automatic Control*, **39**, 2166–2171.

71. Sigman, K. (1990). The stability of open queueing networks. *Stochastic Processes and Applications*, **35**, 11–25.

72. Skorokhod, A. V. (1956). Limit theorems for stochastic processes. *Theory of Probability and Its Applications*, **1**, 261–290.

73. Taylor, L. M., and Williams, R. J. (1993). Existence and uniqueness of semimartingale reflecting Brownian motions in an orthant. *Probability Theory and Related Fields*, **96**, 283–317.

74. Wein, L. M. (1990). Scheduling networks of queues: heavy traffic analysis of a two-station network with controllable inputs. *Operations Research*, **38**, 1065–1078.

75. Wein, L. M. (1992). Scheduling networks of queues: heavy traffic analysis of a multistation network with controllable inputs. *Operations Research*, **40** (suppl.), S312–S334.

76. Whitt, W. (1971). Weak convergence theorems for priority queues: preemptive resume discipline. *J. Applied Probability*, **8**, 74–94.

77. Whitt, W. (1974). Heavy traffic theorems for queues: a survey. In *Mathematical Methods in Queueing Theory*, A. B. Clarke (ed.), Springer-Verlag, New York, 307–350.

78. Whitt, W. (1980). Some useful functions for functional limit theorems. *Mathematics of Operations Research*, **3**, 67–85.

79. Whitt, W. (1993). Large fluctuations in a deterministic multiclass network of queues. *Management Science*, **39**, 1020–1028.

80. Williams, R. J. (1995). Semimartingale reflecting Brownian motions in an orthant. In *Stochastic Networks*, IMA Volumes in Mathematics and Its Applications, F. P. Kelly and R. J. Williams (eds.), **71**, Springer-Verlag, New York, 125–137.

4

The BIGSTEP approach to flow management in stochastic processing networks

J. Michael Harrison

Graduate School of Business, Stanford University

Abstract

Flow management in stochastic networks may involve dynamic routing, dynamic input control, dynamic resource allocation, or combinations thereof. This paper describes a new and very general approach to flow management in which system status is reviewed at discrete points in time, a linear programming problem is solved at each such point, and the processing plan obtained from that linear program is implemented in open-loop fashion over the ensuing period.

Discrete-review policies of the proposed type are motivated by heavy traffic theory for network control problems, in which the loads imposed on at least some processing resources are assumed to approach the capacities of those resources. Striving to construct a discrete-review policy that is asymptotically optimal in the heavy traffic limit, one must first formulate and analyze an approximating Brownian control problem, which typically is much simpler than the original processing network.

In general, data derived from analysis of the Brownian control problem are needed to determine objective function coefficients for the linear programs that are solved at successive review points. For certain important special cases, however, analysis of the Brownian approximation is unnecessary. Moreover, the class of flow management policies described here has great practical appeal, even if one substitutes some crude heuristic for the Brownian analysis.

For networks in heavy traffic, our discretized control policies use review points that are widely spaced in time. Because the proposed method steps through time in big increments, it is called the BIGSTEP method or BIGSTEP approach to dynamic flow management. No attempt is made here to prove asymptotic optimality of BIGSTEP policies, or even to describe the method in precise mathematical terms, but a rigorous analysis of one relatively simple example is described.

1 Introduction

This paper is concerned with dynamic flow control in stochastic processing networks, and more specifically, with the "heavy traffic approach" to network flow control. Roughly speaking, a stochastic processing network is said to be "in

57

heavy traffic" if the loads imposed on one or more of its processing resources, either exogenously or as a consequence of rational endogenous decision making, are approximately equal to the capacities of those resources. Research in the last decade has suggested that a network in heavy traffic can be usefully approximated by a corresponding "Brownian system model," and more specifically, that a near-optimal policy for the original network can be derived by analyzing the Brownian approximation. This general idea was first put forth in [2], and first applied to a concrete problem in [5], but it has been applied, elaborated, rigorized and extended in subsequent work [1, 6, 10, 11, 12, 13, 14, 15, 20, 21, 22, 23].

The heavy traffic approach to network flow control has achieved notable successes, but development of this theory has reached something of an impasse. Broadly speaking, the difficulty is that optimal control policies for Brownian system models, even if they can be determined, do not translate in any obvious way into implementable control policies for the more finely structured processing network models of original interest. This severely restricts the theory's value, whether one is interested in literal computation and practical application or merely in extracting qualitative insights about system behavior and the structure of optimal policies. The successful heavy traffic analyses referred to above have focused primarily on particular examples with just a few processing resources, and the authors' arguments contain *ad hoc* elements depending on the special character of their examples.

The central purpose of this paper is to introduce a class of discrete-review control policies that provide a potential means of mechanically constructing near-optimal controls for processing networks in heavy traffic. In a policy of the proposed type one must solve a linear programming problem at each review point, and in general one needs the solution of an approximating Brownian control problem to determine the objective coefficients of those linear programs. However, construction of the discrete-review policy does not involve any problem-specific ingenuity. Moreover, many problems that have been successfully analyzed in the heavy traffic control literature involve models whose Brownian approximations admit pathwise optimal controls [2], and for those models the discrete-review policies referred to above can be constructed entirely from original problem data.

To achieve near-optimality in heavy traffic when using the proposed method of discretized control, we take the review periods to be long in absolute terms, but short in comparison with a certain scaling parameter. With such a parameter choice, one might say that the proposed method steps through time in big increments, and hence it will be called the BIGSTEP method or BIGSTEP approach to dynamic flow management. The method will be described here in terms of a three-station network example, but using notation and terminology that readily suggest the general version to be developed in future work. (A general description of the method requires a general model formulation and a general definition of heavy traffic, each of which is a subtle and complex topic in its own right.) The three-station example contains issues of dynamic routing, dynamic resource allocation, and dynamic input control, so it represents well the class of flow management problems for which a general method is needed. The three-station example is complex enough to defy complete and explicit solution, and thus it leads one to consider what sort of analysis is possible and desirable for systems of realistic scale. However, a complete exposition of BIGSTEP mechanics for the three-station example would be prohibitively complex, so for that and other purposes a simpler example involving two servers in parallel will be introduced later.

This paper is aimed at readers who are already familiar with the heavy traffic approach to network control problems, and more specifically, with the general approach to formulation of Brownian approximations that was developed in [2]. More recent papers like [5] and [22] have applied and extended that approach in particular contexts, and the recapitulation of the general method contained in those papers may help readers to acquire the intuitive grasp of Brownian approximations that is taken for granted here. In the same vein, [7] provides a very readable survey of heavy traffic theory for networks with discretionary routing, emphasizing qualitative insights about system behavior that one derives from approximating Brownian models, and it too contains a good deal of recapitulation.

None of the papers mentioned in the previous paragraph actually prove, or even state in a precise form, a heavy traffic limit theorem, although each of them expresses a hope or expectation that network control policies derived from Brownian approximations are "asymptotically optimal" in the heavy traffic limit. A major attraction of the BIGSTEP approach lies in the potential framework it provides for proving general results on heavy traffic convergence in a network control context, but the only rigorous results described here are for the relatively simple parallel-server example. There are now a number of papers that rigorously prove limit theorems of the desired type for specially structured examples [10, 11, 12, 14, 15], and review of that work will help readers to appreciate the difficulties involved in developing a limit theory that is both general and rigorous.

Several other aspects of heavy traffic background knowledge, apart from the mechanics of Brownian model formulation, are implicitly assumed in later sections, and misunderstandings may be avoided by highlighting these matters at the outset. First, our discussion of the BIGSTEP method is conducted in terms of two examples whose special distributional assumptions (Poisson arrivals, exponential or deterministic service times) and other fine structure (single-server processing stations, interruptible service times, and so forth) are convenient but actually irrelevant to the main ideas developed. All of heavy traffic theory, at least as that term is used here, is based on the central limit theorem, so only the first two moments of the underlying distributions are relevant, and even the assumption of renewal input processes and IID service time sequences can be weakened. Similarly, many years of experience have shown that Brownian approximations and heavy traffic limits are insensitive to model microstructure, such as single-server stations versus multi-server stations, but no mention of these points will be made hereafter. Also, in the examples discussed later all servers are assumed to be heavily loaded (traffic intensity parameter near unity), but it is a point of faith in heavy traffic theory, borne out by long experience, that such analyses carry over to systems with some lightly loaded stations. The lightly loaded stations, or non-bottleneck stations, are simply irrelevant to gross system behavior in the heavy traffic limit, and hence heavy traffic analysis focuses on the "bottleneck sub-network." All of these points are discussed at some point in [2].

The rest of the paper is structured as follows. Sections 2 and 3 deal with the three-station example referred to earlier and its Brownian approximation, and the BIGSTEP method is described in the context of that example in section 4. To make plausible the conjecture that BIGSTEP controls are asymptotically optimal in heavy traffic, rough heuristic arguments are presented, but several crucial issues are swept under the rug. In particular, we do not address the question of how to choose an interval length between review points so as to achieve asymptotic optimality, nor the question of how to execute BIGSTEP plans on the detailed level. In sections 5 and 6 those matters are discussed in the context of a simpler network model with two servers working in parallel. These sections essentially summarize a complete analysis undertaken in a companion paper, but with enough detail to give readers a clear picture of how BIGSTEP controls actually work in the relatively simple parallel-server system. While we cannot pretend that the analysis carries over directly to more complex environments, readers will see that certain themes have obvious relevance beyond the parallel-server example. In particular, large deviations theory is crucial in determining the proper interval length between review points, and this is a new development in heavy traffic analysis. Finally, sections 7 and 8 give an overview of the BIGSTEP method from two different perspectives, summarizing first the tasks involved in its rigorous justification, and then its potential appeal from a practical standpoint. In the latter regard, we note that BIGSTEP control policies have a three-tiered hierarchical structure, with goal setting at the highest level, flow planning at the intermediate level, and execution of flow plans at the lowest level.

In concluding this introduction, a few words are appropriate about terminology. Systems of the type considered here are often called "queueing networks," but the term "processing network" will be used instead throughout this paper, for two reasons. First, the alternate name emphasizes the positive purpose for which such a system exists, whereas terms like "queue" and "congestion" refer to undesirable phenomena that may be experienced in the course of processing activity. Second, the analytical framework described here is potentially applicable to a broader class of systems than what is usually understood by the term "queueing network." Motivated by manufacturing applications, we call the discrete units that flow through a processing network "jobs" rather than "customers," and the basic unit of time in each of our examples is taken to be hours. In conformance with standard queueing theoretic usage, processing resources are called "servers" or "stations," those two terms being used interchangeably.

2   An illustrative three-station processing network

Consider the system pictured in Fig. 1, having three single-server stations (represented by circles) and three external input processes. As shown in the figure, jobs of type B require a service at station 3, then a service at station 2, and then they exit the system. Similarly, jobs of type C require three services, at stations 3, 1 and then 3 again, before they exit. For jobs of type A there are two processing routes available: if a type A job is directed to the lower route, it requires just a single service at station 3, but if it is directed to the upper route then two services are required, at stations 1 and 2 in that order. For concreteness, let us suppose that the routing decision for each type A job must be made irrevocably at the moment of its arrival.

In the usual fashion, we define a different "job class" for each combination of job type and stage of completion, including one class for type A jobs seeking their first service on the upper route and another class for type A jobs following the lower route. There are eight such classes in total, and their numbering is indicated in the open-ended rectangles in Fig. 1. (These rectangles represent infinite-capacity buffers in which the various job classes reside.) Because individual servers are responsible for more than one job class, the system portrayed in Fig. 1 is a "multi-class" processing network, and decisions can be made dynamically as to the sequence in which job classes are served at each station.

In addition to the dynamic routing and sequencing decisions referred to above, new jobs of any type can be rejected at the moment of arrival, but there is an economic penalty for doing so, because a fixed reward (perhaps depending on type) is received each time a new arrival is accepted. On the other hand there are linear holding costs that provide a motivation for rejecting new inputs when queues are long.

For concreteness, let us suppose that jobs of types A, B and C arrive according to independent Poisson processes at average rates of 0.50, 0.25 and 0.25 jobs per hour, respectively. Also, jobs of each class k have exponentially distributed service times with mean m_k where

$$m = (m_1, \ldots, m_8) = (2, 3, 1, 1, 1, 1, 2, 1). \tag{2.1}$$

Let us denote by $\sigma(k)$ the (unique) server responsible for processing job class k, so

$$\sigma(1) = \sigma(7) = 1, \quad \sigma(2) = \sigma(5) = 2, \quad \text{and} \quad \sigma(3) = \sigma(4) = \sigma(6) = \sigma(8) = 3.$$

Finally, it will make things easiest to assume that services can be interrupted at any desired time and then resumed later by the responsible server. The rather special assumptions enunciated in this paragraph are inessential, but they make for a simple and familiar setting, so attention is not distracted from the main ideas introduced later.

To facilitate connection with existing heavy traffic literature, let us first suppose that inputs are uncontrollable (that is, arriving jobs cannot be turned away) and routing decisions for type A arrivals are settled by independent coin flips.

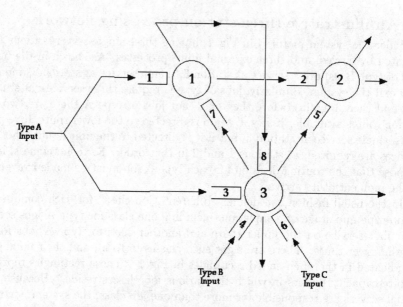

FIG. 1. A three-station network

Then external arrivals of classes 1, 3, 4 and 6 are independent Poisson processes, each with an average arrival rate of 0.25 jobs per hour. Since there are no other external arrivals, the vector of average external arrival rates into the various classes is

$$\lambda = (\lambda_1, \dots, \lambda_8) = \left(\frac{1}{4}, \ 0, \ \frac{1}{4}, \ \frac{1}{4}, \ 0, \ \frac{1}{4}, \ 0, \ 0 \right). \qquad (2.2)$$

(Later, when matrix–vector relationships are written, λ will be treated as a column vector. In such relationships readers should associate with vectors whatever orientation is required for the equation to make sense.)

With input control and routing decisions eliminated from the problem, only the issue of dynamic resource allocation remains. That is, a dynamic control strategy is expressed by an eight-dimensional vector process $T = \{T(t), t \geq 0\}$ with components $T_1(t), \dots, T_8(t)$, where $T_j(t)$ represents the total amount of time that server $\sigma(j)$ devotes to processing class j over the interval $[0, t]$. To introduce some useful generic language, one may say that eight different *activities* are available, with server $\sigma(j)$ uniquely responsible for activity j. Following the notational conventions introduced in [2], we associate with the network control problem (of course, no objective has been stated yet) an 8×8 *input–output matrix* $R = (R_{kj})$ and a 3×8 *resource consumption matrix* $A = (A_{ij})$ defined as follows. First, R_{kj} is the average rate at which activity j depletes buffer k (that is, depletes the queue of class k jobs) when server $\sigma(j)$ is devoted entirely to that activity. With the job routing and numerical data introduced above we have

$$R = \begin{bmatrix} \frac{1}{2} & & & & & & & \\ -\frac{1}{2} & \frac{1}{3} & & & & & & \\ & & 1 & & & & & \\ & & & 1 & & & & \\ & & & -1 & 1 & & & \\ & & & & & 1 & & \\ & & & & & -1 & \frac{1}{2} & \\ & & & & & & -\frac{1}{2} & 1 \end{bmatrix}. \tag{2.3}$$

Second, $A_{ij} = 1$ if $\sigma(j) = i$ (that is, server i is responsible for activity j), and $A_{ij} = 0$ otherwise. For our illustrative three-station network, then,

$$A = \begin{bmatrix} 1 & & & & 1 & & & \\ & 1 & & 1 & & & & \\ & & 1 & 1 & & 1 & & 1 \end{bmatrix}. \tag{2.4}$$

To complete the problem formulation with input control and routing eliminated, let us suppose that a large integer n is given (see below), cost is continuously incurred at rate h_k/n for each class k job in existence (thus we have a
problem with *linear holding costs*), the planning horizon is infinite, and costs are
continuously discounted at interest rate γ/n, where $\gamma > 0$. Later, a heavy traffic
limit will be achieved by letting $n \to \infty$ with all other problem data fixed. One
may interpret n as a unit of time that is natural for the problem under consideration: by assumption, the interest rate per natural time unit remains fixed at
γ as n grows large, and the holding cost per natural time unit similarly remains
fixed at h_k for class k jobs. An alternative interpretation will be provided at the
end of this section.

Denoting by $Q_k(t)$ the number of class k jobs in existence at time t, either
waiting or being served, the objective is thus to

$$\text{minimize} \quad f = E\left\{ \int_0^\infty e^{-\gamma t/n} \frac{1}{n} h \cdot Q(t) \, dt \right\}, \tag{2.5}$$

where $h \cdot Q(t) = \sum h_k Q_k(t)$ as usual. In the language of [2], this initial version
of our dynamic control problem is one of *pure sequencing*. That is, the only
questions that arise in dynamic flow management concern the sequence in which
various servers will process the jobs queued at their stations. Because services are
assumed to be interruptible and exponentially distributed, we have a continuous-
time Markov decision process whose state at time t is summarized by the eight-
dimensional queue length vector $Q(t)$. The dynamic control process or server
allocation process $T = \{T(t), t \geq 0\}$ introduced earlier must satisfy certain
constraints, of course, but it is not necessary for our purposes to write them out.

The pure sequencing problem described above is of the general type considered in [2], and now we verify that "heavy traffic conditions" prevail as
$n \to \infty$. Deviating slightly from the expositional sequence in [2], we seek a
vector $\beta = (\beta_1, \ldots, \beta_8)$ of *nominal activity rates* that satisfies

$$R\beta = \lambda, \quad A\beta \le e \quad \text{and} \quad \beta \ge 0, \tag{2.6}$$

where e denotes the three-vector of ones. (The same letter e will be used later to represent a vector of ones having different dimension, but the appropriate dimension will always be clear from context.) Readers should interpret β_j as the long-run fraction of time that server $\sigma(j)$ devotes to activity j. Thus the equation $R\beta = \lambda$ is a *material balance constraint*, requiring that the long-run average rates at which jobs of each class arrive, due either to external inputs or internal transitions, be equal to the long-run average rates at which they are processed or consumed. The constraint $A\beta \le e$ says that no more than 100% of each server's time can be allocated (these are *capacity constraints*), and the non-negativity constraints in (2.6) are obviously needed as well. Given the data (2.2)–(2.4), the unique solution of (2.6) is

$$\beta = \left(\frac{1}{2}, \frac{3}{4}, \frac{1}{4}, \frac{1}{4}, \frac{1}{4}, \frac{1}{4}, \frac{1}{2}, \frac{1}{4} \right). \tag{2.7}$$

This vector of nominal allocations satisfies $A\beta = e$, which means that the processing capacities of the three servers are *perfectly balanced* with the external arrival rates. In the usual language of queueing theory, each server has traffic intensity parameter equal to one, and thus we know from classical theory that the expected total number of jobs in the system at time t must diverge as $t \to \infty$ under any control policy. If n is held fixed in the discounted objective (2.5), it is easy to show that f remains finite for any policy, but even the best achievable f value diverges as $n \to \infty$, and roughly speaking, the objective in heavy traffic analysis is to find a control policy under which f diverges as slowly as possible as $n \to \infty$. The situation considered here, involving a single perfectly balanced system and only the interest rate and holding cost rates changing with n, is the simplest example of a heavy traffic limit. More generally one can allow the mean service times m_k and external arrival rates λ_k to depend on n in such a way that perfect balance is approached as $n \to \infty$, but generalizing the problem setting in that way would add nothing to the basic ideas explained below.

Having discussed the simplified problem of pure sequencing, let us now extend the formulation to allow dynamic routing of type A arrivals (recall that, by assumption, the routing decision must be made at the instant an A job arrives) and further suppose that any input process can be turned off (or equivalently, jobs of that type can be rejected upon arrival) at any time if doing so is economically advantageous. An economic motivation for accepting new arrivals will be introduced shortly. In extending the previous formulation it is conceptually easiest to speak in terms of three fictional "input servers" numbered 4, 5 and 6 who generate arrivals of types A, B and C respectively. We associate with the input servers four new "activities," as follows:

activity   9 = creation of class 1 jobs   (by server 4),
activity 10 = creation of class 3 jobs   (by server 4),

activity 11 = creation of class 4 jobs   (by server 5),
activity 12 = creation of class 6 jobs   (by server 6).

Hereafter when we say that "server 4 is engaged in activity 9" this is understood to mean that the type A input process is turned on and any resulting arrival will be routed to buffer 1, so the instantaneous Poisson arrival rates to buffers 1 and 3 are $\lambda_1 = 0.50$ and $\lambda_3 = 0$, and similarly for activities 10, 11 and 12. For concreteness, we allow the input servers to work at less than full capacity and server 4 to divide its time between activities 9 and 10 if doing so is deemed desirable. Such actions correspond to randomized acceptance of new arrivals and randomized routing to type A arrivals, respectively. Similarly, servers 1, 2 and 3 are allowed to divide their time among activities available to them, processing several job classes simultaneously, provided that the total rate of effort allocation does not exceed 100%.

In the problem conceptualization outlined above, one takes the view that all job flows are the consequence of some "activity" undertaken by one of the system's six "servers." In the obvious way, the 8×8 input–output matrix R defined by (2.3) is extended to an 8×12 matrix (one column for each of the new activities) as follows:

$$R = \begin{bmatrix} \frac{1}{2} & & & & & & -\frac{1}{2} & & \\ -\frac{1}{2} & \frac{1}{3} & & & & & & & \\ & & 1 & & & & & -\frac{1}{2} & \\ & & & 1 & & & & & -\frac{1}{4} \\ & & -1 & 1 & & & & & \\ & & & & 1 & & & & -\frac{1}{4} \\ & & & & -1 & \frac{1}{2} & & & \\ & & & & & -\frac{1}{2} & 1 & & \end{bmatrix}. \qquad (2.8)$$

Similarly, the 3×8 capacity consumption matrix A defined by (2.4) is extended to the following 6×12 matrix, whose last three rows correspond to the new input servers and last four columns correspond to the new input activities:

$$A = \begin{bmatrix} 1 & & & & & 1 & & & & & \\ & 1 & & 1 & & & & & & & \\ & & 1 & 1 & & 1 & & 1 & & & \\ & & & & & & & 1 & 1 & & \\ & & & & & & & & & 1 & \\ & & & & & & & & & & 1 \end{bmatrix}. \qquad (2.9)$$

Given a positive economic incentive for accepting new arrivals (see below), and given that enough processing capacity exists on average to accommodate all arrivals, it is natural to extend the vector β of nominal activity rates defined by (2.7) as follows (the last four entries correspond to accepting all arrivals and splitting A jobs equally between the upper and lower routes):

$$\beta = \left(\frac{1}{2}, \frac{3}{4}, \frac{1}{4}, \frac{1}{4}, \frac{1}{4}, \frac{1}{4}, \frac{1}{2}, \frac{1}{4}, \frac{1}{2}, \frac{1}{2}, 1, 1 \right). \qquad (2.10)$$

To state precisely the objective in our dynamic control problem, let us denote by $I_i(t)$ the cumulative idleness experienced by server i up to time t (under a given policy), where "idleness" for an input server simply means time when the corresponding input process is turned off. In the obvious way, $I = \{I(t), t \geq 0\}$ is the six-dimensional process with components $I_i(t)$. A control policy now takes the form of a twelve-dimensional cumulative allocation process $T = \{T(t), t \geq 0\}$ with component processes $T_j(t)$. The objective is to find a control T which minimizes

$$f = E \left\{ \int_0^\infty e^{-\gamma t/n} \left[\frac{1}{n} \, h \cdot Q(t) \, dt + r \cdot dI(t) \right] \right\}, \qquad (2.11)$$

where

$$r = (0, 0, 0, r_4, r_5, r_6) \qquad (2.12)$$

and $r \cdot dI(t) = \sum r_i dI_i(t)$. Here r_4, r_5 and r_6 are the rewards associated with accepting jobs of types A, B and C, respectively. The minimand f defined by (2.11) includes both expected discounted holding costs and expected discounted revenue *losses* that result from turning off inputs (that is, rejecting arrivals).

The obvious missing elements in the model formulation presented thus far are (a) a specification of system dynamics, showing how the multi-dimensional queue length process $Q(t)$ and idleness process $I(t)$ are determined by our probabilistic primitives and the chosen control $T(t)$, and (b) a precise articulation of constraints on the control process T. To fill these gaps would require a good deal of new notation, and similar developments can be found in [2], so we shall proceed directly to heavy traffic scaling and development of the approximating Brownian system model. Viewing n as a large but fixed integer for the time being, let

$$Z_k(t) = n^{-\frac{1}{2}} Q_k(nt) \quad \text{and} \quad U_i(t) = n^{-\frac{1}{2}} I_i(nt), \qquad (2.13)$$

and define multi-dimensional scaled processes $Z(t)$ and $U(t)$, eight-dimensional and six-dimensional respectively, in the obvious way. These definitions involve a compression of the time scale by a factor of n and a compression of spatial dimensions by a factor of $\sqrt{n}$. We now apply a similar scaling to the controls $T_j(t)$, after first re-expressing each such cumulative allocation as a deviation from the nominal cumulative allocation $\beta_j t$ given by (2.10). Let

$$Y_j(t) = n^{-\frac{1}{2}} [\beta_j nt - T_j(nt)] \quad \text{for } j = 1, \ldots, 12, \qquad (2.14)$$

with the vector process $Y(t)$ defined in the obvious way. From their definitions, we have the following link between the vector Y of scaled deviation controls and the vector U of scaled cumulative idleness processes:

$$U(t) = A\,Y(t). \qquad (2.15)$$

With one difference to be noted, readers will find that (2.13)–(2.15) follow the same logical pattern and use the same notation that was established in [2] to treat pure sequencing problems, and was subsequently extended in work surveyed by Kelly and Laws [7] to treat problems with an additional aspect of dynamic routing. The one significant difference is that Wein [22] and Kelly and Laws [7] used two matrices to represent information which is here incorporated in a single input–output matrix R. Arguing exactly as in the work cited above, one may express system dynamics in terms of scaled quantities through the matrix–vector equation

$$Z(t) = X(t) + R\,Y(t), \tag{2.16}$$

where $X = \{X(t), t \geq 0\}$ is an eight-dimensional scaled process whose definition involves the original dynamic allocation processes $T_j(t)$. It was argued in [2] that the process X, whose definition we delete in the interest of brevity, converges as $n \to \infty$, under *any* interesting control policy, to a multi-dimensional Brownian motion whose drift vector and covariance matrix depend only on original problem data. In the current context the drift vector is zero because we treat a single perfectly balanced system. There will be no need to write out the covariance matrix Γ, but that could easily be done. Now let

$$g = n^{-\frac{1}{2}} f, \tag{2.17}$$

where f is the objective quantity (to be minimized) defined by (2.11). Making the change of variable $s = t/n$ in (2.11), readers may verify (this is not a trivial task) that

$$g = E\left\{ \int_0^\infty e^{-\gamma s}[h \cdot Z(s)\ ds + r \cdot dU(s)] \right\}. \tag{2.18}$$

The change of variable $s = t/n$ suggests an alternative interpretation for the large parameter n that indexes our sequence of network control problems. One may suppose that the time scale remains the same in each problem, with γ the fixed interest rate and h the fixed vector of holding cost rates, but all arrival processes and service processes are speeded up by a factor of n. The crucial assumption is that the natural unit of time for measuring interarrival and service times be small compared with the natural unit of time for purposes of discounting and holding cost measurement.

3 The Brownian approximation

Proceeding exactly as in [2], we now describe a "Brownian analog" of the dynamic control problem introduced above, reusing the notation of section 2 to emphasize the intended parallelism. The data of our Brownian approximation are the interest rate $\gamma > 0$, the input–output matrix R defined by (2.10), the capacity consumption matrix A defined by (2.11), the vector h of holding costs, the vector r of acceptance rewards, and a covariance matrix Γ determined by the external arrival rates, mean service times and transition structure of our processing network model. (With general interarrival and service time distributions, elements of Γ would also depend on the variances of those distributions.) Taken as primitive in the Brownian system model is an eight-dimensional Brownian motion $X = \{X(t), t \geq 0\}$ with zero drift, covariance matrix Γ, and $X(0) = 0$ (assuming that the original processing network model was initially empty). In the Brownian model a system manager must choose a twelve-dimensional control process $Y = \{Y(t), t \geq 0\}$ which is adapted to X and further satisfies constraints (3.3)–(3.4) below. Vector processes Z and U are defined in terms of X and the chosen control Y via

$$Z(t) = X(t) + R\, Y(t), \quad \text{and} \tag{3.1}$$

$$U(t) = A\, Y(t), \tag{3.2}$$

all of these processes being interpreted as in the preceding section. The control Y is said to be *admissible* if it is adapted to X and further satisfies

$$Z(t) \geq 0 \ \text{ for all } t \geq 0, \text{ and} \tag{3.3}$$

$$U(\cdot) \text{ is non-decreasing with } U(0) = 0. \tag{3.4}$$

These constraints follow naturally from the interpretation of Z as a queue length process and U as a cumulative idleness process. As in section 2, the objective is to find an admissible control (or flow management policy) Y which minimizes

$$g = E\left\{ \int_0^\infty e^{-\gamma t}\, [h \cdot Z(t)\, dt + r \cdot dU(t)] \right\}. \tag{3.5}$$

Roughly speaking, the assertion in heavy traffic theory is that Brownian control problems like the one described above provide good approximations to the dynamic flow control problems they replace, assuming that parameter values of the original system fall in the heavy traffic domain. To make that statement precise, one would like to prove a heavy traffic limit theorem, considering a sequence of dynamic control problems indexed by $n = 1, 2, \ldots$ whose associated parameters vary with n so as to approach a heavy traffic limit. But what, exactly, is to be proved? The ideal result would be one that directly reveals a sequence of policies (one for each system n) that are "asymptotically optimal" in the heavy traffic limit, meaning that the objective value achieved by the nth policy cannot be improved by more than $\epsilon(n)$ percent in the nth system, where $\epsilon(n) \to 0$ as $n \to \infty$. A more modest goal would be to show that the optimal (scaled) objective value g for the nth system approaches the optimal g value for the approximating Brownian control problem, but such a result does not provide a guide to action in the original problem of interest and hence is of no direct value from a system management perspective.

To repeat, the ultimate objective in heavy traffic analysis of network control problems must be construction of asymptotically optimal policies, or at least construction of implementable policies which are "good" under heavy traffic conditions. Unfortunately, as stated in section 1, the optimal control policy for our approximating Brownian system model does not translate in any obvious way into an implementable policy for the original network. To understand why one must look more carefully at the Brownian approximation. The analysis undertaken in the remainder of this section, summarizing results proved in [4], shows that the Brownian control problem (3.1)–(3.5) is far simpler than the original problem described in section 2, which is both good news and bad. On the positive side of the ledger, there is reason to believe that Brownian system models capture the essential features of network control problems while supressing inessential fine structure. As Kelly and Laws [7] observed, in heavy traffic analysis "the important features of good control policies are displayed in sharpest relief." On the negative side of the ledger, interpretation of such insights in terms of our original, finely structured processing network requires that a gap be bridged, and we shall turn to that problem in section 4.

In the Brownian control problem (3.1)–(3.5) we have an eight-dimensional state vector $Z(t)$ which is confined to the non-negative orthant by means of a twelve-dimensional control process $Y(t)$. The term "Brownian network" was used in [2] to describe stochastic control systems having dynamics of the general kind specified in (3.1)–(3.4). Past analyses have shown that such control problems admit an "equivalent workload formulation" whose state descriptor, usually called the "workload process," is of lower dimension than Z. For the problem at hand, it is shown in [4] that one suitable definition of workload (other equivalent definitions are possible) is $W(t) = MZ(t)$, where

$$M = \begin{bmatrix} 2 & 0 & 2 & 2 & 0 & 6 & 4 & 2 \\ 3 & 3 & 3 & 4 & 1 & 6 & 3 & 3 \end{bmatrix}. \tag{3.6}$$

With the additional definitions

$$G = \begin{bmatrix} 1 & 0 & 2 & -1 & -1/2 & -3/2 \\ 0 & 1 & 3 & -3/2 & -1 & -3/2 \end{bmatrix}, \tag{3.7}$$

and

$$\xi(t) = MX(t), \quad t \geq 0, \tag{3.8}$$

the system dynamics for the equivalent workload formulation are as follows:

$$W(t) = \xi(t) + GU(t), \tag{3.9}$$

$$U(\cdot) \text{ is non-decreasing with } U(0) = 0, \text{ and} \tag{3.10}$$

$$W(t) = MZ(t), \quad \text{where} \tag{3.11}$$

$$Z(t) \geq 0. \tag{3.12}$$

That is, in the equivalent workload formulation we take as primitive a two-dimensional Brownian motion ξ and choose a six-dimensional, non-decreasing control $U(\cdot)$ so as to maximize the discounted quantity g defined by (3.5), subject to (3.9)–(3.12). Obviously, (3.11) and (3.12) demand that U be chosen to keep W within the state space

$$S = \{w \in \mathbb{R}^2 : w = Mz, \ z \geq 0\}. \tag{3.13}$$

The positive cone S is pictured in Fig. 2, along with the directions of control that are available by increasing various components of U.

It is certainly not obvious that our specification (3.1)–(3.4) of system dynam-

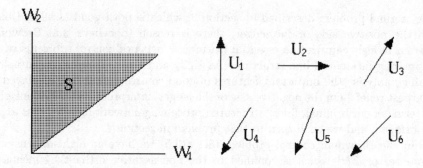

FIG. 2. Graphical representation of equivalent workload formulation

ics for the Brownian control problem is equivalent to (3.9)–(3.12). A proof of that equivalence is given in [4], where a general theory of equivalent workload formulations is developed and this same example is discussed in detail. Because it is rather peripheral to the concerns of this paper, the argument will not be repeated here, but a few more words may serve as an aid to intuition. First, it can be verified that $MR = GA$, so equation (3.9) is obtained by pre-multiplying both sides of (3.1) by M. Thus it is obvious that any pair (Z, U) which is achievable using an admissible control Y in the original system model (3.1)–(3.4) is also achievable in the workload formulation (3.9)–(3.12). The converse is not at all obvious, but exact analogs of this result are discussed by Kelly and Laws [7] in their survey of Brownian network models with alternate routing.

The components of $W(t)$ are two, different, positive linear combinations of the eight-dimensional (scaled) queue length process $Z(t)$. In the end we find that only these two quantities need be recorded to maintain an adequate summary of system status, because in the Brownian system model any two Z vectors that yield the same W vector can be instantaneously converted one for the other by applying a jump control $\Delta Y = y$ which does not affect the cumulative idleness process U (that is, $Ay = 0$). In the absence of server idleness, the workload $W(t)$ evolves as the Brownian motion $\xi(t) = MX(t)$, regardless of how servers distribute their effort among available activities, and increases in the cumulative idleness of various servers displace the workload vector W in the directions pictured in Fig. 2. Given a sample path of ξ, idleness must be inserted in quantities sufficient to keep W within the cone S, which means that W must at all times be consistent with some non-negative choice of the vector Z. Costs are associated with the Z vector chosen and with increases in certain of the idleness processes, and the system controller must typically address trade-offs between these two types of cost. If there exists a control strategy which minimizes both cumulative holding cost and cumulative idleness of all servers for every realization of ξ, we call this a *pathwise solution*. An example of a Brownian control problem having a pathwise solution will be presented later, but the example currently under discussion does not fall in that category.

As explained in [4], the two components of W can be interpreted in terms of two overlapping, *resource pools*, one composed of servers 1 and 3, the other composed of servers 2 and 3. When properly understood, the equivalent workload formulation of the approximating Brownian control problem gives deep insight into the structure of our original network control problem, even before one sets out to compute the optimal solution. But this representation of the original system is far removed from the intuitive understanding that one brings to the problem initially, and the optimal controls that emerge from a complete analysis can be even harder to interpret. For example, a particular choice of holding costs h_k and acceptance rewards r_i leads to an optimal policy with the following characteristics: the only arrivals that are ever rejected are those of type A, the only queues that are ever positive are those containing type A jobs (classes 1, 2 and 3), servers 1 and 2 are never idle, and server 3 is idle only when the workload process W is on the boundary of the cone S. This prescription cannot be taken literally, of course.

In earlier work like [5] and [22], similar optimal policies for Brownian control problems have been interpreted as *threshold policies* in the original network context, but such interpretations are derived from the authors' intuitive understanding of the special network structures under study, and their ability to derive an explicit analytical solution for the approximating Brownian control problem. For more complicated examples like the one under discussion here, where numerical methods must eventually be used to determine an optimal control for the Brownian problem, even in its simplified workload formulation, existing theory gives no hint as to how one may use that solution in the original network context.

Before addressing that problem, it will be useful to note a few properties of the Brownian control problem in equivalent workload form, with discounted objective (3.5) and system dynamics specified by (3.9)–(3.12). First, let us extend the problem to allow a general initial state $W(0) = x \in S$, denoting by $v(x)$ the infimum of the objective value g in (3.5) over admissible control policies. It is quite easy to show that the optimal return function $v(\cdot)$ is convex on S (the key observation is that if U and U' are admissible controls starting from x and x' respectively, then $\frac{1}{2}U + \frac{1}{2}U'$ is an admissible control starting from $\frac{1}{2}x + \frac{1}{2}x'$), and hence we can define the generalized gradient function $\nabla v(x), x \in S$. In the BIGSTEP method there is a need to estimate the impact of changes in system state on expected discounted cost, and the function $\nabla v(\cdot)$ provides a means of doing that, but one must take care about scaling.

Given an initial queue length vector $Q(0) = q$, one may estimate the corresponding minimum achievable expected discounted cost as follows. First, defining the two-dimensional workload vector $w = Mq$, we note that the scaled workload process $W(\cdot) = MZ(\cdot)$ defined in our analysis of the Brownian system model has initial state

$$x = W(0) = MZ(0) = n^{-\frac{1}{2}}MQ(0) = n^{-\frac{1}{2}}Mq = n^{-\frac{1}{2}}w.$$

The scaled objective value associated with this initial state in the Brownian model is $v(x)$, but taking account of the relationship (2.19) between scaled and original objective functions we have the following estimate: minimum discounted cost beginning with unscaled workload vector $w \simeq n^{1/2} v(n^{-1/2} w)$. If an action is taken which changes the unscaled initial workload vector to $w + \delta$, say, then the obvious first-order estimate of this action's effect on economic performance is as follows: increase in minimum expected discounted cost due to change in unscaled initial workload vector from w to $w + \delta \simeq \delta \cdot \pi(w)$, where

$$\pi(w) = \nabla v(n^{-1/2} w). \tag{3.14}$$

4   BIGSTEP control of the three-station network

Returning now to the network control problem described in section 2, let us consider a discrete-review policy with review periods of length l. Throughout most of the following discussion n will be viewed as a fixed large integer, but when it becomes necessary to indicate the dependence of a policy parameter on n, that will be done by attaching a functional argument of n to notation established earlier. From a theoretical perspective, we wish to consider a sequence of problems indexed by $n = 1, 2, \ldots$. The interest rate for discounting in the nth problem is γ/n, so increasingly long spans of time become relevant to the economic objective as $n \to \infty$, and one might describe the parameter n as a natural unit of time (see section 2). For reasons explained later, we restrict attention to discrete-review policies such that

$$l(n) \to \infty \quad \text{but} \quad \frac{l(n)}{n} \to 0 \quad \text{as } n \to \infty. \tag{4.1}$$

To construct a BIGSTEP control policy one must specify not only the period length l but also a non-negative vector $\theta = (\theta_1, \ldots, \theta_8)$ of *threshold parameters*. These parameters may be interpreted as *planned safety stocks* for the various job classes, as we shall see shortly. For purposes of heavy traffic limit theory, one must allow θ to vary with n, and to achieve asymptotic optimality it may be necessary to have $\theta_k(n) \to \infty$ as $n \to \infty$ for some job classes k, but an essential restriction is that

$$n^{-1/2} \theta_k(n) \to 0 \quad \text{as} \quad n \to \infty \quad \text{for all} \quad k = 1, \ldots, 8. \tag{4.2}$$

Now with n fixed, and given a choice of l and θ, consider a policy that reviews system status at times $t = 0, l, 2l, \ldots$. At each such t we observe the current queue length vector $q = Q(t)$, compute the corresponding workload vector $w = Mq$, and then determine $\pi = \pi(w)$ via (3.14). As a final preliminary we define a six-vector p of *penalty rates*, with one component for each server, via (here π is viewed as a row vector)

$$p = \pi G. \tag{4.3}$$

Using these data we solve a *BIGSTEP planning problem* for the upcoming review period. This is a two-stage linear program with the following decision variables: x_j is the amount of time that server $\sigma(j)$ will devote to activity j over the coming period, u_i is the amount of time that server i will be idle, and z_k is the expected or planned inventory of class k jobs at the end of the period. These variables specify a "processing plan" for the coming period, and the question of how to execute that plan will be taken up later. The two stages of the BIGSTEP planning problem can be simultaneously expressed in the following form (this is sometimes called a linear program with lexicographic criterion): denoting by ϵ a strictly positive but arbitrarily small scalar, choose $x = (x_1, \ldots, x_{12})$ so as to

$$\text{minimize } (r + p) \cdot u + \epsilon \, h \cdot z, \qquad (4.4)$$

subject to the constraints

$$z = b + l\lambda - Rx, \quad u = le - Ax, \quad z \geq 0, \quad u \geq 0 \quad \text{and} \quad x \geq 0. \qquad (4.5)$$

The lexicographic objective (4.4) will be explained and justified later in this section. With regard to the constraints (4.5), readers should first note that the obvious physical constraint $Ax \leq le$, requiring that servers allocate no more than 100% of their time, is actually implied by $u = le - Ax$ and $u \geq 0$. Also, the constraint $z \geq \theta$ imposes a lower bound on each of the planned ending inventories, which justifies the previous characterization of θ_k as a "safety stock" parameter of the BIGSTEP planning problem.

The linear programming problem (4.4)–(4.5) involves decision variables x_j which are natural and intuitive in our network control setting. To relate this problem to the approximating Brownian control problem described in section 3, we define a twelve-vector

$$y = l\beta - x$$

of translated decision variables, corresponding to the process Y of deviation controls introduced in section 2. That is, y_j expresses the time to be allocated to activity j as a deviation, which can be either positive or negative in general, from the nominal allocation or long-run average allocation $l\beta_j$. Recalling from section 2 that $R\beta = \lambda$ and $A\beta = e$, we substitute the definition of y into (4.5) to get the following alternate form of the BIGSTEP planning problem: choose y so as to minimize the objective (4.4) subject to

$$z = b + Ry, \quad u = Ay, \quad z \geq \theta, \quad u \geq 0 \quad \text{and} \quad y \leq l\beta. \qquad (4.6)$$

The constraints (4.6) are formally analogous to the system equations (3.1)–(3.4) that define our Brownian system model, except for three factors. First, of course, our BIGSTEP planning problem treats the processing network as if its evolution over the upcoming period were deterministic, so the driftless Brownian motion X that appears in (3.1) is simply suppressed in our deterministic system equation $z = b + Ry$. Second, given the deterministic idealization implicit in (4.6), we protect against unplanned server idleness in the BIGSTEP method by imposing planned safety stock constraints $z \geq \theta$, whereas the Brownian analog (3.3) requires only that inventories remain non-negative. Finally, the upper bound $y \leq l\beta$ on deviation controls in the BIGSTEP planning problem has no analog in our Brownian system model, but an argument originally advanced in section

5 (page 167) of [2] to motivate or justify the approximating Brownian control problem suggests that those upper bound constraints are almost always inactive in heavy traffic.

Condition (4.1) expresses the requirement that review periods for the BIG-STEP control policy be long in absolute terms but small on the Brownian time scale. To be more specific, the length of a BIGSTEP planning period decreases toward zero on the Brownian time scale as $n \to \infty$, and thus the influence of stochastic variability on changes in (scaled) system state over such an interval is small compared with the influence of controls applied. With this argument, of course, we are striving to justify the deterministic approximation of system dynamics embodied in (4.5), or equivalently, (4.6), but there is one important caveat. When available stocks of one or more job classes are low, relatively small amounts of stochastic variability may cause comparable amounts of unplanned server idleness, but even small amounts of idleness may have a significant influence on system behavior under heavy traffic conditions. Enforcement of the planned safety stock requirement $z \geq \theta$ is intended to eliminate that concern, but then one worries that the cost of holding safety stock may itself become economically significant, and our requirement (4.2) on the magnitude of safety stock parameters assures that such holding costs are asymptotically negligible, because only stocks of order $\sqrt{n}$ or larger remain significant under Brownian scaling. The remaining question is whether adequate protection against unplanned idleness can be afforded by safety stocks required to satisfy (4.2), and in the next section it will be shown for a simpler example that the answer is affirmative.

Turning now to the lexicographic objective function (4.4), the rationale is as follows. First recall that h/n is the vector of inventory holding cost rates. We estimate the average inventory vector for the coming period by $\frac{1}{2}b + \frac{1}{2}z$ (more will be said about this estimate later), so the corresponding estimate of inventory holding costs for the period is $\frac{1}{2n}h \cdot (b + z)$. The direct cost of server idleness is $r \cdot u$, but one must also consider the effect of server idleness on the expected discounted cost achievable in *future* periods. As explained earlier, analysis of the Brownian model suggests that only changes in the two-dimensional system workload vector w are relevant, and the text immediately preceding (3.14) gives a first-order estimate of the increase in future expected discounted cost due to a change of δ in the workload vector. In the BIGSTEP planning problem a vector x of time allocations changes the queue length or inventory vector from b to z, and hence changes the workload vector from $w = Mb$ to $w' = Mz$. Because $z = b + Ry$, we then have $\delta = w' - w = M(b - z) = MRy$. But recall from section 3 that $MR = GA$, where G is the 2×6 matrix defined in (3.7). Moreover, $Ay = u$ as a matter of definition, so we have the compact expression $\delta = Gu$. Defining $\pi = \pi(w)$ in accordance with (3.14), and recalling from (4.3) that $p = \pi G$ by definition, our first-order estimate of the increase in future expected discounted cost becomes $\delta \cdot \pi = (Gu) \cdot \pi = p \cdot u$. Adding this to the other cost elements described earlier, we arrive at the following objective function for our BIGSTEP planning problem:

$$\text{minimize } (r + p) \cdot u + \frac{l}{2n} \, h \cdot (b + z).$$ (4.7)

Discarding the constant term involving b, one sees that (4.7) is equivalent to (4.4) with $\epsilon = l/2n$. Of course, $\epsilon \to 0$ as $n \to \infty$ by (4.1). Combining this with the identities $u = Ay$ and $z = b + Ry$, one sees that the first term in (4.7) dominates in the heavy traffic regime where n is large, which leads to the two-stage optimization logic described above.

This justification is obviously crude, and one important direction for future work is to prove rigorously that the BIGSTEP policy parameters l and θ can be chosen to give asymptotic optimality in the heavy traffic limit. (It is natural to start by investigating special model structures, of course.) In the absence of rigorous proofs, a few comments may help to bolster readers' confidence in the BIGSTEP approach. First, the two-stage planning logic described above parallels precisely the hierarchical view of system control that emerges from analysis of Brownian models: at the higher level one chooses cumulative idleness processes so as to optimally "steer" a multi-dimensional system workload, appropriately defined; and at the lower level one must choose a queue length or inventory vector at each point in time which is consistent with the resulting workload, the lower-level objective being to "store" work in a form that minimizes associated holding costs. A second comment concerns the estimate of average inventory levels during a planning period that was used in deriving the BIGSTEP objective function (4.4). The estimate used above was $\frac{1}{2}(b + z)$, but readers may verify that any other convex combination of b and z leads to the same final conclusion. Finally, some readers may be dismayed at the extreme character of BIGSTEP processing plans, especially with regard to resource utilization. To be specific, when the gradient vector $\pi(w)$ is such as to make *some* idleness of a given server attractive, the BIGSTEP method may lead to idling that server for an *entire* planning period. However, because (4.1) demands that planning periods be small on the Brownian time scale, there is reason to believe that such "overshoot" will be asymptotically negligible in the heavy traffic limit.

With regard to implementation, the question is when within the planning period servers are to undertake the various activities for which they are responsible, and the concern is whether they will actually have on hand the jobs that are needed in those activities. That is, we must concern ourselves with the fine-scale coordination of servers' activities, because an input activity or processing activity at one station may provide jobs that are needed for a processing activity at another station, and we must concern ourselves with the effect of stochastic variability on a plan formulated by means of a deterministic optimization model. However, one can choose safety stock parameters θ_k in (4.7) large enough to render these concerns irrelevant. To be more specific, one can set $\theta_1, \ldots, \theta_8$ in such a way that, with very high probability, each server will be able to execute the activities that it is assigned during each review period working only on jobs that are on hand at the start of the period. This eliminates all need for coordination of servers' activities within a period and enables the servers to execute their

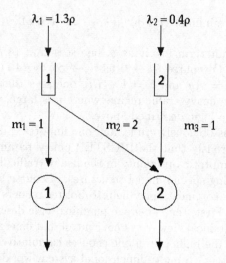

FIG. 3. Two servers working in parallel

assigned activities in any order they may wish.

If there is to be any hope of asymptotic optimality, of course, it must be that the planned safety stocks are negligible in comparison with $\sqrt{n}$ as $n \to \infty$, and that is the content of (4.2). In the next section it will be shown in the context of a simpler example that adequate protection against server idleness can in fact be achieved with safety stock values that satisfy (4.2). Finally, it should be said that the safety stock parameters θ_k which one would adopt in a specific situation to get the best practical performance might be quite different from the values one would choose to prove theoretical properties of the BIGSTEP method, and careful coordination of servers' activities within periods may facilitate reduction of safety stock levels.

5   A network with two servers working in parallel

To enable a more detailed discussion of issues that have been glossed over in the preceding account of BIGSTEP control policies, let us consider now the simple processing network pictured in Fig. 3. There are two independent Poisson processes that deliver jobs of class 1 and class 2 at rates λ_1 and λ_2 respectively, and there are two servers. As indicated in the figure, the external arrival rates (expressed in jobs per hour) are taken to be $\lambda_1 = 1.3\rho$ and $\lambda_2 = 0.4\rho$, where ρ is a parameter that will be allowed to vary. Each job requires a single service before it departs, and for class 2 jobs that service can only be provided by server 2. Class 1 jobs, in contrast, can be processed by either server. In the three-station example discussed earlier, it was required that routing decisions be made at the moment of arrival, but for variety let us now suppose that routing decisions can be postponed. That is, jobs of class 1 enter buffer 1 and remain there until they are selected for processing by either server 1 or server 2.

Also for contrast with the previous example, we assume that services must be completed without interruption once they have been started. There are three processing activities, as follows:

activity 1 = processing of class 1 jobs (by server 1),
activity 2 = processing of class 1 jobs (by server 2),
activity 3 = processing of class 2 jobs (by server 2).

In general one can allow a different service time distribution for each activity. To make things as simple as possible, let us assume that the service time for each activity j is deterministically of length m_j, where $m_1 = m_3 = 1$ and $m_2 = 2$ as shown in Fig. 3. That is, it takes exactly one hour for server 1 to process a class 1 job or server 2 to process a class 2 job, but it takes two hours for server 2 to process a class 1 job.

If ρ is near one, the parallel-server system is "in heavy traffic," meaning that its overall input rate and overall processing capacity are approximately equal. To see that this is true, consider the case $\rho = 1$. Devoting all of its time to class 1, server 1 can handle only one of the 1.3 class 1 jobs that arrive per hour on average, and server 2 must spend 60% of its time in processing the class 1 jobs that are left over (0.3 jobs per hour × 2 hours per job = 0.6). Server 2 has 40% of its time left for class 2 jobs, which is just adequate to handle the arrival rate of $\lambda_2 = 0.4$. Thus the nominal activity rates

$$\beta_1 = 1, \quad \beta_2 = 0.6 \quad \text{and} \quad \beta_3 = 0.4 \tag{5.1}$$

give a total output rate from each buffer equal to its input rate, with the capacity of each server fully utilized. That is, defining the 2×3 input–output matrix (one row for each buffer or job class, and one column for each activity)

$$R = \begin{bmatrix} 1 & \frac{1}{2} \\ & & 1 \end{bmatrix} \tag{5.2}$$

and the 2×3 resource consumption matrix (one row for each server and one column for each activity)

$$A = \begin{bmatrix} 1 \\ & 1 & 1 \end{bmatrix}, \tag{5.3}$$

we have

$$R\beta = \lambda \quad \text{and} \quad A\beta = e. \tag{5.4}$$

To complete the problem formulation, let us assume linear holding costs at rate h_k/n for each job class k, with the specific numerical data

$$h_1 = 3 \quad \text{and} \quad h_2 = 1, \tag{5.5}$$

and continuous discounting of costs at

$$\text{interest rate} = \gamma/n, \tag{5.6}$$

where $\gamma > 0$ is fixed. Letting $n \to \infty$ as in earlier discussion of the three-station example, one may imagine a sequence of systems indexed by n, each having the structure described above, where the only system parameter changing with n is ρ. To achieve a heavy traffic limit it is necessary to have $\rho(n) \to 1$, and more specifically, $\rho(n)$ must approach unity fast enough that

$$\sqrt{n}[1 - \rho(n)] \to \kappa \text{ (a finite constant)} \quad \text{as } n \to \infty. \tag{5.7}$$

Following standard practice in queueing theory, let us denote by $\mu_j = 1/m_j$ the average service rate for activity j, and further define

$$c_1 = h_1 = 3, \quad c_2 = h_1 = 3, \quad \text{and} \quad c_3 = h_2 = 1. \tag{5.8}$$

Thus c_j/n is the holding cost rate for the job class that is served in activity j. When the system contains q_1 jobs of class 1 and q_2 jobs of class 2, holding costs are continuously incurred at a total rate of $(h_1 q_1 + h_2 q_2)/n$, and activity j decreases this total at an average rate of $c_j \mu_j/n$ per hour spent in the activity. Guided by the classical $c\mu$ rule of scheduling theory, one might plausibly adopt the following *greedy scheduling heuristic*. First, server 1 should obviously spend as much time as it can processing class 1, going idle only when buffer 1 is empty. Second, when server 2 completes the processing of a job and finds new jobs waiting in both buffer 1 and buffer 2, it should choose among activities 2 and 3 so that the chosen activity j maximizes $c_j \mu_j$. (If there are waiting jobs in only one buffer, the server will presumably take one of them, and if both buffers are empty it must go idle.) Since $c_2 \mu_2 = 3/2$ and $c_3 \mu_3 = 1$, this means that server 2 will always choose to serve class 1 when confronted with a choice.

For the case $\rho = 0.95$, simulation results presented in [3] show that our parallel-server system is actually unstable under the greedy scheduling heuristic described above. To be more specific, the backlog of class 2 jobs grows in roughly linear fashion as time passes, and so the greedy heuristic is presumably far from optimal. The problem is that this scheduling rule causes server 2 to spend about 71% of its time on class 1 jobs, which does not leave that server with enough residual capacity to handle the demands of class 2. (Simultaneously, server 1 spends about 13% of its time idle under the greedy heuristic.) So we see that this problem of dynamic flow management, although simple, does involve some subtlety, and it will be interesting to see the control policy obtained by mechanical application of the BIGSTEP method.

As a preliminary to the discussion of BIGSTEP controls, it will be useful to recapitulate insights derived from the approximating Brownian system model, whose construction and analysis for the case at hand are undertaken in [3]. Dynamics of the Brownian model are exactly as in (3.1)–(3.4), but R is now defined by (5.2), A is defined by (5.3) and X is a two-dimensional Brownian motion whose drift vector and covariance matrix can be written out in terms of basic system parameters. Here Z approximates a two-dimensional scaled queue length process, U approximates a two-dimensional scaled idleness process, and Y approximates a three-dimensional process of scaled deviation controls, each defined exactly as in section 2. In the current problem there are no input control decisions, so by analogy with (3.5), the objective is to choose a control Y so as to

$$\text{minimize} \quad g = E\left\{ \int_0^\infty e^{-\gamma t}\, h \cdot Z(t)\, dt \right\}. \tag{5.9}$$

Moving now to the equivalent workload formulation, appropriate analogs of the matrices M and G defined in (3.6) and (3.7) are found to be (here M and G are 1×2 matrices that happen to be identical)

$$M = G = [2,\ 1]. \tag{5.10}$$

Defining the one-dimensional Brownian motion $\xi(t) = MX(t)$ exactly as in (3.8), we can state the equivalent workload formulation as follows: choose a two-dimensional control U to minimize the discounted objective (5.13), subject to

$$W(t) = \xi(t) + GU(t), \tag{5.11}$$

$$U(\cdot) \text{ is non-decreasing with } U(0) = 0, \tag{5.12}$$

$$W(t) = MZ(t), \tag{5.13}$$

$$Z(t) \geq 0. \tag{5.14}$$

The system equations (5.11)–(5.14) are identical to (3.9)–(3.12). For the problem at hand, this equivalent workload formulation is so simple that one can essentially solve for an optimal control strategy by inspection, as explained below.

A relatively complete analysis and interpretation of the equivalent workload formulation is given in [3], but here we shall mention just three of its salient properties. First, for any given workload level $W(t) = w$, one obviously wants to choose a queue length vector $Z(t) = z$ so as to minimize the instantaneous cost rate $h \cdot z = 3z_1 + z_2$ in (5.9), subject to the constraints $2z_1 + z_2 = w$ and $z \geq 0$, which are imposed by (5.13) and (5.14) respectively. Given a value $W(t) > 0$, then, we solve this simple linear program (it is a knapsack problem with continuous decision variables) to obtain the optimal queue lengths

$$Z_1^*(t) = 0 \quad \text{and} \quad Z_2^*(t) = W(t). \tag{5.15}$$

The corresponding instantaneous cost rate is $h \cdot Z^*(t) = W(t)$, so a smaller workload level is always preferable to a larger one. Defining

$$V(t) = GU(t) = 2U_1(t) + U_2(t), \tag{5.16}$$

we see from (5.12) that $V(\cdot)$ must be non-decreasing with $V(0) = 0$, and (5.11) obviously reduces to $W(t) = \xi(t) + V(t)$. Thus, to minimize $W(\cdot)$ we choose (here it is assumed for simplicity that $Z(0) = 0$, hence $W(0) = \xi(0) = 0$)

$$V^*(t) = - \inf_{0 \leq s \leq t} \xi(s), \tag{5.17}$$

which is the least non-decreasing process $V(\cdot)$ such that $W(\cdot) = \xi(\cdot) + V(\cdot)$ remains non-negative as implied by (5.14). The optimal control V^* defined by

(5.17) increases only at time points t such that $W^*(t) = \xi(t) + V^*(t) = 0$, which correspond to times at which our processing system is empty. To achieve this solution we can choose any non-decreasing control U^* such that $2U_1^* + U_2^* = V^*$, and for concreteness let us take

$$U_1^*(t) = 0 \quad \text{and} \quad U_2^*(t) = V^*(t). \tag{5.18}$$

This control achieves the workload process W^*, which is a pathwise lower bound for any other workload process W achievable with an admissible control U. That is, $W^*(t) \leq W(t)$ for every time t and every realization of the underlying Brownian motion ξ. Given the remarks above, it follows that $h \cdot Z^*(t) \leq h \cdot Z(t)$ for every achievable queue length process Z, every $t \geq 0$, and every realization of ξ. Thus the control policy defined by (5.18), (5.17) and (5.15) is what was earlier called a *pathwise solution* for the Brownian control problem. Tying this to the original Brownian formulation with system equations identical to (3.1)–(3.4), readers may verify that (U^*, Z^*) is achieved by the following process Y^* of scaled deviation controls:

$$Y_1^*(t) = 0, \quad Y_3^*(t) = W^*(t) - X_2(t) \quad \text{and} \quad Y_2^*(t) = V^*(t) - Y_3^*(t). \tag{5.19}$$

In a sense it is easy to "interpret" this optimal control policy: processing resources should be allocated to jobs on hand in such a way that server 1 is never idle ($Y_1^* = U_1^* = 0$), yet no class 1 jobs are ever held in inventory ($Z_1^* = 0$), and server 2 is idle only when no jobs of either class are present ($U_2^* = V^*$, which only increases when $W^* = 0$). Such ideal system behavior is not literally achievable, of course, because phrases like "server never idle" and "no jobs present" must be understood to mean "server idleness is negligible" in some sense, and "few if any jobs are present." The Brownian system model corresponds to what might be called a *frictionless ideal*, and the "insights" derived from this model must be understood as limiting statements, or statements about relative magnitudes. The big question is how to construct an implementable control policy under which the ideal system behavior described above is approached in the heavy traffic limit, and that is the subject of the next section.

As a final note, let us consider the Brownian control problem in equivalent workload form (5.11)–(5.14), allowing a general initial state $W(0) = w \geq 0$ and denoting by $v(w)$ the optimal return function. As explained in [3], one can write out $v(\cdot)$ as an explicit formula for the one-dimensional problem at hand, but that is not necessary for current purposes. However, defining

$$\pi(w) = v'(n^{-1/2}w), \quad w \geq 0, \tag{5.20}$$

as in (3.14), it *is* relevant that $\pi(\cdot) \geq 0$ over its whole range (this reflects the fact that increasing initial workload can only increase minimum achievable discounted cost). Adapting to the problem at hand the description of the BIGSTEP method in section 4, one must determine the workload $w = MQ(t)$ at each review point, then determine $\pi = \pi(w)$, and then compute a two-dimensional penalty-rate vector

$$p = \pi G = (2\pi, \pi) \tag{5.21}$$

for use in the objective function of the BIGSTEP planning problem for the period to follow.

6 BIGSTEP control of the parallel-server system

Proceeding exactly as in section 3, we choose a length l for the planning periods, choose a positive vector $\theta = (\theta_1, \theta_2)$ of threshold parameters, and review system status at times $t = 0, l, 2l, \ldots$. At each such review point t we observe the current backlog $b = Q(t)$, compute the corresponding workload level $w = Mb$, and then compute the penalty-rate vector $p = \pi G = (2\pi, \pi)$ as in (5.21). Here π is a positive number whose value remains unspecified for the time being. Now a processing plan (x, u, z) for the coming period is formulated by solving the following two-stage linear program:

$$\text{minimize } p \cdot u + \epsilon \, h \cdot z, \tag{6.1}$$

subject to

$$z = b + l\lambda - Rx, \quad u = le - Ax, \quad z \geq \theta, \quad u \geq 0 \text{ and } x \geq 0, \tag{6.2}$$

where ϵ is arbitrarily small. This parallels precisely problem (4.4)–(4.5) for the earlier three-station example. For example, x_j is the number of hours that server $\sigma(j)$ will devote to activity j over the period, and z_k is the expected number of class k jobs remaining at the end of the period.

Given the specific numerical data above, and for any positive values of ρ, π, l and θ, it is shown in [3] that the optimal solution of (6.1)–(6.2) is as follows (here we express the solution in terms of the service times m_j for which numerical values have been fixed, to facilitate intuitive understanding):

$$x_1 = [m_1(b_1 + l\lambda_1 - \theta_1)^+] \wedge l, \tag{6.3}$$

$$x_2 = [m_2(b_1 + l\lambda_1 - \theta_1 - x_1/m_1)^+] \wedge l, \tag{6.4}$$

$$x_3 = [m_3(b_2 + l\lambda_2 - \theta_2)^+] \wedge (l - x_2). \tag{6.5}$$

Interested readers will find that this general solution is actually quite easy to verify, once the following verbal interpretation is understood. First, (6.3) says that server 1 will devote as much time as possible to class 1 given the constraint $z_1 \geq \theta_1$ on planned ending inventory. Then (6.4) says that server 2 will devote as much time as possible to class 1 given that same lower bound constraint *and* the allocation that has already been determined for server 1. Finally, (6.5) says that server 2 will devote as much of its remaining time as possible to class 2, given the planned ending inventory constraint $z_2 \geq \theta_2$.

A notable feature of this solution is that it remains optimal for any positive number π determining the penalty-rate vector $p = (2\pi, \pi)$. This occurs for the same reason that a pathwise optimal solution exists for the approximating Brownian control problem described in section 5, namely in this problem there is no trade-off between minimizing holding costs and minimizing idleness. If there exists any feasible solution with $u_1 = u_2 = 0$, then there exists an optimal solution with that property. The objective coefficients p_1 and p_2 only matter if our threshold constraints $z \geq \theta$ force either u_1 or u_2 or both to be strictly positive, and then it is only their relative magnitudes that matter. To be more specific, if there is a forced choice between idling server 1 and idling server 2, the former action is strictly preferable if and only if $p_1 > 2p_2$. Because we have $p_1 = 2p_2$, one optimal solution can always be obtained by maximizing utilization of server 1, and that is the solution displayed above.

To explain how the time allocations (6.3)–(6.5) are implemented, and how the ideal system behavior described in section 5 is approached in the heavy traffic limit, the following choice of BIGSTEP policy parameters is particularly convenient: we shall restrict attention to values of l that are even positive integers, and further take

$$\theta_1 = l\lambda_1 \quad \text{and} \quad \theta_2 = l\lambda_2. \tag{6.6}$$

Substituting (6.10) into (6.7)–(6.9) gives

$$x_1 = b_1 \wedge l, \tag{6.7}$$

$$x_2 = 2(b_1 - x_1) \wedge l, \tag{6.8}$$

$$x_3 = b_2 \wedge (l - x_2). \tag{6.9}$$

Let us assume that the system is initially empty, so $b = 0$ at the initial review point $t = 0$. Then (6.6)–(6.9) reduce to $x_1 = x_2 = x_3 = 0$, meaning that both servers are idle through the first review period (which is l hours long). At every subsequent review point, because we have chosen $\theta = l\lambda$ in (6.6), the ending inventory constraint $z \geq \theta$ and basic system equation $z = b + l\lambda - Rx$ combine to imply $Rx \leq b$. That is, because the planned safety stock requirement θ has been chosen equal to expected arrivals for one period, servers are restricted to work throughout the period on jobs which are on hand at its beginning. At the end of the first period, whatever backlog vector b may be observed, the values of x_1 and x_3 dictated by (6.7) and (6.9) are integers, and the value of x_2 dictated by (6.10) is an even integer, because only even integer values of l are considered. Recalling that all service times are deterministic, with $m_1 = m_3 = 1$ and $m_2 = 2$, we then observe that each of the time allocations x_j translates into an integer number of jobs to be processed during the review period. Thus no services will be in process when the period ends, and the same story is repeated in every subsequent review period.

Combining (6.7) with the constraint $u = le - Ax$, one arrives at the following simple expression for the idleness u_1 to be experienced by server 1 during a period that begins with backlog b:

$$u_1 = l - x_1 = (b_1 - l)^-. \tag{6.10}$$

This formula is obvious from the fact that b_1 represents the number of hours of potential work for server 1 embodied in the current backlog, and server 1 is given first priority in the assignment of that work.

The discussion above applies to a period starting at any of the review points $l, 2l, \dots$. Fixing attention on any one such review point, let us denote by N_1 the number of class 1 jobs that arrived during the preceding period. As noted earlier, servers never work on jobs during the period in which they arrive, so $b_1 \geq N_1$ and then (6.10) implies

$$u_1 \leq (N_1 - l)^-. \tag{6.11}$$

The random variable N_1 has a Poisson distribution with mean $l\lambda_1 = 1.3l\rho$. As an application of elementary large deviations theory, it is shown in [3] that

$$P(N_1 \leq l) \leq \exp[-\psi(\lambda_1)l] \tag{6.12}$$

where $\psi(\cdot)$ is continuous, strictly positive and strictly increasing on the interval $(1, \infty)$. Combining (6.11) and (6.12), we have that

$$P(u_1 > 0) \leq \exp[-\psi(\lambda_1)l], \tag{6.13}$$

and hence, because $u_1 \leq l$,

$$E(u_1) \leq l \exp[-\psi(\lambda_1)l]. \tag{6.14}$$

Recall that $I_1(t)$ denotes total idleness by server 1 up to time t. The first t hours of operation are contained within the first $[t/l] + 1$ review periods (here square brackets represent the integer part operator), and because (6.14) gives a bound on the expected idleness in any one of those periods, we have

$$E[I_1(t)] \leq (t + l) \exp[-\psi(\lambda_1)l]. \tag{6.15}$$

Returning now to the heavy traffic theme, but viewing n as a fixed large integer for the moment, we assume that ρ is near 1 and hence λ_1 is near 1.3. Defining

$$c = \frac{1}{\psi(\lambda_1)} > 0, \tag{6.16}$$

suppose that review periods are taken to be of length

$$l = c \log n. \tag{6.17}$$

Then (6.15) becomes

$$E[I_1(t)] \leq (t + l)n^{-1}. \tag{6.18}$$

Finally, recalling the definition $U_1(t) = n^{-\frac{1}{2}} I_1(nt)$, we have from (6.18) that the scaled idleness process U_1 satisfies

$$E[U_1(t)] \leq n^{-\frac{1}{2}}(t + l/n). \tag{6.19}$$

Thus, given a sequence of systems indexed by n with $\rho(n) \to 1$ fast enough to satisfy (5.7), it follows that

$$E[U_1(t)] \to 0 \quad \text{as } n \to \infty \tag{6.20}$$

for each fixed $t > 0$. (Note that the constant c involved in our choice $l = c \log n$ is itself a function of n, but $c \to 1/\psi(1.3) > 0$ as $n \to \infty$.) Using a very similar argument, it is shown in [3] that the scaled queue length process $Z_1(t) = n^{-\frac{1}{2}} Q_1(nt)$ satisfies, for each fixed $t > 0$,

$$E[Z_1(t)] \to 0 \quad \text{as} \quad n \to \infty. \tag{6.21}$$

A crucial fact in the proof of (6.21) is that the length of a review period specified by (6.17) is small compared with the spatial scaling factor $\sqrt{n}$ as n gets large. Thus, because planned safety stocks are of order l with our parameter choice (6.6), they too are negligible in the heavy traffic limit.

Given a sequence of systems approaching heavy traffic, the parameter choices (6.17) and (6.6) define a corresponding sequence of BIGSTEP control policies. It is shown in [3] that the scaled processes (U, Z) associated with those control policies converge in an appropriate sense to the pair (U^*, Z^*) that define an optimal control for the Brownian system model described in section 5, and that the sequence of BIGSTEP policies is asymptotically optimal. Apart from standard weak convergence arguments based on Donsker's theorem and the continuous mapping theorem, which have not even been mentioned in this paper but are ubiquitous in heavy traffic theory, the essential features of the proof are the large deviations bound (6.12) and the choice of review period length (6.17) that it suggests.

The limiting statements (6.20) and (6.21) correspond to two seemingly paradoxical properties of the Brownian system model described in section 5, namely that there exists an optimal control strategy for that idealized system model under which server 1 is never idle and yet buffer 1 is always empty. The third property of the optimal Brownian control policy described in section 5 was that server 2 is never idle unless the system is empty, and the appropriate interpretation of that statement is as follows: under the BIGSTEP policy described above, server 2 experiences idleness only during periods that begin with too little work in the system to keep both servers busy for a full l hours, and that happens only when both beginning queue lengths are small compared with the spatial scaling factor of $\sqrt{n}$.

Although the parameter choices (6.17) and (6.6), together with the rigid implementation rules described earlier, are particularly convenient for proving asymptotic optimality, one would never make these choices in a practical situation. For example, given the special structure of our parallel-server system, one should simply take $\theta_2 = 0$, because there is no motivation to hold a safety stock of class 2 jobs. Similarly, when "implementing" a set of computed time allocations during a review period, one would never allow a server to go idle when there is work for it to do in either buffer. Making these and other obvious pragmatic adjustments within the basic BIGSTEP framework, one is led to an appealing family of control policies characterized by two parameters: the period length l and threshold parameter θ_1. For the case $\rho = 0.95$ and interest rate $= 0.0001$, a simulation study has been conducted to determine which values of l and θ_1 minimize expected present value of total cost. Results presented in [3] show that there are many parameter combinations achieving nearly identical performance, and that performance is in turn nearly identical to the best achievable within a one-parameter family of "threshold policies" analogous to those studied in [5].

7 An agenda for mathematical research

This paper has described for each of two illustrative processing network models a family of discretized control policies, called BIGSTEP policies, and has implicitly advanced the following two conjectures. First, the BIGSTEP method is broadly applicable, meaning that similar discretized control policies can be defined or constructed for a very general class of stochastic processing networks. Second, the parameters of a BIGSTEP policy can be chosen in such a way as to achieve asymptotic optimality in the heavy traffic limit. In this section we describe the mathematical tasks that must be undertaken to prove the conjecture of asymptotic optimality. Section 8 then discusses the potential practical appeal of the BIGSTEP method, quite apart from its theoretical foundations.

The mathematical research program must begin by defining a stochastic processing network in general terms. An ideal model formulation would include not only the structural features allowed in complex queueing networks, but also unconventional features like simultaneous resource requirements, whereby two or more servers must work together to complete certain processing tasks. With regard to probabilistic assumptions, we know from long experience that the arrival processes and service time sequences that are taken as primitive in a network model must satisfy a functional central limit theorem. Analysis of the parallel-server example in [3], summarized in section 6 of this paper, suggests that some of the arrival processes and service time sequences must further satisfy a large deviations principle, but finer probabilistic assumptions are almost certainly unnecessary for a satisfactory heavy traffic theory. A general formulation should include aspects of dynamic routing, dynamic input control, and dynamic resource allocation (sequencing).

A final and crucial concern in formulating a general network model is its cost structure, or global objective. In the three-station example described in section 2, we credited the system manager with a fixed reward for each job of a given class that was accepted for processing, which seems quite natural in problems where input acceptance is discretionary. On the other hand, the linear holding costs assumed in both our examples are not very realistic, however common such formulations may be in the mathematical theory of optimal network control. In real applications it is invariably difficult to quantify "congestion costs," but still one can say with some confidence that linear holding costs seldom provide a satisfactory representation. A more satisfactory representation would express congestion cost as a convex increasing function of the total delay experienced by a job in the network (one such function for each arrival type). Such formulations are considered hopelessly intractable by most mathematical theorists, but an important recent paper [20] has shown that convex delay penalties do not cause serious difficulties in heavy traffic analysis, because an application of Reiman's "snapshot principle" [16, 18, 19] allows one to represent the total delay experienced by a job as a linear combination of the queue lengths seen by the job upon arrival. (This is a statement about representation of total job delay in the limiting Brownian control problem.)

Continuing this same line of thought, one realizes that *rigid constraints on total delay* can be incorporated in heavy traffic formulations of network control problems, although such constraints make no sense in conventional formulations. That is, an upper bound constraint on the total delay experienced by arriving jobs of a given type can be represented in the limiting Brownian control problem as an upper bound constraint on a certain positive linear combination of queue lengths. The system manager then strives to maximize rewards earned through acceptance of new arrivals, subject to a set of linear constraints on the system's various queue lengths. This is a very appealing type of formulation, because maximum tolerable delays are often quite easy to articulate, whereas "congestion costs" are usually nebulous.

Having formulated a general model of a processing network and its associated flow control problem, a mathematical theorist must next define what is meant by "heavy traffic." While this is a relatively clear-cut matter in networks without alternate routing opportunities [2], the work of Laws [13] shows that alternate routing makes the definition of heavy traffic very complicated, and no more will be said about that matter in this paper.

The next phase of the BIGSTEP research program involves identification and characterization of the "limiting Brownian control problem" for a sequence of stochastic networks in heavy traffic. As explained in section 4, the linear programming problems to be solved as part of the BIGSTEP method involve objective function coefficients derived from the gradient $\pi(\cdot)$ of the optimal return function for an appropriate Brownian system model. One must specify the Brownian model, derive its "equivalent workload formulation," and show that the gradient function $\pi(\cdot)$ is well defined as a preliminary to defining a sequence of BIGSTEP policies for the sequence of network control problems. Of course, practical implementation of the BIGSTEP method further requires numerical evaluation of the gradient function (see below), but existence of $\pi(\cdot)$ and perhaps some regularity are the key issues in a proof of asymptotic optimality.

A next step in formulating the desired heavy traffic limit theorem is to specify exactly how the BIGSTEP policy parameters l (length of review periods) and θ (a vector of threshold parameters or planned safety stocks) are to be chosen for each network control problem in the sequence. One plausible conjecture, suggested by the analysis summarized in section 6, is that it suffices for asymptotic optimality to take $l(n) = c \log n$ for a properly chosen constant c, and to have components of θ grow at rate $\log n$ as well. As the final step in specifying a sequence of BIGSTEP control policies, one must explain exactly how server time allocations x_j, derived by solving successive linear programs, are to be implemented within each planning period. In general, of course, these time allocations cannot be enforced exactly (they might be characterized as guides to lower level flow management), but conformance to the BIGSTEP flow plan must be increasingly precise as the heavy traffic limit is approached.

All of the steps enumerated above are necessary preliminaries to a precise statement of a heavy traffic limit theorem that would provide theoretical justification for the BIGSTEP approach. The last item on the theoretical agenda, obviously, is to prove the limit theorem, showing that the specified sequence of discrete-review control policies is asymptotically optimal as one approaches heavy traffic.

8 Practical appeal of BIGSTEP policies

Let us consider now the practical appeal of BIGSTEP control policies, starting with a few obvious comments. From a system manager's perspective, the review period length l and threshold vector θ are policy parameters that can be set and dynamically readjusted through a combination of intuition, accumulated operating experience, simulation studies, and simple considerations of physical convenience. In a manufacturing environment, for example, one naturally thinks of l as an integer number of shifts or days, and initial high values for the threshold parameters θ might be gradually adjusted downward until unacceptable levels of resource idleness suggest that a practical limit has been reached. The key point is that there are relatively few of these policy parameters, compared to the staggeringly complex array of control strategies that is potentially available, and the parameter values need not be set in concrete. In a data communications context, these parameter choices might be imbedded in hardware designs, and hence difficult to change, in which case one would expect simulation studies to play a larger role.

With regard to the method of implementing BIGSTEP processing plans within individual periods, the idea behind BIGSTEP is to make safety stocks large enough relative to the planning period length so that details of implementation do not matter very much. In manufacturing contexts there are organizational benefits to be gained by giving workers and supervisors discretionary control over hour-to-hour decision making, provided that first-order considerations of total system performance are addressed by an aggregate processing plan in the form of BIGSTEP time allocations.

Without doubt, the most sensitive aspect of the BIGSTEP method from a practical perspective is the apparent need to solve a Brownian control problem in order to determine the penalty-rate vector p appearing in the objective function of the linear program to be solved each period (see section 4). Research by Kushner and his co-workers [8, 9] suggests that practical numerical methods may in fact be achievable, but before such highly technical issues are addressed one wants to think hard about simpler approaches. For concreteness, consider again the three-station example discussed in sections 2–4, and recall that analysis of the approximating Brownian control problem begins with the identification of a two-dimensional workload process $W(t) = MZ(t)$. As explained in [20], the 2×8

matrix M is determined by ordinary linear algebra, using the input–output matrix R and resource consumption matrix A as data, and $W(t)$ gives an adequate summary of current system status for certain purposes. In particular, we know that the BIGSTEP penalty-rate vector p is determined by $w = MQ(t)$ via the equation $p = \pi(w)G$. In a practical situation it is a simple matter to calculate M, and its row dimension (the number of components in a workload vector) is never larger than the number of processing resources, which is typically much smaller than the number of queues or buffers that figures in a full-blown description of system status. Thus, knowing that the proper choice of a penalty-rate vector p for BIGSTEP planning purposes depends only on the current system workload vector w, it becomes more plausible that one might be able to choose p by some direct method without a serious negative effect on system performance. Also, heavy traffic theory suggests that workload changes relatively slowly, so p values do not need to be frequently updated for BIGSTEP planning purposes.

The idea of directly assessing the penalty-rate vector p, without actually attempting to solve a Brownian control problem, will not be discussed further here, but still it is advisable to say a bit more about the penalty-rate vector and the role it plays in BIGSTEP system management. If one simply sets $p = 0$ in a planning problem objective function like (4.4), the solution that emerges is *myopic* in the sense that single-period costs are minimized without regard to costs in future periods. At the gross level of system representation that a Brownian model provides, current decisions only affect future costs through the server idleness vector u that they imply, because system workload is completely determined by idleness and uncontrollable stochastic factors. Thus myopic cost minimization is fine unless it forces us into avoidable resource idleness. (Such idleness causes at least some components of the system workload vector to be higher in future periods than might otherwise be the case, which may result in higher future costs.) The role of the penalty-rate vector p in a linear program like (4.4)–(4.5) is to quantify the impact of current idleness on future costs, thus allowing evaluation of trade-offs between immediate cost and immediate resource utilization. In other words, all reasoning about *dynamic* system behavior in the BIGSTEP method is embodied in the penalty-rate vector p, and the influence of stochastic variability (specifically, the covariance matrix Γ that captures variability information in the Brownian system model) is expressed entirely through penalty-rate vectors as well.

Reviewing the two examples described in this paper, and the discussion immediately above, one is led to characterize the BIGSTEP method in terms of a three-tiered hierarchy of system management activities. At the highest level there is a *goal-setting function* that results in assessment of both direct costs and the penalty-rate vectors that appear in the objective functions of BIGSTEP planning problems. As discussed in this section, the quantification of direct costs involves a good deal of modeling discretion, although these are treated as "problem data" in the formal mathematical development, and in principle the penalty-rate vector is determined by analysis of an approximating Brownian control problem in its low dimensional "equivalent workload formulation."

At the intermediate level there is a *flow planning function*, accomplished through solution of a linear programming problem at the beginning of each review period. Flow planning output takes the form of server time allocations for successive review periods, and these flow plans or processing plans are executed as nearly as possible in the lowest level function of *detailed flow management*. A distinctive feature of the BIGSTEP method, alluded to in the abstract and opening paragraphs of this paper, is that penalty vector calculations derived from a highly aggregated Brownian system model are translated into server time allocations at the level of job classes, which are operationally meaningful, by repeated application of linear programming. In any application of realistic scale, the linear programming problems will have many variables and many constraints, because their level of detail matches that of the processing network under study, but in this crucial intermediate function the BIGSTEP method is exploiting the most highly refined of all optimization techniques, so the high dimensionality would not appear to be a practical problem.

Bibliography

1. P. B. Chevalier and L. M. Wein. (1993). Scheduling networks of queues: heavy traffic analysis of a multistation closed network. *Operations Research*, **41**, 743–758.

2. J. M. Harrison. (1988). Brownian models of queueing networks with heterogeneous customer populations. In *Stochastic Differential Systems, Stochastic Control Theory and Applications*, IMA Vol. 10, eds. W. Fleming and P.-L. Lions. Springer, New York, 147–186.

3. J. M. Harrison. BIGSTEP control of a processing network with two servers working in parallel. In preparation.

4. J. M. Harrison and J. A. Van Mieghem. Dynamic control of Brownian networks: state space collapse and equivalent workload formulations. In preparation.

5. J. M. Harrison and L. M. Wein. (1989). Scheduling networks of queues: heavy traffic analysis of a simple open network. *Queueing Systems*, **5**, 265–280.

6. J. M. Harrison and L. M. Wein. (1990). Scheduling networks of queues: heavy traffic analysis of a two-station closed network. *Operations Research*, **38**, 1052–1064.

7. F. P. Kelly and C. N. Laws. (1993). Dynamic routing in open queueing networks: Brownian models, cut constraints and resource pooling. *Queueing Systems*. **13**, 47–86.

8. H. J. Kushner and P. Dupuis. (1992). *Numerical Methods for Stochastic Control Problems in Continuous Time*. Springer, New York and Berlin.

9. H. J. Kushner and L. F. Martins. (1991). Numerical methods for stochastic singular control problems. *SIAM Journal on Control and Optimization*, **29**, 1443–1475.

10. H. J. Kushner and L. F. Martins. (1993). Limit theorems for pathwise average cost per unit time problems for controlled queues in heavy traffic. *Stochastics and Stochastics Reports,* **42**, 25–51.

11. H. J. Kushner and L. F. Martins. (1994). Heavy traffic analysis of a controlled multi-class queueing network via weak convergence methods. Technical Report, Brown University, Lefschetz Center for Dynamical Systems: Division of Applied Math., September.

12. H. J. Kushner and K. M. Ramachandran. (1989). Optimal and approximately optimal control policies for queues in heavy traffic. *SIAM Journal on Control and Optimization,* **27**, 1293–1318.

13. C. N. Laws. (1992). Resource pooling in queueing networks with dynamic routing. *Advances in Applied Probability,* **24**, 699–726.

14. L. F. Martins and H. J. Kushner. (1990). Routing and singular control for queueing networks in heavy traffic. *SIAM Journal on Control and Optimization,* **28**, 1209–1233.

15. L. F. Martins, S. E. Shreve, and H. M. Soner. (1994). Heavy traffic convergence of a controlled multi-class queueing system. Technical Report 94-SA002, Carnegie Mellon University Center for Nonlinear Analysis.

16. M. I. Reiman. (1982). The heavy traffic diffusion approximation for sojourn times in Jackson networks. In *Applied Probability-Computer Science: The Interface,* Vol. 2, eds. R. L. Disney and T. J. Ott. Birkhäuser, Boston, 409–422.

17. M. I. Reiman. (1983). Some diffusion approximations with state space collapse. In *Proc. Int. Seminar on Modelling and Performance Evaluation Methodology,* Lecture Notes in Control and Informational Sciences 60, eds. F. Baccelli and G. Fayolle. Springer, Berlin, 209–240.

18. M. I. Reiman. (1984). Open queueing networks in heavy traffic, *Mathematics of Operational Research,* **9**, 441–458.

19. M. I. Reiman and B. Simon. (1990). A network of priority queues in heavy traffic: one bottleneck station. *Queueing Systems,* **6**, 33–58.

20. J. Van Mieghem. (1995). Dynamic scheduling with convex delay costs: the generalized $c\mu$ rule. *Annals of Applied Probability,* **5**, 809–833.

21. L. M. Wein. (1991). Scheduling networks of queues: heavy traffic analysis of a two-station network with controllable inputs. *Operations Research,* **39**, 322–340.

22. L. M. Wein. (1990). Brownian networks with discretionary routing. *Operations Research,* **38**, 1065–1078.

23. L. M. Wein. (1992). Scheduling networks of queues: heavy traffic analysis of a multistation network with controllable inputs. *Operations Research,* **40**, S312–S334.

5

Queue lengths and departures at single-server resources

Neil O'Connell

BRIMS, Hewlett-Packard Laboratories, Bristol

Abstract

In this paper I will review and illustrate some large deviation results for queues with interacting traffic, both for shared buffer and shared capacity models. These results are examples of a general scheme which can be applied to an endless variety of network problems where the goal is to establish probability approximations for aspects of a system (such as queue lengths) under very general ergodicity and mixing assumptions about the network inputs.

1 Introduction

In this paper I will review and illustrate some large deviation results for queues with interacting traffic, both for shared buffer and shared capacity models. These results are examples of a general scheme which can be applied to an endless variety of network problems where the goal is to establish probability approximations for aspects of a system (such as queue lengths) under very general ergodicity and mixing assumptions about the network inputs. I will begin by motivating such a scheme and briefly describing how it works.

We will suppose that the inputs to a network can be represented by a sequence of random variables (X_k) in $\mathbb{R}^d$, and that the (sequence of) objects of interest, (O_n), can be expressed as a continuous function of the partial sums process corresponding to X. To make this more precise, for $t \geq 0$ set

$$S_n(t) = \frac{1}{n} \sum_{k=1}^{[nt]} X_k, \qquad (1.1)$$

and write $\tilde{S}_n$ for the polygonal approximation to S_n:

$$\tilde{S}_n(t) = S_n(t) + \left(t - \frac{[nt]}{n}\right)\left(S_n\left(\frac{[nt]+1}{n}\right) - S_n\left(\frac{[nt]}{n}\right)\right). \qquad (1.2)$$

For $\mu \in \mathbb{R}^d$, denote by $\mathcal{A}_\mu(\mathbb{R}_+)$ the space of absolutely continuous paths $\phi : \mathbb{R}_+ \to \mathbb{R}^d$, with $\phi(0) = 0$ and limits

$$\lim_{t\to\infty} \frac{\phi(t)}{1+t} = \mu,$$

equipped with the topology induced by the norm

$$\|\phi\|_u = \sup_t \left| \frac{\phi(t)}{1+t} - \mu \right|. \tag{1.3}$$

Our supposition is that there exists a continuous function $f : \mathcal{A}_\mu(\mathbb{R}_+) \to \mathcal{X}$, for some Hausdorff topological space $\mathcal{X}$, such that $O_n = f(\tilde{S}_n)$, for each n. (Note that we are also implicitly assuming that $\tilde{S}_n \in \mathcal{A}_\mu(\mathbb{R}_+)$, for each n.)

For example, suppose $d = 1$ and X_k is the amount of work arriving at time $-k$ at a single-server queue with constant service capacity $c > 0$. Suppose also that the limit

$$\mu := \lim_{n\to\infty} \sum_{k=1}^{n} X_k/n$$

exists almost surely and is less than c. The queue length at time zero is given by

$$Q_0 = \sup_{n\geq 0} \sum_{k=0}^{n} (X_k - c), \tag{1.4}$$

or, equivalently, $Q_0/n = f(\tilde{S}_n)$, where $f : \mathcal{A}_\mu(\mathbb{R}_+) \to \mathbb{R}_+$ is defined by

$$f(\phi) = \sup_{t>0} [\phi(t) - ct]. \tag{1.5}$$

It is easy to check that f is a continuous function.

Why is this a useful supposition? To answer this, we need to introduce some large deviation theory.

Let $\mathcal{X}$ be a Hausdorff topological space with Borel σ-algebra $\mathcal{B}$, and let μ_n be a sequence of probability measures on $(\mathcal{X}, \mathcal{B})$. We say that μ_n satisfies the *large deviation principle* (LDP) with rate function I, if for all $B \in \mathcal{B}$,

$$- \inf_{x\in B^\circ} I(x) \leq \liminf_n \frac{1}{n} \log \mu_n(B) \leq \limsup_n \frac{1}{n} \log \mu_n(B) \leq - \inf_{x\in \bar{B}} I(x); \tag{1.6}$$

if, for each n, Z_n is a realisation of μ_n, it is sometimes convenient to say that the sequence Z_n satisfies the LDP. A rate function is *good* if its level sets are compact.

A useful tool in large deviation theory is the *contraction principle*. This states that if Z_n satisfies the LDP in a Hausdorff topological space $\mathcal{X}$ with good rate function I, and f is a continuous mapping from $\mathcal{X}$ into another Hausdorff topological space $\mathcal{Y}$, then the sequence $f(Z_n)$ satisfies the LDP in $\mathcal{Y}$ with good rate function given by

$$J(y) = \inf\{I(x) : f(x) = y\}.$$

Now consider the partial sums process $\tilde{S}_n$. Denote by $\tilde{S}_n[0,1]$ the restriction of $\tilde{S}_n$ to the unit interval, by $C[0,1]$ the space of continuous functions on $[0,1]$, equipped with the uniform topology, and by $\mathcal{A}[0,1]$ the subspace of absolutely continuous functions on $[0,1]$ with $\phi(0) = 0$. Dembo and Zajic (1995) establish quite general conditions for which $\tilde{S}_n[0,1]$ satisfies the LDP in $\mathcal{A}[0,1]$ with good convex rate function given by

$$I(\phi) = \begin{cases} \int_0^1 \Lambda^*(\dot{\phi})ds & \phi \in \mathcal{A}[0,1] \\ \infty & \text{otherwise,} \end{cases} \tag{1.7}$$

where Λ^* is the Fenchel–Legendre transform of the scaled cumulant generating function

$$\Lambda(\lambda) = \lim_{n\to\infty} \frac{1}{n} \log E e^{n\lambda \cdot S_n(1)}, \tag{1.8}$$

which is assumed to exist for each $\lambda \in \mathbb{R}^{d+1}$ as an extended real number. For such an LDP to hold in the i.i.d. case, it is sufficient that the moment generating function $E e^{\lambda \cdot X_1}$ exists and is finite everywhere; this is a classical result, due to Varadhan (1966) and Mogulskii (1976). This is usually extended to the space $C(\mathbb{R}_+)$ (of continuous functions on $\mathbb{R}_+$), via the Dawson–Gärtner theorem for projective limits. However, the projective limit topology (the topology of uniform convergence on compact intervals) is not strong enough for many applications; in particular, the function f defined by (1.5) is not continuous in this topology on any supporting subspace, and so the contraction principle does not apply. This has motivated the consideration of stronger topologies by Dobrushin and Pechersky (1995) and O'Connell (1996). In the latter it is proved that if the LDP holds in $C[0,1]$ and Λ is differentiable at the origin with $\nabla\Lambda(0) = \mu$, then the LDP holds in the space $\mathcal{A}_\mu(\mathbb{R}_+)$ with the topology induced by the norm (1.3), and with good rate function given by

$$I(\phi) = \int_0^\infty \Lambda^*(\dot{\phi})ds.$$

As we remarked earlier, the function f defined by (1.5) is continuous in this topology, provided $\mu < c$.

Getting back to our network problem we see that under very general conditions on the input process, if the objects of interest can be written as $O_n = f(\tilde{S}_n)$, for some continuous f, we have an LDP for O_n with rate function given by

$$J(y) = \inf \left\{ \int_0^\infty \Lambda^*(\dot{\phi})ds : f(\phi) = y \right\}. \tag{1.9}$$

This will provide probability approximations for O_n. However, for it to be useful, we must first simplify the rate function J (as it stands, it is an infinite-dimensional optimisation problem). This is where we use the convexity of Λ^*: combined with Jensen's inequality it allows us to restrict our consideration to a set of piecewise linear paths that depends on f and the problem becomes finite dimensional.

To illustrate this, consider the single-server queue with arrivals process (X_k)

and constant capacity $c > 0$: if $\tilde{S}_n$ satisfies the LDP in $\mathcal{A}_\mu(\mathbb{R}_+)$ with good convex rate function given by

$$I(\phi) = \int_0^\infty \Lambda^*(\dot{\phi})ds,$$

then the normalised queue length at time zero, Q_0/n, satisfies the LDP in $\mathbb{R}_+$ with good rate function

$$
\begin{aligned}
J(q) &= \inf\left\{\int_0^\infty \Lambda^*(\dot{\phi})ds : \sup_{t>0}[\phi(t) - ct] = q\right\} \\
&= \inf_{\tau>0}\inf\left\{\int_0^\tau \Lambda^*(\dot{\phi})ds : \phi(\tau) - c\tau = q\right\} \\
&= \inf_{\tau>0}\tau\Lambda^*(c + q/\tau).
\end{aligned}
$$

This fact has previously been demonstrated by several authors, under similar conditions (Chang, 1994; de Veciana *et al.*, 1993; Duffield and O'Connell, 1995; Glynn and Whitt, 1995).

Finally, why is all this potentially useful? Because it is very general, and rate functions can (in principle) be estimated from real traffic observations: see, for example, Courcoubetis *et al.* (1995) or Duffield *et al.* (1995) for more about the estimation problem.

The outline of the paper is as follows. In Section 2, we present the LDP for departures of traffic streams from an initially empty shared buffer with stochastic service capacity; in Section 3 we present an equilibrium version of this result, along with an LDP for the state of the system in equilibrium. In Section 4 we consider a system with dedicated buffers, served with weighted priority by a single server; in Section 5 we consider the problem of optimal resource allocation in such a system, and present some surprising results.

We will adopt the following convention throughout the paper: if x is a vector-valued object, denote by x^i the components of x and by $\hat{x}$ the sum of the components of x.

2 Departures from a shared buffer

Suppose we have d arrival streams $X = (X^1, \ldots, X^d)$ sharing an infinite buffer, initially empty, according to an FCFS policy with stochastic service rate C: we will begin by making this statement precise. For the moment, the only assumption is that $X^1, \ldots, X^d$ and C are non-negative sequences of random variables, indexed by the positive integers. For each n, set

$$A_n = \sum_{k=1}^n X_k, \qquad B_n = \sum_{k=1}^n C_k. \tag{2.1}$$

The total amount of work in the queue at time n is given by the recursion $(Q_0 = 0)$

$$Q_n = (Q_{n-1} + X_n - C_n)^+, \tag{2.2}$$

and the total departures (amount of work serviced) up to time n are given by

$$D_n^{(t)} = \hat{A}_n - Q_n, \tag{2.3}$$

or, equivalently,

$$D_n^{(t)} = \inf_{0 \le k \le n} (\hat{A}_k - B_k) + B_n. \tag{2.4}$$

It remains to specify the quantities of interest, namely the amounts of work, $D_n = (D_n^1, \ldots, D_n^d)$, serviced from each input stream by time n. To do this we set

$$T_n = \sup\{k \le n : \hat{A}_k \le D_n^{(t)}\}, \tag{2.5}$$

$$D_n^i = A_{T_n}^i + (D_n^{(t)} - \hat{A}_{T_n})X_{T_n+1}^i / \hat{X}_{T_n+1}. \tag{2.6}$$

Note that $D_n^{(t)} = \hat{D}_n = D_n^1 + \cdots + D_n^d$. In words, work is serviced in the order received and simultaneous arrivals from each source are thoroughly mixed in the queue.

For $0 \le t \le 1$, set

$$S_n(t) = \left(\frac{1}{n}A_{[nt]}, \frac{1}{n}B_{[nt]}\right), \tag{2.7}$$

and write $\tilde{S}_n$ for the polygonal approximation to S_n. The following is a slight modification of Corollary 2.2 in O'Connell (1994a).

Theorem 2.1 *Suppose the sequence of partial sums $\tilde{S}_n$ satisfies the LDP in $A_\mu(\mathbb{R}_+)$ with good convex rate function given by*

$$I(\phi) = \int_0^1 \Lambda^*(\dot{\phi})ds, \tag{2.8}$$

where Λ^* is the Fenchel–Legendre transform of the scaled cumulant generating function

$$\Lambda(\lambda) = \lim_{n \to \infty} \frac{1}{n} \log E e^{n\lambda \cdot S_n(1)}, \tag{2.9}$$

which is assumed to exist for each $\lambda \in \mathbb{R}^{d+1}$ as an extended real number, and $\nabla\Lambda(0) = \mu$. Suppose also that Λ^* is of the form

$$\Lambda^*(x, c) = \Lambda_a^*(x) + \Lambda_b^*(c), \tag{2.10}$$

for $(x, c) \in \mathbb{R}^d \times \mathbb{R}$. Then D_n/n satisfies the LDP in $\mathbb{R}_+^d$ with good rate function given by

$$\Lambda_d^*(z) = \inf\left\{\beta\Lambda_a^*(x/\beta) + \sigma\Lambda_a^*\left(\frac{z-x}{\sigma}\right) + \beta\Lambda_b^*(c) + (1-\beta)\Lambda_b^*\left(\frac{\hat{z}-\hat{x}}{1-\beta}\right) : \right.$$
$$\left. \beta, \sigma \in [0, 1], \ c \in \mathbb{R}, \ \beta + \sigma \le 1, \ x \in \mathbb{R}_+^d, \ \hat{x} \le \beta c\right\}. \tag{2.11}$$

De Veciana *et al.* (1993) showed that under similar hypotheses, with the arrivals assumed to be independent $(\Lambda_a^*(z) = \Lambda_1^*(z^1) + \cdots + \Lambda_d^*(z^d)$, say) and service assumed to be constant ($C_n = c$, say), the sequence of departures corresponding to the first stream (D_n^1/n) satisfies the LDP in $\mathbb{R}_+$ with rate function $\Lambda_{D^1}^*$ which is equal to Λ_1^* on the interval $[\mu_1, c - \mu_2 - \cdots - \mu_d]$, where $\mu_i = \Lambda_i'(0)$: a full description of $\Lambda_{D^1}^*$ can be obtained from (2.11) by taking an infimum over the second and subsequent variables.

Theorem 2.1 also generalises the one-dimensional results of De Veciana *et al.* (1993), Chang *et al.* (1994) and Duffield and O'Connell (1994a).

A natural question to ask, if one were hoping to consider the departure process as an arrival process at a subsequent queue and iterate the results, is whether the departure process satisfies the same hypotheses as the arrival process. The answer is that this is not generally the case (Ganesh and O'Connell, 1996).

3   Equilibrium results for a shared buffer

In the last section we assumed that the buffer was initially empty. In the case of a single input ($d = 1$), Chang and Zajic (1995) prove a stationary version of Theorem 2.1 and make the important observation that the rate function for the departures in the stationary case is generally different from the 'transient' case when the service rate is stochastic (otherwise it is the same); the difference stems from the fact that a large (positive) deviation in the departures can be encouraged by starting with a very long queue. In this section we present the LDP for departures from a shared buffer when the system is assumed to be initially in equilibrium; note, however, that to describe the state of the system in equilibrium requires more than just a single queue length, or even d queue lengths.

We will begin by setting up a stationary version of the system described in the previous section. Suppose $\{(X_k, C_k) : k \in \mathbb{Z}\}$ is a stationary, ergodic sequence in $\mathbb{R}_+^d \times \mathbb{R}_+$, with $E\hat{X}_1 < EC_1$ for stability. It is convenient to define cumulative arrivals and service on intervals: set

$$A_{k,l} = \sum_{j=k+1}^{l} X_j, \qquad\qquad B_{k,l} = \sum_{j=k+1}^{l} C_j; \qquad (3.1)$$

we will write A_n for $A_{0,n}$ and B_n for $B_{0,n}$. As before, for $0 \le t \le 1$ we set

$$S_n(t) = \left(\frac{1}{n} A_{[nt]}, \frac{1}{n} B_{[nt]} \right), \qquad (3.2)$$

and write $\tilde{S}_n$ for the polygonal approximation to S_n. The (total) amount of work in the queue at time $n \in \mathbb{Z}$ is given by

$$Q_n^{(t)} = \sup_{k \ge 0} (\hat{A}_{n-k,n} - B_{n-k,n}). \qquad (3.3)$$

The (total) departures during the interval $(k, l]$ are given by

$$D_{k,l}^{(t)} = \hat{A}_{k,l} + Q_k^{(t)} - Q_l^{(t)}. \qquad (3.4)$$

Set
$$K = \inf\{k \geq 0: Q^{(t)}_{-k} = 0\}, \tag{3.5}$$

and note that $Q^{(t)}_0 = \hat{A}_{-K,0} - B_{-K,0}$.

Just as in the previous set-up, we need to specify the quantities of interest, and this requires an assumption about how service is distributed between inputs. Let
$$L = -\sup\{k \leq 0: \hat{A}_{-K,k} \leq D^{(t)}_{-K,0}\}$$

and define the departures $D_{-K,0} = (D^1_{-K,0}, \dots, D^d_{-K,0})$ from the respective inputs on the interval $(-K, 0]$ to be
$$D^i_{-K,0} = A^i_{-K,-L} + \epsilon^i,$$

where
$$\epsilon^i = (D^{(t)}_{-K,0} - \hat{A}_{-K,-L}) X^i_{1-L} / \hat{X}_{1-L}.$$

Note that $\hat{D}_{-K,0} = D^1_{-K,0} + \cdots + D^d_{-K,0} = D^{(t)}_{-K,0}$.

To describe the state of the system at time 0, we consider the following $\mathbb{R}^d_+$-valued process: set $N(L) = A_{-L,0} - \epsilon$, and for $k = 1, \dots, L-1$ set $N(k) = A_{-k,0}$. Note that $N^i(k)$ is the amount of work of type i that has been waiting in the queue for at most k units of time; $\hat{N}(L) = Q_0$, and $N^i(L)$ is the amount of work of type i in the queue at time zero. For clarity we will write Q_0 for $N(L)$; Q_n, the respective amounts of work in the queue at any other time n, is defined similarly. Write $\tilde{O}_n$ for the polygonal approximation to $\{A_{-nt,0}/n, \ t \geq 0\}$ on the interval $[0, \tilde{L}]$, where $n\tilde{L} = L - \epsilon/X_{1-L}$. Note that $\tilde{O}_n(\tilde{L}) = Q_0/n$. To state the LDP for $\tilde{O}_n$, we need to define a suitable path space. For each positive integer k denote by $\mathcal{L}^k_\tau$ the subspace of paths in $L_\infty([0, \tau])^k$ with non-decreasing components, by $\mathcal{C}^k_\tau \subset \mathcal{L}^k_\tau$ the subspace of continuous paths starting at zero, and by $\mathcal{A}^k_\tau \subset \mathcal{C}^k_\tau$ the set of those paths with absolutely continuous components; now set
$$\mathcal{B}_k = \{\theta \in \mathcal{C}^k_\tau: \ \tau > 0\}, \tag{3.6}$$

and equip $\mathcal{B}_k$ with the topology defined by the metric
$$d(\theta_1, \theta_2) = \sup_{0 \leq t \leq \tau_1 \wedge \tau_2} \sum_{i=1}^k |\theta^i_1(t) - \theta^i_2(t)| + \sum_{i=1}^k |\theta^i_1(\tau_1) - \theta^i_2(\tau_2)|, \tag{3.7}$$

for $\theta_1 \in \mathcal{C}^k_{\tau_1}$, $\theta_2 \in \mathcal{C}^k_{\tau_2}$.

Theorem 3.1 *Under the hypotheses of Theorem 2.1, $\tilde{O}_n$ satisfies the LDP in $\mathcal{B}_d$ with good rate function given by*

$$K(\theta) = \inf\left\{\rho\Lambda^*(x, c) + \int_0^\beta \Lambda^*(\dot{\theta}, \dot{\phi}): \right.$$
$$\left. x \in \mathbb{R}^d_+, \ \beta, \rho, c \in \mathbb{R}_+, \ \phi \in \mathcal{A}_\beta, \ \tau(\hat{x} - c) = \phi(\beta)\right\}.$$

Corollary 3.2 *Under the hypotheses of Theorem 2.1, Q_0/n satisfies the LDP in $\mathbb{R}_+^d$ with good rate function*

$$L(q) = \inf\{\rho\Lambda^*(x,c) + \beta\Lambda^*(q/\beta, \rho(\hat{x} - c)/\beta) : \ x \in \mathbb{R}_+^d, \ \beta, \rho, c \in \mathbb{R}_+\}. \quad (3.8)$$

To state the LDP for the departures from an equilibrium system we define the cumulative departures from respective inputs up to time n, by

$$D_n = Q_0 + A_n - Q_n. \quad (3.9)$$

Theorem 3.3 *Under the hypotheses of Theorem 2.1, D_n/n satisfies the LDP in $\mathbb{R}_+^d$ with good rate function given by*

$$\begin{aligned}
\tilde{\Lambda}_d^*(z) &= \inf\left\{ L(q) + \beta_1\Lambda^*\left(\frac{z_1 - q}{\beta_1}, \frac{\hat{z}_1}{\beta_1}\right) + \beta_2\Lambda^*(z_2/\beta_2, c_2) \right. \\
&\quad + \tau\Lambda_a^*\left(\frac{z - z_1 - z_2}{\tau}\right) + (1 - \beta_1 - \beta_2)\Lambda_b^*\left(\frac{\hat{z} - \hat{z}_1 - \hat{z}_2}{1 - \beta_1 - \beta_2}\right) : \\
&\quad q, z_1, z_2 \in \mathbb{R}_+^d, \ c_2, \beta_1, \beta_2, \tau \in \mathbb{R}_+, \ \beta_1 + \beta_2 + \tau \le 1, \\
&\quad \left. \beta_2 c_2 \ge \hat{z}_2 \right\}.
\end{aligned}$$

Proofs of the above results can be found in O'Connell (1994a).

Again, a natural question to ask here is whether the departure process satisfies the same hypotheses assumed to hold for the arrivals; again, generally not is the answer (Ganesh and O'Connell, 1996). There is, however, one situation where it is the case, namely if the arrival processes are independent Poisson processes and the service times are exponential; then the departure processes are also independent Poisson processes.

4   Queue lengths at a system with dedicated buffers and shared service capacity

Consider a single-server queue with two inputs (X_n^1) and (X_n^2) and constant service capacity c shared between the inputs according to a weighted priority scheme with weights $p_1 + p_2 = 1$. To be more precise, X_n^1 and X_n^2 are sequences of non-negative random variables and, starting with an empty system, the respective queue lengths at time n are defined recursively by the equations

$$\begin{aligned}
Q_n^1 &= (Q_{n-1}^1 + X_n^1 - \max(c - Q_{n-1}^2 - X_n^2, p_1 c))^+ \\
Q_n^2 &= (Q_{n-1}^2 + X_n^2 - \max(c - Q_{n-1}^1 - X_n^1, p_2 c))^+
\end{aligned}$$

with $Q_0^1 = Q_0^2 = 0$. We will write $Q_n = (Q_n^1, Q_n^2)$.

For each n define a path $S_n : [0,1] \to \mathbb{R}_+^2$ by

$$S_n(t) = \left(\frac{1}{n}\sum_{k=1}^{[nt]} X_k^1, \frac{1}{n}\sum_{k=1}^{[nt]} X_k^2\right), \quad (4.1)$$

and denote its polygonal approximation by $\tilde{S}_n$. For $\lambda \in \mathbb{R}^2$ set

$$\Lambda(\lambda) = \lim_{n\to\infty} \frac{1}{n} \log E e^{\lambda \cdot S_n(1)}, \tag{4.2}$$

whenever this limit exists. Write Λ^* for the convex dual of Λ. Assuming $\tilde{S}_n$ satisfies the LDP in $\mathcal{A}_\mu(\mathbb{R}_+)$, where $\mu = \nabla\Lambda(0)$, with good rate function given by

$$I(\phi) = \int_0^\infty \Lambda^*(\dot\phi)ds,$$

and we can write

$$Q_n/n = \Pi(\tilde{S}_n),$$

for some continuous function Π, we have by the contraction principle that the sequence Q_n/n satisfies the LDP in $\mathbb{R}_+^2$ with good rate function given by

$$J(q) = \inf\{I(\phi): \ \Pi(\phi) = q\}, \tag{4.3}$$

and the first queue length Q_n^1/n satisfies the LDP in $\mathbb{R}_+$ with good rate function given by

$$L(q) = \inf\{I(\phi): \ \Pi(\phi)^1 = q\}. \tag{4.4}$$

The mapping Π is formally defined in O'Connell (1994b), where the following simplifications of J and L are obtained for the case where the inputs are assumed to be independent:

$$\Lambda(\lambda) = \Lambda_1(\lambda_1) + \Lambda_2(\lambda_2).$$

Set $\mu_i = \Lambda_i'(0)$.

Theorem 4.1 *In the above setting:*

(a) If $\mu_i \leq p_i c$ $(i = 1, 2)$,

$$J(a) = \inf\{\tau\Lambda^*(x,y) + \tau'\Lambda^*(x',y') : \ (\tau, \tau', x, x', y, y') \in E(a)\}, \tag{4.5}$$

where $E(a) = E_1(a) \cup E_2(a)$ and

$$\begin{aligned} E_1(a) &= \{(\tau, \tau', x, x', y, y') \in \mathbb{R}_+^6 : \ \tau + \tau' \leq 1, \ y \leq p_2 c, \\ &\quad \tau(x + y - c) + \tau'(x' - p_1 c) = a_1, \ \tau'(y' - p_2 c) = a_2\}, \end{aligned}$$

$$\begin{aligned} E_2(a) &= \{(\tau, \tau', x, x', y, y') \in \mathbb{R}_+^6 : \ \tau + \tau' \leq 1, \ x \leq p_1 c, \\ &\quad \tau(x + y - c) + \tau'(y' - p_2 c) = a_2, \ \tau'(x' - p_1 c) = a_1\}. \end{aligned}$$

(b) If $\mu_1 + \mu_2 \leq c$ and $\mu_2 \leq p_2 c$ then

$$L(a) = \inf\{\tau\Lambda^*(c - y + a/\tau, y) : \ 0 \leq \tau \leq 1, \ y \leq p_2 c\}.$$

(c) If $\mu_1 + \mu_2 \leq c$ and $\mu_2 \geq p_2 c$ then

$$L(a) = \inf_{0 \leq \tau \leq 1} \tau\Lambda_1^*(p_1 c + a/\tau).$$

This complements and extends results of de Veciana and Kesidis (1993) and Bertsimas *et al.* (1995), where the tail asymptotics for the limiting distribution of Q_n^1 are obtained in the ergodic case; of Weber (1995) on the large deviation principle for queue lengths in a similar system with state-dependent service; and of Ignatyuk *et al.* (1993), and Borovkov and Mogulskii (1995), on the large deviation behaviour of random walks in a two-dimensional quadrant. See also Dupuis and Ellis (1995), and references therein, for related work.

This can also be extended to the equilibrium case, where the LDP holds with rate functions given by the expressions in Theorem 4.1 without the restrictions $\tau + \tau' \le 1$ for case (a) and $\tau \le 1$ for cases (b) and (c).

5 Resource allocation

Suppose we have two buffers of sizes an and $(1-a)n$ (n is large and $0 < a < 1$) and service capacity c per unit time distributed between the buffers with respective priority weights p and $1 - p$. The two input streams are independent, and are characterised by their rate functions Λ_1^* and Λ_2^*. How should we allocate service capacity and buffer space—that is, how should we choose p and a—in order to minimise the overall frequency of buffer overflow? Well, we can approximate the overall frequency of overflow by

$$P(Q^1 > an \text{ or } Q^2 > (1 - a)n),$$

where Q^1 and Q^2 are the queue lengths at an infinite buffer version of the system in equilibrium. Applying the principle of the largest term and an equilibrium version of Theorem 4.1, we have

$$P(Q^1 > an \text{ or } Q^2 > (1 - a)n) \approx e^{-\delta(a,p)n},$$

where $\delta(a, p) = [a\delta_1(p)] \wedge [(1 - a)\delta_2(p)]$ and

$$\delta_1(p) = \inf_{\tau \ge 0, x_2 \ge 0} \tau[\Lambda_1^*(1/\tau + (pc) \vee (c - x_2)) + \Lambda_2^*(x_2)],$$

$$\delta_2(p) = \inf_{\tau \ge 0, x_1 \ge 0} \tau[\Lambda_2^*(1/\tau + ((1 - p)c) \vee (c - x_1)) + \Lambda_1^*(x_1)].$$

The problem of minimising the overall frequency of overflow is thus approximately equivalent to the problem of maximising $\delta(a, p)$ with respect to a and p. For fixed p, this is achieved by setting

$$a = a^*(p) = \frac{\delta_1(p)}{\delta_1(p) + \delta_2(p)},$$

yielding

$$\delta(a^*(p), p) = \frac{\delta_1(p)\delta_2(p)}{\delta_1(p) + \delta_2(p)};$$

thus, to maximise $\delta(a, p)$ we should choose $p = p^*$ to minimise

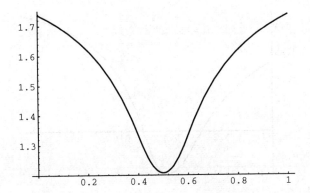

FIG. 1. Plot of $\delta(a^*(p), p)$ against p for the parameter values $\mu_1 = \mu_2 = 0.4$, $\sigma_1^2 = \sigma_2^2 = 0.1$, $c = 1$. The optimal policy has $p = 1$ and $a = 0.13$.

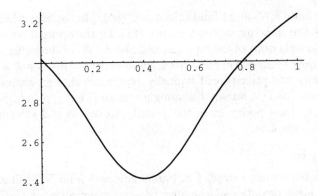

FIG. 2. Plot of $\delta(a^*(p), p)$ against p for the parameter values $\mu_1 = 0.2$, $\mu_2 = 0.4$, $\sigma_1^2 = \sigma_2^2 = 0.1$, $c = 1$. The optimal policy has $p = 1$ and $a = 0.075$.

$$\frac{\delta_1(p)\delta_2(p)}{\delta_1(p) + \delta_2(p)}$$

and set $a^* = a^*(p^*)$.

For example, suppose $\Lambda_i^*(x) = (x - \mu_i)^2/2\sigma_i^2$, $\mu_1 + \mu_2 < c$. Then, after some straightforward calculations, we get

$$\delta_2(p) = \begin{cases} \frac{2(c-\mu_1-\mu_2)}{\sigma_1^2+\sigma_2^2} & pc > \frac{2\sigma_1^2(c-\mu_1-\mu_2)}{\sigma_1^2+\sigma_2^2} + \mu_1 \\[2ex] 2\frac{(1-p)c-\mu_2}{\sigma_2^2} & pc \le \mu_1 \\[2ex] \frac{(1-p)c-\mu_2}{\sigma_2^2} + \frac{1}{\sigma_2}\sqrt{\left(\frac{(1-p)c-\mu_2}{\sigma_2}\right)^2 + \left(\frac{pc-\mu_1}{\sigma_1}\right)^2} & \text{otherwise} \end{cases}$$

and a similar expression for $\delta_1(p)$. Figures 1–3 are plots of $\delta(a^*(p), p)$ against p for different parameter values. To interpret these, recall that the optimal policy, if one wishes to minimise the overall frequency of overflow, is to choose p in

Neil O'Connell

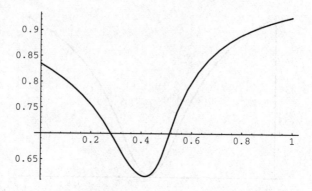

FIG. 3. Plot of $\delta(a^*(p), p)$ against p for the parameter values $\mu_1 = \mu_2 = 0.4$, $\sigma_1^2 = 0.1$, $\sigma_2^2 = 0.3$, $c = 1$. The optimal policy has $p = 1$ and $a = 0.19$.

order to maximise $\delta(a^*(p), p)$ (and take $a = a^*(p^*)$). In the case where the input streams have the same mean and variance (Fig. 1), the optimal policy is to give top priority to either one of the streams, and about 87% of the buffer space to the other. This may seem surprising. Note, however, that this is not a fair policy: the stream with top priority will typically experience shorter waiting times. In the cases where the first stream has a higher mean (Fig. 2), or a higher variance (Fig. 3), the optimal policy gives top priority to the second stream and more buffer space to the first.

Bibliography

1. Dimitris Bertsimas, Ioannis Ch. Paschalidis and John N. Tsitsiklis (1994). On the large deviations behaviour of acyclic networks of G/G/1 queues. LIDS Report: LIDS-P-2278.

2. Dimitris Bertsimas, Ioannis Ch. Paschalidis and John N. Tsitsiklis (1995). RSS Research Workshop in Stochastic Networks, Edinburgh, August 1995.

3. A.A. Borovkov and A.A. Mogulskii (1995). Large deviations for stationary Markov chains in a quarter plane. Preprint.

4. Cheng-Shang Chang (1994). Stability, queue length and delay of deterministic and stochastic queueing networks. *IEEE Trans. Autom. Control* 39:913–931.

5. Cheng-Shang Chang. Approximations of ATM networks: effective bandwidths and traffic descriptors. Submitted.

6. Cheng-Shang Chang, Philip Heidelberger, Sandeep Juneja and Perwez Shahabuddin (1994). Effective bandwidth and fast simulation of ATM intree networks. *Performance Eval.* 20:45–66.

7. C.-S. Chang and T. Zajic (1995). Effective bandwidths of departure processes from queues with time varying capacities. INFOCOM, 1995.

8. C. Courcoubetis, G. Kesidis, A. Ridder, J. Walrand and R. Weber (1995).

Admission control and routing in ATM networks using inferences from measured buffer occupancy. To appear in *IEEE Trans. Commun.*

9. Amir Dembo and Tim Zajic (1995). Large deviations: from empirical mean and measure to partial sums process. *Stoch. Proc. Appl.* 57:191–224.

10. Amir Dembo and Ofer Zeitouni (1992). *Large Deviations Techniques and Applications.* Jones and Bartlett, London.

11. G. de Veciana, C. Courcoubetis and J. Walrand (1993). Decoupling bandwidths for networks: a decomposition approach to resource management. Memorandum No. UCB/ERL M93/50, University of California.

12. G. de Veciana and G. Kesidis (1993). Bandwidth allocation for multiple qualities of service using generalised processor sharing. Preprint.

13. R.L. Dobrushin and E.A. Pechersky (1995). Large deviations for random processes with independent increments on infinite intervals. Preprint.

14. N.G. Duffield, J.T. Lewis, Neil O'Connell, Raymond Russell and Fergal Toomey (1995). Entropy of ATM traffic streams: a tool for estimating QoS parameters. *IEEE J. Sel. Areas Commun.* 13(6):981–990.

15. N.G. Duffield and Neil O'Connell (1994). Large deviations for arrivals, departures, and overflow in some queues of interacting traffic. *Proceedings of the 11th IEE Teletraffic Symposium,* Cambridge, March 1994.

16. N.G. Duffield and Neil O'Connell (1995). Large deviations and overflow probabilities for the general single server queue, with applications. *Proc. Cambridge Philos. Soc.* 118(1).

17. Paul Dupuis and Richard S. Ellis. The large deviation principle for a general class of queueing systems, I. *Trans. Am. Math. Soc.,* to appear.

18. Paul Dupuis and Richard S. Ellis (1995). Large deviation analysis of queueing systems. In *Stochastic Networks,* IMA Volumes in Mathematics and Its Applications, F. P. Kelly and R. J. Williams (eds.), 71, Springer-Verlag, New York, 347–365.

19. A. Ganesh and Neil O'Connell (1996). The linear geodesic property is not generally preserved by a FIFO queue. Technical Report HPL-BRIMS-96-006, BRIMS, Hewlett-Packard, Bristol.

20. Peter W. Glynn and Ward Whitt (1995). Logarithmic asymptotics for steady-state tail probabilities in a single-server queue. *J. Appl. Prob.,* to appear.

21. I.A. Ignatyuk, V. Malyshev and V.V. Scherbakov (1993). Boundary effects in large deviation problems. Preprint.

22. F.P. Kelly and P.B. Key (1994). Dimensioning playout buffers from an ATM network. *Proceedings of the 11th IEE Teletraffic Symposium,* Cambridge, March 1994.

23. A.A. Mogulskii (1976). Large deviations for trajectories of multi dimensional random walks. *Th. Prob. Appl.* 21:300–315.

24. Neil O'Connell (1994a). Large deviations for departures from a shared

buffer. *J. Appl. Prob.*, to appear. (Revised version of 'Large deviations in queueing networks', DIAS Technical Report DIAS-APG-9413.)

25. Neil O'Connell (1994b). Large deviations for queue lengths at a multi-buffered resource. *J. Appl. Prob.*, to appear.

26. Neil O'Connell (1996). Stronger topologies for sample path large deviations in Euclidean space. BRIMS Technical Report: HPL-BRIMS-96-005.

27. Shyam Parekh and Jean Walrand (1989). A quick simulation method for excessive backlogs in networks of queues. *IEEE Trans. Autom. Control* 34:54–66.

28. S.R.S. Varadhan (1966). Asymptotic probabilities and differential equations. *Comm. Pure Appl. Math.* 19:261–286.

29. Richard Weber (1995). Estimation of overflow probabilities for state-dependent service of traffic streams with dedicated buffers. RSS Research Workshop in Stochastic Networks, Edinburgh, August 1995.

6

Large deviations of stationary reflected Brownian motions

Kurt Majewski

Ludwig-Maximilians-Universität München

Abstract

Let W be a stationary reflected Brownian motion with a reflection matrix of the form $(I+Q)^{-1}$, where I is the identity matrix and Q is lower diagonal with non-negative entries. We derive a large deviation principle for the sequence $(W(0)/u)_{u>0}$ and thus characterize the asymptotic logarithmic decay rates of the tail probabilities of the distribution of $W(0)$.

We also give an example of an explicit decay rate calculation.

1 Introduction

Large deviation principles (LDPs) for queue length distributions of queueing systems are of great interest for the analysis of tail probabilities and thus for the analysis of the performance of these systems. So far, for queueing systems with more than one queue LDPs are only available in particular cases; see Bertsimas *et al.* (1994), Dupuis and Ellis (1995) and O'Connell (1996).

Working in a heavy traffic regime, stochastic queueing networks can be approximated by multidimensional reflected Brownian motions (RBMs); see Harrison and Nguyen (1993). In particular, for multiclass feedforward queueing systems Peterson (1991) proved the weak convergence of the properly scaled workload process to an RBM, when the traffic intensity converges to one with a certain speed. Here, a feedforward queueing system is a queueing system in which the nodes can be numbered in such a way that each customer is routed along nodes with increasing numbers. The resulting limit RBMs, which we call feedforward RBMs, have a reflection matrix of the form $(I + Q)^{-1}$, where I is the identity matrix and Q is a lower diagonal matrix with non-negative entries. In this paper, we prove an LDP for the steady-state distribution of feedforward RBMs. We also give an example of an explicit calculation of a certain large deviation rate.

The question arises if this LDP for RBMs can provide additional insight into the large deviation behaviour of steady-state queue length distributions. We can

---

Work supported by Siemens Corp. Research and Development Munich.

positively answer this question in at least three regards. First, the abstract form of the variational formula in our LDP seems to be valid for stationary queue length distributions, too; compare Ramanan and Dupuis (1995). Its structure reveals an intimate relation between time-inhomogeneous fluid models and large deviations and gives additional information on how large queue lengths build up. Second, some of the analytical tools we developed to derive the LDP carry over to derive analogous LDPs for queue length distributions; see Majewski (1996). And last but not least, in some cases the large deviation rates which we calculated from the RBM model are good approximations for the large deviation rates of the approximated queue length distribution.

In what follows, we omit some of the less important proofs to keep this paper to a reasonable length. Missing details can be found in Majewski (1995).

2 Feedforward reflected Brownian motions

Let $k \in I\!\!N$. We denote the set of all continuous mappings from $I\!\!R_+ := [0, \infty[$ resp. $I\!\!R$ to $I\!\!R^k$ by $\mathcal{C}^k$ resp. $\overline{\mathcal{C}}^k$. (All overlined symbols refer to the time interval $I\!\!R$.) Let p be the canonical projection of $\overline{\mathcal{C}}^k$ onto $\mathcal{C}^k$. For $T \geq 0$ the supremum semi-norm $\| \cdot \|_{\infty,T}$ on $\mathcal{C}^k$ is defined by $\|\omega\|_{\infty,T} := \sup_{0 \leq t \leq T} \|\omega(t)\|_\infty$, where $\| \cdot \|_\infty$ is the usual supremum norm for vectors. We will often consider the non-Hausdorff topology induced on $\mathcal{C}^k$ by one of these semi-norms. Let θ_s resp. $\overline{\theta}_s$ be the *shift mapping* on $\mathcal{C}^k$ resp. on $\overline{\mathcal{C}}^k$ defined by $(\theta_s \omega)(t) := \omega(t + s)$ for $s, t \in I\!\!R_+$ resp. $(\overline{\theta}_s \omega)(t) := \omega(t + s)$ for $s, t \in I\!\!R$.

Define the *reflection mapping* $\psi \colon \mathcal{C}^1 \to \mathcal{C}^1$ resp. $\overline{\psi} \colon \overline{\mathcal{C}}^1 \to \overline{\mathcal{C}}^1$ by setting for $\omega \in \mathcal{C}^1$ and $t \in I\!\!R_+$

$$\psi(\omega)(t) \;:=\; \omega(t) \;-\; \min\left\{ 0, \inf_{0 \leq s \leq t} \omega(s) \right\} \tag{2.1}$$

resp. for $\omega \in \overline{\mathcal{C}}^1$ and $t \in I\!\!R$

$$\overline{\psi}(\omega)(t) \;:=\; \begin{cases} \omega(t) \;-\; \inf_{s \leq t} \omega(s) & \text{if } \inf_{s \leq 0} \omega(s) \in I\!\!R \\ 0 & \text{otherwise.} \end{cases} \tag{2.2}$$

The reflection mapping possesses a certain *loss-of-memory property*, which is partially captured in the two equations $\theta_s \psi(\omega) = \psi(\psi(\omega)(s) - \omega(s) + \theta_s(\omega))$ for $s \geq 0$ and $\omega \in \mathcal{C}^1$ and $p(\overline{\psi}(\omega)) = \psi(\overline{\psi}(\omega)(0) - \omega(0) + p(\omega))$ for $\omega \in \overline{\mathcal{C}}^1$ with $\inf_{s \leq 0} \omega(s) \in I\!\!R$.

A one-dimensional RBM can be constructed as the image of a one-dimensional Brownian motion (possibly with drift) under the mapping ψ. This can be generalized to construct multidimensional RBMs: fix $n \in I\!\!N$ and let $Q = (Q_{i,j})_{i,j \leq n}$ be a non-negative lower diagonal matrix, that is $Q_{i,j} \geq 0$ for $i < j$ and $Q_{i,j} = 0$ for $i \geq j$. In Theorem 2.1, the matrix $R := (I + Q)^{-1}$, where I is the identity matrix, will be the reflection matrix of certain RBMs. Define the mapping $\Psi \colon \mathcal{C}^n \to \mathcal{C}^n$ by $\Psi((\omega_1, \ldots, \omega_n)^T) := (\eta_1, \ldots, \eta_n)^T$, where $\eta_k \in \mathcal{C}^1$ is recursively defined for $k = 1, \ldots, n$ by

$$\eta_k := \psi\left(\omega_k - \sum_{i=1}^{k-1} Q_{k,i}(\eta_i - \eta_i(0))\right). \tag{2.3}$$

Define similarly the mapping $\overline{\Psi}\colon \overline{C}^n \to \overline{C}^n$ by replacing in eqn 2.3 ψ by $\overline{\psi}$ and taking ω from $\overline{C}^n$ instead of C^n. The mapping Ψ inherits several properties from ψ: for example, there is a $b > 0$ such that for all $T \geq 0$ and $\omega, \eta \in C^n$ we have $\|\Psi(\omega) - \Psi(\eta)\|_{\infty,T} \leq b\|\omega - \eta\|_{\infty,T}$. Also the first loss-of-memory equation stated above remains valid if ψ is replaced by Ψ and ω is taken from C^n instead of C^1. Furthermore, the image of a process adapted to some filtration under Ψ is again adapted to this filtration.

We now proceed with the construction of feedforward RBMs. Fix $m \in \mathbb{N}$ and let $B = (B_1, \ldots, B_m)^T$ and $B' = (B'_1, \ldots, B'_m)^T$ be two independent m-dimensional vectors of independent standard Brownian motions defined on some probability space $(\Omega, \mathcal{F}, P)$. Set for $t \in \mathbb{R}$

$$\overline{B}(t) \quad := \quad \begin{cases} B(t) & \text{for } t \geq 0 \\ B'(-t) & \text{otherwise.} \end{cases} \tag{2.4}$$

Thus the process $\overline{B} := (\overline{B}(t))_{t \in \mathbb{R}}$ is an m-dimensional continuous process with stationary increments. Define for $t \geq 0$ the σ-algebra $\mathcal{F}_t \subset \mathcal{F}$ as the one generated by the collection of random variables $\{\overline{B}(s)\colon s \in \,]-\infty, t]\}$. Clearly B is adapted to the filtration $(\mathcal{F}_t)_{t \in \mathbb{R}_+}$ and is independent of $\mathcal{F}_0$.

Let $M = (M_{i,j})_{i \leq n,\, j \leq m}$ be an $n \times m$ matrix, set $\Sigma := MM^T$ and let $d = (d_1, \ldots, d_n)^T \in \mathbb{R}^n$. Define the processes $Y = (Y_1, \ldots, Y_n)^T$ and $\overline{Y}$ by

$$Y := MB - D \quad \text{resp.} \quad \overline{Y} := M\overline{B} - \overline{D}, \tag{2.5}$$

where $D := (td)_{t \in \mathbb{R}_+}$ and $\overline{D} := (td)_{t \in \mathbb{R}}$. Clearly Y is an m-dimensional Brownian motion with covariance matrix Σ and drift d adapted to the filtration $(\mathcal{F}_t)_{t \in \mathbb{R}_+}$ and independent of $\mathcal{F}_0$. The process $\overline{Y}$ has stationary increments.

Let $V = (V_1, \ldots, V_n)^T \geq 0$ be an n-dimensional $\mathcal{F}_0$-measurable random variable. Define the process $W^V = (W_1^V, \ldots, W_n^V)^T$ by

$$W^V \quad := \quad \Psi(V + Y). \tag{2.6}$$

The process W^V is adapted to the filtration $(\mathcal{F}_t)_{t \geq 0}$.

Theorem 2.1 *On the filtrated probability space* $(\Omega, \mathcal{F}, (\mathcal{F}_t)_{t \geq 0}, P)$, *the process* W^V *is a semimartingale reflected Brownian motion with reflection matrix* R, *covariance matrix* $R\Sigma R^T$, *drift* $-Rd$ *and initial distribution* V *in the sense of Taylor and Williams* (1993) *Definition 1.1.*

Proof Define recursively for $k = 1, \ldots, n$ the processes $J_k = (J_k(t))_{t \geq 0}$ by

$$J_k \quad := \quad W_k^V - V_k - Y_k + \sum_{i=1}^{k-1} Q_{k,i}(W_i^V - V_i). \tag{2.7}$$

The process $J := (J_1, \ldots, J_n)^T$ is non-decreasing, $J(0) = 0$ and a component J_k

of J can only increase at times where $W_k^V = 0$. Furthermore we have $(I + Q)W^V = (I + Q)V + Y + J$. Multiplying this equation with R gives

$$W^V \;=\; V + RY + RJ. \tag{2.8}$$

As V and B are independent we get that the process $V + RY + RD = V + RMB$ is a martingale with respect to the filtration $(\mathcal{F}_t)_{t\geq 0}$. Therefore, W^V is a semimartingale reflected Brownian motion with reflection matrix R, covariance matrix $R\Sigma R^T$, drift $-Rd$ and initial distribution V. $\square$

We call RBMs with the particular kind of reflection matrix R defined above *feedforward reflected Brownian motions*. What can be said about stationary feedforward RBMs? To get an answer define the process $\overline{W} = (\overline{W}_1, \ldots, \overline{W}_n)^T := \overline{\Psi}(\overline{Y})$ and its restriction $W := (\overline{W}(t))_{t\geq 0}$. The process W is adapted to the filtration $(\mathcal{F}_t)_{t\in\mathbb{R}_+}$.

Lemma 2.2 *The processes W and $\overline{W}$ are stationary.*

Proof It can be shown that $\overline{\Psi}$ commutes with the shift mappings $\overline{\theta}_s$ for all $s \in \mathbb{R}$. Furthermore $\overline{\Psi}$ is invariant under additions of constants. Due to the stationary increments of the process $\overline{Y}$ the stationarity of $\overline{W}$ and in particular of W follows. $\square$

Define $\overline{Z}_k := \overline{Y}_k - \sum_{i=1}^{k-1} Q_{k,i}(\overline{W}_i - \overline{W}_i(0))$ for $k = 1, \ldots, n$. This implies $\overline{W}_k = \overline{\psi}(\overline{Z}_k)$.

Lemma 2.3 *If all components of d are strictly positive, the processes $W^{W(0)}$ and W are indistinguishable and the process $W^{W(0)}$ is a stationary RBM.*

Proof Suppose d has only strictly positive components. We will show in Lemma 3.2 that $P(\inf_{s\leq 0} \overline{Z}_k(s) = -\infty) = 0$. Therefore, the loss-of-memory property of $\overline{\psi}$ and the equation $\overline{W}_k = \overline{\psi}(\overline{Z}_k)$ imply that the processes $W_k^{W(0)}$ and W_k are indistinguishable for all $k = 1, \ldots, n$. $\square$

Suppose from now on that the vector d is strictly positive in every component. The interest of the rest of this paper is to derive an LDP for the sequence of scaled stationary states $(W(0)/u)_{u>0}$.

3   Exponential tightness

Most proofs in this section will be omitted. They are based on the principle of the largest term (see Lemma 1.2.15 of Dembo and Zeitouni (1993)) and can be found in Majewski (1995).

Let $k \in \mathbb{N}$. A sequence of $\mathbb{R}^k$-valued random variables $(X_u)_{u>0}$ is called *exponentially tight*, if for every $c \geq 0$ there is a compact set $K \subset \mathbb{R}^k$ with

$$\limsup_{u\to\infty} \frac{1}{u} \log P(X_u \notin K) \;\leq\; -c. \tag{3.1}$$

The main topic of this section is to derive the exponential tightness of the sequence $(W(0)/u)_{u>0}$. To this end we introduce two other notions of exponential tightness. Let $X = (X(t))_{t \in \mathbb{R}}$ be an $\mathbb{R}$-valued continuous process. We call the process X *locally exponentially tight* (l.e.t.), if for all $c \geq 0$ we can find an $\epsilon > 0$ such that

$$\forall s \leq 0: \quad \limsup_{u \to \infty} \frac{1}{u} \log P\left(\sup_{s-\epsilon \leq t \leq s} |X(su) - X(tu)| \geq u \right) \leq -c. \quad (3.2)$$

We call the process X *exponentially reflection tight* (e.r.t.), if

$$\limsup_{u \to \infty} \frac{1}{u} \log P\left(\inf_{t \leq -u} X(t) \leq X(0) \right) < 0. \quad (3.3)$$

One motivation for these definitions is

Lemma 3.1 *Let $X = (X(t))_{t \in \mathbb{R}}$ be an $\mathbb{R}$-valued continuous, locally exponentially tight and exponentially reflection tight process. Then $P(\inf_{t \leq 0} X(t) = -\infty) = 0$ and the sequence $(\overline{\psi}(X)(0)/u)_{u>0}$ is exponentially tight.*

It is not difficult to show that for $a > 0$ the process $(X(t) - at)_{t \in \mathbb{R}}$, where X is a linear combination of components of $\overline{B}$, is l.e.t. and e.r.t. Furthermore, linear combinations with non-negative coefficients of l.e.t. and e.r.t. processes are again l.e.t. and e.r.t. Also the image of an l.e.t. process under the mapping $\overline{\psi}$ is again l.e.t. Finally, if the process $X = (X(t))_{t \in \mathbb{R}}$ is non-negative, stationary and l.e.t. and if $(X(0)/u)_{u>0}$ is exponentially tight, then for all $a > 0$ the process $(-X(t) - at)_{t \in \mathbb{R}}$ is e.r.t. Put together, these statements lead to

Lemma 3.2 *For all $k = 1, \ldots, n$ we have $P(\inf_{t \leq 0} \overline{Z}_k(t) = -\infty) = 0$ and the sequence $(W_k(0)/u)_{u>0}$ is exponentially tight. In particular $(W(0)/u)_{u>0}$ is exponentially tight.*

Proof The statement follows from Lemma 3.1 once we show that for $k = 1, \ldots, n$ the process $\overline{Z}_k$ is l.e.t. and e.r.t. This can be proved by induction over k using that $\overline{Z}_k$ is a linear combination with positive coefficients of the l.e.t. and e.r.t. processes $\sum_{j=1}^{m} M_{k,j} \overline{B}_j - \overline{D}_k/k$ and $Q_{k,i}(\overline{W}_i(0) - \overline{W}_i) - \overline{D}_k/k$ for $i = 1, \ldots, k-1$. $\square$

We conclude this section with two statements useful to bound large deviations rates. These statements will be important in the example of the last section.

Proposition 3.3 *Let $X = (X(t))_{t \in \mathbb{R}}$ be a càdlàg process, which is locally exponentially tight and exponentially reflection tight. Then for every mapping $f :]0, \infty[\to \mathcal{F}$ with $f(u) \subset f(v)$ if $u \leq v$*

$$\limsup_{u \to \infty} \frac{1}{u} \log P(f(u) \cap \{\psi(X)(0) \geq u\})$$

$$\leq \inf_{c>0} \limsup_{u \to \infty} \frac{1}{u} \log P(f(u) \cap \{X(0) - X(-cu) \geq u\}). \quad (3.4)$$

Lemma 3.4 *Let $k \in \mathbb{N}$ and for $i = 1, \ldots, k$ and $u > 0$ let $Z_i(u)$ be random variables such that the sequence $(Z(u)/u)_{u>0}$ is exponentially tight. Then for every mapping $f :]0, \infty[\to \mathcal{F}$ and $\delta \in]0, 1[$*

$$\limsup_{u\to\infty} \frac{1}{u} \log P\left(f(u) \;\cap\; \left\{ \sum_{i=1}^{k} Z_i(u) \geq u \right\} \right)$$

$$\leq \inf_{(q_1,\dots,q_k)^T \in E_\delta^k} \limsup_{u\to\infty} \frac{1}{u} \log P\left(f(u) \cap \bigcap_{i=1}^{k} \{ Z_i(u) \geq q_i u \} \right), \quad (3.5)$$

where $E_\delta^k = \{(q_1,\dots,q_k)^T \in I\!\!R^k \colon \sum_{i=1}^{k} q_i = \delta\}$.

4 The variational formula

Let $\mathcal{H}^m$ be the Sobolev space of absolutely continuous functions $\omega \colon I\!\!R_+ \to I\!\!R^m$ with $\omega(0) = 0$ and square integrable derivative on bounded intervals. For $\omega \in \mathcal{H}^m$ let $\dot\omega$ be (a version of) its derivative. For $T > 0$ we define the *rate function* $I_T \colon \mathcal{H}^m \to I\!\!R_+$ by

$$I_T(\omega) \;:=\; \frac{1}{2} \int_0^T \|\dot\omega(t)\|_2^2 \, dt, \quad (4.1)$$

where $\|\cdot\|_2$ is the usual Euclidean norm for real vectors. The convexity of the square norm implies that $\frac{1}{2T}\|\omega\|_{\infty,T}^2 \leq I_T(\omega)$ for $\omega \in \mathcal{H}^m$.

For $u > 0$ let the *scaling mapping* $\Gamma_u \colon \mathcal{C}^m \to \mathcal{C}^m$ be defined by $(\Gamma_u \omega)(t) := \omega(ut)/u$ for $t \in I\!\!R_+$. Note that $\Gamma_u \circ \Psi = \Psi \circ \Gamma_u$ for all $u > 0$. Recall that B is an m-dimensional standard Brownian motion. Here is a version of Schilder's theorem adapted to our purposes.

Theorem 4.1. (Schilder) *Let $T > 0$ and let A^c resp. A^o be a closed resp. an open subset of $\mathcal{C}^m$ in the topology induced by the semi-norm $\|\cdot\|_{\infty,T}$. Then*

$$\limsup_{u\to\infty} \frac{1}{u} \log P(\Gamma_u B \in A^c) \;\leq\; -\inf_{\omega \in \mathcal{H}^m \cap A^c} I_T(\omega) \quad (4.2)$$

and

$$\liminf_{u\to\infty} \frac{1}{u} \log P(\Gamma_u B \in A^o) \;\geq\; -\inf_{\omega \in \mathcal{H}^m \cap A^o} I_T(\omega).$$

Proof First note that for all $u > 0$

$$P(\Gamma_u B \in A^c) \;=\; P(\sqrt{u} B_{[0,T]} \in A_{[0,T]}^c), \quad (4.3)$$

where $B_{[0,T]}$ is the Brownian motion B restricted to the time interval $[0,T]$ and similarly $A_{[0,T]}^c$ is the image of A^c under the mapping, which restricts every element of $\mathcal{C}^m$ onto the time interval $[0,T]$. The same holds when A^c is replaced by A^o. Therefore, the statement of this theorem is a consequence of Schilder's theorem; see Theorem 5.2.3 and Exercise 5.2.8 of Dembo and Zeitouni (1993). □

The difficulties in deriving an LDP for the sequence $(W(0)/u)_{u>0}$ stem from the fact that the mapping $\overline{\Psi}$ is not continuous with respect to the topology of uniform convergence on compact intervals. So we cannot simply use Schilder's theorem and apply the contraction principle. Instead we will use the following two lemmas.

Lemma 4.2 *For every $c, r \geq 0$ there is a $T > 0$ such that for all $x \in \mathbb{R}_+^n$ with $\|x\|_\infty \leq r$ and $\omega \in \mathcal{H}^m$ with $\Psi(x + M\omega - D)(t) \neq 0$ for all $t \in [0, T]$ we have $I_T(\omega) \geq c$.*

Proof Due to the strict positivity of the components of d, there is a constant $N > 0$ such that for all $x \in \mathbb{R}_+^n$ and $t \geq N\|x\|_\infty$ we have $\Psi(x - D)(t) = 0$. Let $b > 0$ such that for all $T > 0$, $x \in \mathbb{R}_+^n$ and $\omega, \eta \in \mathcal{C}^m$ we have $\|\Psi(x + M\omega - D) - \Psi(x + M\eta - D)\|_{\infty, T} \leq b\|\omega - \eta\|_{\infty, T}$. Fix a $T > Nr$ such that $(T - rN)^2 \geq 2Tcb^2N^2$.

Let $x \in \mathbb{R}_+^n$ with $\|x\|_\infty \leq r$ and $\omega \in \mathcal{H}^m$ be such that $\eta := \Psi(x + M\omega - D)$ does not vanish on $[0, T]$. As η is continuous there is an $\epsilon > 0$ such that $\|\eta(t)\|_\infty \geq \epsilon$ for all $t \in [0, T]$. Define recursively the time points $t_1 := 0$ and $t_{k+1} := t_k + N\|\eta(t_k)\|_\infty$ and set $\Delta_k := t_{k+1} - t_k$ for $k \in \mathbb{N}$. There exists an $l \in \mathbb{N}$ such that $t_l \leq T < t_{l+1}$, because $\|\eta(t_k)\|_\infty \geq \epsilon$ as long as $t_k \leq T$. By the loss-of-memory property of Ψ we have

$$\eta(t_{k+1}) = \Psi(\eta(t_k) + M(\theta_{t_k}\omega - \omega(t_k)) - D)(t_{k+1}). \tag{4.4}$$

On the other hand $\Psi(\eta(t_k) - D)(t_{k+1}) = 0$. By the definition of b it follows that $\|\eta(t_{k+1})\|_\infty \leq b\|\theta_{t_k}\omega - \omega(t_k)\|_{\infty, \Delta_k}$. Therefore

$$I_T(\omega) \geq \sum_{k=1}^{l} I_T(\theta_{t_k}\omega - \omega(t_k))$$

$$\geq \sum_{k=1}^{l} \frac{1}{2\Delta_k}\|\theta_{t_k}\omega - \omega(t_k)\|_{\infty, \Delta_k}^2$$

$$\geq \sum_{k=1}^{l} \frac{1}{2b^2\Delta_k}\|\eta(t_{k+1})\|_\infty^2$$

$$= \sum_{k=1}^{l} \frac{\Delta_k}{2b^2N^2}\left(\frac{\Delta_{k+1}}{\Delta_k}\right)^2$$

$$\geq \frac{1}{2b^2N^2\sum_{k=1}^{l}\Delta_k}\left(\sum_{k=2}^{l+1}\Delta_k\right)^2$$

$$\geq \frac{(T - rN)^2}{2Tb^2N^2}$$

$$\geq c. \tag{4.5}$$

In the fifth step we applied Jensen's inequality. This proves the lemma. $\square$

Define for every $T > 0$ the *quasi-potential* $L_T \colon \mathbb{R}_+^n \to [0, \infty]$ by setting

$$L_T(y) := \inf_{\omega \in \mathcal{H}^m,\ \Psi(M\omega - D)(T) = y} I_T(\omega). \tag{4.6}$$

Lemma 4.3 *For every $c > 0$ and compact set $K \subset \mathbb{R}^n_+$ there exists a $T > 0$ such that for all $y \in \mathbb{R}^n_+$*

$$\inf_{\omega \in \mathcal{H}^m, \; \exists x \in K: \; \Psi(x + M\omega - D) = y} I_T(\omega) \geq \min \left\{ c, \inf_{S \geq 0} L_S(y) \right\}. \qquad (4.7)$$

Proof Let $r := \sup_{x \in K} \|x\|_\infty$. Choose T according to Lemma 4.2. Let $x \in K$ and $\omega \in \mathcal{H}^m$. Set $\eta := \Psi(x + M\omega - D)$ and suppose $\eta(T) = y$.

Case one: there is an $s \in [0, T]$ with $\eta(s) = 0$. Define $\omega' := \theta_s \omega - \omega(s)$ and $\eta' := \Psi(M\omega' - D)$. We have by the loss-of-memory property $\eta'(T - s) = y$ and also $\inf_{S \geq 0} L_S(y) \leq I_{T-s}(\omega') \leq I_T(\omega)$.

Case two: there is no $s \in [0, T]$ with $\eta(s) = 0$. Then $I_T(\omega) \geq c$ by Lemma 4.2. Both cases together give the claimed inequality. $\qquad \square$

We can now prove an LDP for the sequence $(W(0)/u)_{u>0}$:

Theorem 4.4. (Variational formula) *For every measurable $A \subset \mathbb{R}^n$*

$$\limsup_{u \to \infty} \frac{1}{u} \log P(W(0)/u \in A) \leq - \inf_{T > 0, \; y \in A^c} L_T(y) \qquad (4.8)$$

and

$$\liminf_{u \to \infty} \frac{1}{u} \log P(W(0)/u \in A) \geq - \inf_{T > 0, \; y \in A^\circ} L_T(y), \qquad (4.9)$$

where A^c resp. A° is the closed hull resp. open interior of A.

Proof *The upper bound*: Let $c > 0$. The sequence $(W(0)/u)_{u>0}$ is exponentially tight by Lemma 3.2 and therefore we can find a compact set $K \subset \mathbb{R}^n_+$ such that $\limsup_{u \to \infty} \frac{1}{u} \log P(W(0) \notin Ku) \leq -c$. Let $T > 0$ such that the statement of Lemma 4.3 is satisfied for the constant c and the compact set K. Then we have by stationarity of W and Lemma 1.2.15 of Dembo and Zeitouni (1993)

$$\limsup_{u \to \infty} \frac{1}{u} \log P(W(0)/u \in A)$$

$$\leq \max \left\{ -c, \limsup_{u \to \infty} \frac{1}{u} \log P(W(0)/u \in K, \; W(Tu)/u \in A) \right\}. \qquad (4.10)$$

The indistinguishability of W and $W^{W(0)}$ (see Lemma 2.3) gives

$$\begin{aligned}
P(W(0)/u \in K, \; W(Tu)/u \in A) \\
= \; & P(W(0)/u \in K, \; \Psi(W(0) + MB - D)(Tu)/u \in A) \\
\leq \; & P(\exists x \in K: \; \Psi(xu + MB - D)(Tu)/u \in A^c) \\
= \; & P(\exists x \in K: \; \Psi(x + M\Gamma_u B - D)(T) \in A^c). \qquad (4.11)
\end{aligned}$$

Note that the set $\{ \omega \in \mathcal{C}^m: \; \exists x \in K \; \Psi(x + M\omega - D)(T) \in A^c \}$ is closed in the topology induced on $\mathcal{C}^m$ by the semi-norm $\| \cdot \|_{\infty, T}$. If we use this estimate and our version of Schilder's theorem 4.1 to continue inequality 4.10, we get

$$\limsup_{u \to \infty} \frac{1}{u} \log P(W(0)/u \in A)$$

$$\leq -\min \left\{ c, \inf_{\omega \in \mathcal{H}^m, \; \exists x \in K: \; \Psi(x + M\omega - D)(T) \in A^c} I_T(\omega) \right\}. \quad (4.12)$$

Using Lemma 4.3 and letting c tend to infinity, we get the upper bound.

The lower bound: Set $\mathcal{I} := \inf_{T>0, \; y \in A^o} L_T(y)$ and suppose without loss of generality that $\mathcal{I} < \infty$. Let $\epsilon > 0$ and choose a $T > 0$ and a $y \in A^o$ such that $L_T(y) \leq \mathcal{I} + \epsilon/2$. Fix an $\eta \in \mathcal{H}^m$ such that $\Psi(M\eta - D)(T) = y$ and $I_T(\eta) \leq L_T(y) + \epsilon/2 \leq \mathcal{I} + \epsilon$. Define for $\delta > 0$ the set $E_\delta := \{\omega \in \mathcal{C}^m : \|\omega - \eta\|_{\infty,T} < \delta\}$. Because of the openness of A^o and the Lipschitz continuity of the mapping Ψ with respect to the semi-norm $\|\cdot\|_{\infty,T}$ we can find $\zeta, \xi > 0$ such that for all $x \in \mathbb{R}^n_+$ with $\|x\|_\infty < \xi$ and $\omega \in E_\zeta$ we have $\Psi(x + M\omega - D)(T) \in A^o$. The set E_ζ is open in the topology induced on $\mathcal{C}^m$ by the semi-norm $\|\cdot\|_{\infty,T}$. We deduce from the above considerations the inclusion

$$\{W(Tu)/u \in A\} \supset \{\|W(0)/u\|_\infty < \xi, \; \Gamma_u B \in E_\zeta\}. \quad (4.13)$$

The events $\{\|W(0)/u\|_\infty < \xi\}$ and $\{\Gamma_u B \in E_\zeta\}$ are independent. The stationarity of W, the above inclusion and this independence yield the estimate

$$\liminf_{u \to \infty} \frac{1}{u} \log P(W(0)/u \in A)$$

$$= \liminf_{u \to \infty} \frac{1}{u} \log P(W(Tu)/u \in A)$$

$$\geq \liminf_{u \to \infty} \frac{1}{u} \log P(\|W(0)/u\|_\infty < \xi, \; \Gamma_u B \in E_\zeta)$$

$$\geq \liminf_{u \to \infty} \frac{1}{u} \log P(\|W(0)/u\|_\infty < \xi) + \liminf_{u \to \infty} \frac{1}{u} \log P(\Gamma_u B \in E_\zeta). \quad (4.14)$$

The first summand on the right-hand side of this inequality is zero. To calculate the value of the second summand we can use our version of Schilder's theorem 4.1. This gives

$$\liminf_{u \to \infty} \frac{1}{u} \log P(W(0)/u \in A) \geq \liminf_{u \to \infty} \frac{1}{u} \log P(\Gamma_u B \in E_\zeta)$$

$$\geq -\inf_{\omega \in \mathcal{H}^m \cap E_\zeta} I_T(\omega)$$

$$\geq -I_T(\eta)$$

$$\geq -\mathcal{I} - \epsilon. \quad (4.15)$$

Letting ϵ tend to zero, we get the claimed lower bound. $\qquad \square$

5   Example: queues in series

Let W be an n-dimensional stationary RBM with reflection matrix R, covariance matrix $R\Sigma R^T$ and drift $-Rd$, where R, Σ and d are given by

$$R := \begin{pmatrix} 1 & 0 & 0 & \cdots & 0 \\ -\alpha & 1 & 0 & \ddots & \vdots \\ -\alpha\beta & -\alpha & 1 & \ddots & 0 \\ \vdots & \ddots & \ddots & \ddots & 0 \\ -\alpha\beta^{n-2} & \cdots & -\alpha\beta & -\alpha & 1 \end{pmatrix}, \tag{5.1}$$

$$\Sigma := \begin{pmatrix} \sigma_1 & & 0 \\ & \ddots & \\ 0 & & \sigma_n \end{pmatrix}, \tag{5.2}$$

$$\text{and} \quad d := \begin{pmatrix} a_1 \\ \vdots \\ a_n \end{pmatrix}, \tag{5.3}$$

with $\sigma_1, \ldots, \sigma_n, a_1, \ldots, a_n > 0$, $\alpha \in \,]0,1[$ and $\beta := 1 - \alpha$. The inverse of R is $I + Q$, where

$$Q := \begin{pmatrix} 0 & 0 & \cdots & 0 \\ \alpha & 0 & \ddots & \vdots \\ \vdots & \ddots & \ddots & 0 \\ \alpha & \cdots & \alpha & 0 \end{pmatrix}. \tag{5.4}$$

An $n \times n$ matrix M with $\Sigma = MM^T$ is

$$M := \begin{pmatrix} \sqrt{\sigma_1} & & 0 \\ & \ddots & \\ 0 & & \sqrt{\sigma_n} \end{pmatrix}. \tag{5.5}$$

The RBM with these parameters represents the heavy traffic approximation of a series of n deterministic first-come first-served nodes, which have to be traversed by a constant bitrate traffic stream C. The n nodes have to cope with additional bursty background traffic. The background arrival times at a node are independent of the background arrival times at the other nodes and of the arrival times of the stream C. Background traffic leaves the net after completion of its first service. For $i = 1, \ldots, n$ the parameter σ_i is the coefficient of dispersion of the background stream at node i. The parameter a_i can be interpreted as one minus the traffic intensity at the ith node and α as the fraction of the traffic C at the nodes.

Note that by the *snapshot principle* (see Harrison and Nguyen (1993)), the value of $\sum_{i=1}^{n} W_i(0)$ gives the heavy traffic approximation of the sojourn time of a customer of the stream C, which arrives at the network at time 0. Be aware that the distribution of $W(0)$ is *not* of product form. This is a consequence of Theorem 3.1 in Harrison and Williams (1992).

Proposition 5.1 *For $l = 1, \ldots, n$ we have*

$$\lim_{u \to \infty} \frac{1}{u} \log P\left(\sum_{i=1}^{l} W_i(0) \geq u\right) \;=\; -\max_{i \leq l} \frac{2a_i}{\sigma_i}. \tag{5.6}$$

Proof Define the stationary random processes $S_k := \sum_{i=1}^{k} W_i$ for $k = 1, \ldots, n$. We prove the proposition by induction over l. For $l = 1$ the statement follows from the well-known fact that the stationary distribution of a one-dimensional Brownian motion with drift $-a_1$ and variance σ_1 reflected at zero is exponential with parameter $2a_1/\sigma_1$.

Suppose we have already shown that eqn 5.6 holds for $l = k - 1$ with $1 < k \leq n$. We are going to show that eqn 5.6 also holds for $l = k$.

Upper bound: By Lemma 3.4 and Proposition 3.3 we have for all $\delta \in \,]0,1[$

$$\limsup_{u \to \infty} \frac{1}{\delta u} \log P(S_k(0) \geq u)$$

$$= \limsup_{u \to \infty} \frac{1}{u} \log P(\delta S_k(0) \geq u)$$

$$\leq \inf_{p \in \mathbb{R}} \limsup_{u \to \infty} \frac{1}{u} \log P(W_k(0) \geq pu, \; S_{k-1}(0) \geq (1-p)u)$$

$$= \inf_{p \in [0,1]} \limsup_{u \to \infty} \frac{1}{u} \log P(W_k(0) \geq pu, \; S_{k-1}(0) \geq (1-p)u)$$

$$\leq \inf_{c>0,\, p \in [0,1]} \limsup_{u \to \infty} \frac{1}{u} \log P\Big(-Y_k(-cu) + \alpha S_{k-1}(-cu) - \alpha S_{k-1}(0) \geq pu,$$

$$S_{k-1}(0) \geq (1-p)u\Big). \tag{5.7}$$

From the induction hypothesis and the stationarity and non-negativity of the process S_{k-1} it follows that the sequence $(S_{k-1}(-cu)/u - S_{k-1}(0)/u)_{u>0}$ is exponentially tight. Also the sequence $(-Y_k(cu))_{u>0}$ is exponentially tight. Therefore, we can again apply Lemma 3.4 and obtain

$$\limsup_{u \to \infty} \frac{1}{\delta u} \log P(S_k(0) \geq u)$$

$$\leq \inf_{c>0,\, p \in [0,1],\, q \in \mathbb{R}} \limsup_{u \to \infty} \frac{1}{u} \log P\Big(-Y_k(-cu) \geq qp\delta u,$$

$$S_{k-1}(-cu) - S_{k-1}(0) \geq \frac{(1-q)p}{\alpha}\delta u, \; S_{k-1}(0) \geq (1-p)u\Big)$$

$$\leq \inf_{c>0,\, p \in [0,1],\, q \in \mathbb{R}} \limsup_{u \to \infty} \frac{1}{u} \log P\Big(-\sqrt{\sigma_k}B_k(-cu) \geq (qp\delta + ca_k)u,$$

$$S_{k-1}(-cu) \geq \frac{(1-q)p\delta + \alpha(1-p)}{\alpha}u, \; S_{k-1}(0) \geq (1-p)u\Big)$$

$$\leq \inf_{c>0,\, p \in [0,1],\, q \in \mathbb{R}} \left\{ \limsup_{u \to \infty} \frac{1}{u} \log P(-\sqrt{\sigma_k}B_k(-cu) \geq (qp\delta + ca_k)u) \right.$$

$$+ \limsup_{u\to\infty} \frac{1}{u} \log P\Big(S_{k-1}(-cu) \geq \frac{(1-q)p\delta + \alpha(1-p)}{\alpha}u,$$

$$S_{k-1}(0) \geq (1-p)u\Big)\Big\}. \tag{5.8}$$

In the last step, we have used the independence of the processes S_{k-1} and $\sqrt{\sigma_k}B_k$, coming from the diagonality of matrix M.

The first summand in the infimum on the right side of estimate 5.8 can be calculated by our version of Schilder's theorem 4.1 (or by elementary calculations because the random variable $-\sqrt{\sigma_n}B_k(-cu)$ is Gauss distributed with variance $c\sigma_k u$ for $c, u > 0$). This gives

$$\limsup_{u\to\infty} \frac{1}{u} \log P(-\sqrt{\sigma_k}B_k(-cu) \geq (qp\delta + ca_k)u) = \frac{(qp\delta + ca_k)^2}{2c\sigma_k}. \tag{5.9}$$

We can deduce from the induction hypothesis together with the stationarity of S_{k-1} and a simple scaling argument that $\limsup_{u\to\infty} \frac{1}{u} \log P(S_{k-1}(yu) \geq xu) = -xr_k$ for all $x \geq 0$ and $y \in \mathbb{R}$, where $r_k := -\max_{i<k} 2a_i/\sigma_i$. Therefore, the second summand in the infimum on the right side of estimate 5.8 can be bounded by

$$\limsup_{u\to\infty} \frac{1}{u} \log P\Big(S_{k-1}(-cu) \geq \frac{(1-q)p\delta + \alpha(1-p)}{\alpha}u, \; S_{k-1}(0) \geq (1-p)u\Big)$$

$$\leq \min\Big\{ \limsup_{u\to\infty} \frac{1}{u} \log P\Big(S_{k-1}(-cu) \geq \frac{(1-q)p\delta + \alpha(1-p)}{\alpha}u\Big),$$

$$\limsup_{u\to\infty} \frac{1}{u} \log P(S_{k-1}(0) \geq (1-p)u)\Big\}$$

$$= -r_k \max\Big\{ \frac{(1-q)p\delta + \alpha(1-p)}{\alpha}, 1-p\Big\}. \tag{5.10}$$

These results allow us to continue estimate 5.8 with (recall $\alpha, \delta \in \;]0,1[$)

$$\limsup_{u\to\infty} \frac{1}{\delta u} \log P(S_k(0) \geq u)$$

$$\leq - \inf_{c>0,\; p\in[0,1],\; q\in\mathbb{R}} \Big\{ \frac{(qp\delta + ca_k)^2}{2c\sigma_k} + r_k \max\Big\{ \frac{(1-q)p\delta + \alpha(1-p)}{\alpha}, 1-p\Big\}\Big\}$$

$$\leq - \inf_{p,q\in[0,1]} \Big\{ \frac{2a_k}{\sigma_k} \max\{qp\delta, 0\} + r_k \max\Big\{ \frac{(1-q)p\delta + \alpha(1-p)}{\alpha}, 1-p\Big\}\Big\}$$

$$\leq - \inf_{p,q\in[0,1]} \Big\{ qp\delta \frac{2a_k}{\sigma_k} + \delta r_k((1-q)p + (1-p))\Big\}$$

$$= -\delta \max_{i\leq k} \frac{2a_i}{\sigma_i}. \tag{5.11}$$

Lower bound: Let $j \in \{1, \ldots, k\}$ be such that $a_j/\sigma_j = \min_{i\leq k} a_i/\sigma_i$. Consider the path $\omega = (\omega_1, \ldots, \omega_k)^T \in \mathcal{H}^n$ given by $\omega_i = 0$ if $i \neq j$ and $\omega_j(t) = \frac{2a_j t}{\sqrt{\sigma_j}}$.

Define the image path $\eta := \Psi(M\omega - D)$. It follows that $\eta(0) = 0$ and $\eta(1/a_j)$ is the unit vector with jth component equal to one. Using the lower bound of Theorem 4.4 we obtain for all $\delta \in]0, 1[$

$$
\begin{aligned}
\liminf_{u \to \infty} \frac{\delta}{u} \log P(S_k(0) \geq u) \;\; &\geq \;\; \liminf_{u \to \infty} \frac{1}{u} \log P(S_k(0) > \delta u) \\
&\geq \;\; -I_{1/a_j}(\omega) \\
&= \;\; -\max_{i \leq k} \frac{2a_i}{\sigma_i}.
\end{aligned} \tag{5.12}
$$

Letting δ in inequalities 5.11 and 5.12 tend to one, we obtain eqn 5.6 for $l = k$. By induction we get the statement of the proposition. $\square$

Note that the proof of the upper bound does not make use of the variational formula of Theorem 4.4.

This example shows that the bottleneck node determines the large deviations of sojourn times in the heavy traffic approximation. A similar result for a sequence of queues in tandem was presented by A. J. Ganesh at the Royal Statistical Society workshop with which this volume is associated.

Bibliography

1. Bertsimas, D., Paschalidis, I. Ch. and Tsitsiklis, J. N. (1994). On the large deviations behaviour of acyclic networks of G/G/1 queues. Working paper LIDS-P-2278 of the Massachusetts Institute of Technology. December.

2. Dembo, A. and Zeitouni, O. (1993). *Large Deviations Techniques and Applications*. Jones and Bartlett, London.

3. Dupuis, P. and Ellis, R. S. (1995). Large deviation analysis of queueing systems. In *Stochastic Networks*, Kelly, F. P. and Williams, R. J. (editors), 347–366, Springer, New York.

4. Harrison, J. M. and Nguyen, V. (1993). Brownian models of multiclass queueing networks: current status and open problems. *Queueing Systems*, **13**, 5–40.

5. Harrison, J. M. and Williams, R. J. (1992). Brownian models of feedforward queueing networks: quasireversibility and product form solutions. *The Annals of Applied Probability*, **2(2)**, 263–293.

6. Majewski, K. (1995). Large deviations of reflected Brownian motions. Working paper.

7. Majewski, K. (1996). Large deviations of feedforward queueing networks. Ph.D. thesis (in preparation).

8. O'Connell, N. (1996). Large deviations for departures from a shared buffer. Technical Report HPL-BRIMS-96-003, BRIMS, Hewlett-Packard, Bristol.

9. Peterson, W. P. (1991). A heavy traffic limit theorem for networks of queues with multiple customer types. *Mathematics of Operations Research*, **16(1)**, 90–118.

10. Ramanan, K. and Dupuis, P. (1995). Large deviation properties of data streams that share a large buffer. Report LCDS #95-8. Lefschetz Center for Dynamical Systems and Center for Control Sciences. Division of Applied Mathematics, Brown University, Providence, RI 02912. August.

11. Taylor, L. M. and Williams, R. J. (1993). Existence and uniqueness of semi-martingale reflecting Brownian motions in an orthant. *Probability Theory and Related Fields*, **96**, 283–317.

7

Limit theorems for workload input models

Thomas G. Kurtz

University of Wisconsin–Madison

Abstract

General models for workload input into service systems are considered. Scaling limit theorems appropriate for the formulation of fluid and heavy traffic approximations for systems driven by these inputs are given. Under appropriate assumptions, it is shown that fractional Brownian motion can be obtained as the limiting workload input process. Motivation for these results comes from data on communication network traffic exhibiting scaling properties similar to those for fractional Brownian motion.

1 Discrete source models

We consider models for the input of work into a system from a large number of sources. Each source "turns on" at a random time and inputs work into the system for some period of time. Associated with each active period of a source is a cumulative input process, that is a nondecreasing stochastic process X such that $X(t)$ is the cumulative work input into the system during the first t units of time during the active period. One representation of the process is as follows. Let $N(t)$ denote the number of source activations up to time t, and for the ith activation, let $X_i(s)$ denote the cumulative work input into the system during the first s units of time that the source is on. We will refer to N as the source activation process. The total work input into the system up to time t is then given by

$$U(t) = \int_0^t X_{N(s)}(t - s)dN(s). \qquad (1.1)$$

In some settings, it is useful to model the length of time τ_i that a source remains active separately from X_i. The total work input up to time t then becomes

$$U(t) = \int_0^t X_{N(s)}(\tau_{N(s)} \wedge (t - s))dN(s). \qquad (1.2)$$

Of course these two approaches are equivalent since for the second model, we can always define $\tilde{X}_i(s) = X_i(\tau_i \wedge s)$ and obtain a model of the first form with the same total work input. Note that many sources may be active at the same time, and we let $L(t)$ denote the number of sources active at time t, that is

$$L(t) = \int_0^t I_{[0,\tau_{N(s)})}(t-s)dN(s).$$

We assume that the active periods are identical in the sense that the (X_i, τ_i) are i.i.d. with distribution ν on $D_{\mathbb{R}}[0,\infty) \times [0,\infty)$, although the methods we use can be extended to multiple source classes. Let $\lambda : D_{\mathbb{R}^3}[0,\infty) \to D_{[0,\infty)}[0,\infty)$ be nonanticipating in the sense that $\lambda(v,t) = \lambda(v^t, t)$ for all t where $v^t(s) = v(t \wedge s)$, $v \in D_{\mathbb{R}^3}[0,\infty)$. Assume that N is a counting process with an intensity $\lambda(N,U,L,\cdot)$, that is

$$N(t) - \int_0^t \lambda(N,U,L,s)ds$$

is a martingale with respect to the filtration given by $\mathcal{F}_t = \sigma(N(s), U(s), L(s) : s \le t)$. In particular, $\lambda(N,U,L,t)$ depends on values of N, U, and L only at times $s \le t$.

We can represent the process (N,U,L) as the solution of a stochastic equation involving a Poisson random measure. Recall that a Poisson random measure ξ with mean measure η on a measurable space $(E, \mathcal{B})$ is a random measure on $\mathcal{B}$ such that for each $\Gamma \in \mathcal{B}$, $\xi(\Gamma)$ has a Poisson distribution with mean $\eta(\Gamma)$ and for disjoint $\Gamma_1, \Gamma_2, \ldots$, $\xi(\Gamma_1), \xi(\Gamma_2), \ldots$ are independent.

Let ξ be a Poisson random measure on $[0,\infty) \times D_{\mathbb{R}}[0,\infty) \times [0,\infty)$ with mean measure $\eta = m \times \nu$ (m being Lebesgue measure). ξ can be represented as

$$\xi = \sum_{i=1}^{\infty} \delta_{(S_i, X_i, \tau_i)}$$

where $0 < S_1 < S_2 < \ldots$ are the jump times of a unit Poisson process and the (X_i, τ_i) are as above. Let N, U, and L satisfy the system of equations

$$
\begin{aligned}
N(t) &= \xi(B(t)) \\
U(t) &= \int_{B(t)} u(r \wedge \gamma_t(s))\xi(ds \times du \times dr) \quad\quad (1.3)\\
L(t) &= N(t) - \xi(A(t))
\end{aligned}
$$

where

$$B(t) = \left\{ (s,u,r) : s \le \int_0^t \lambda(N,U,L,z)dz \right\}$$

$$A(t) = \left\{ (s,u,r) : s \le \int_0^{t-r} \lambda(N,U,L,z)dz \right\}$$

$$\int_0^{\beta(t)} \lambda(N,U,L,s)ds = t$$

and

$$\gamma_t(s) = t - \beta(s).$$

Note that $\beta(S_i)$ is the ith activation time, that is the ith jump time of $N(t)$; for $S_i \leq t$, $\gamma_t(S_i)$ is the elapsed time since the ith activation; $T_i = \beta(S_i) + \tau_i$ is the end of the ith active period and satisfies $\tau_i = \gamma_{T_i}(S_i)$. With these interpretations, $X_i(\tau_i \wedge \gamma_t(S_i))$ is the work input by the ith activated source up to time t (giving the equation for U in (1.3)), and $\xi(A(t))$ is the number of active periods that have completed at or before time t (giving the identity for L in (1.3)). It is easy to see that the solution of the above system exists and is unique up to $\sigma_\infty = \inf\{t : N(t) = \infty\}$. This representation is essentially that used in Kurtz (1983) and Solomon (1987) for population models and by Foley (1982) in the particular case of the $M/G/\infty$ queue.

Note that $\eta(B(t)) = \Lambda(t) \equiv \int_0^t \lambda(N, U, L, s)ds$, and the fact that

$$N(t) - \int_0^t \lambda(N, U, L, s)ds = \xi(B(t)) - \eta(B(t))$$

is a martingale follows from the fact that for each $t \geq 0$, $\Lambda(t)$ is a stopping time with respect to the filtration $\{\mathcal{H}_u\}$ given by

$$\mathcal{H}_u = \sigma(\xi(\Gamma) : \Gamma \subset \Gamma_u \equiv [0, u] \times D_{\mathbb{R}}[0, \infty) \times [0, \infty))$$

and that $\xi(\Gamma_u) - u$ is an $\{\mathcal{H}_u\}$-martingale.

Example 1.1 *Queue-dependent arrivals. Consider the simple model for workload processing in which work is processed at rate 1 as long as there is work in the system. The queued work is given by*

$$Q(t) = Q(0) + U(t) - t + \Lambda(t)$$

where Λ denotes the idle time of the processor. Note that Q is uniquely determined by U, so an arrival intensity $\lambda(Q(t))$ is of the form considered above.

Example 1.2 *On/off sources. Let L_0 be a positive integer and set*

$$\lambda(N, U, L, t) = \lambda_0(L_0 - L(t)).$$

This model would correspond to L_0 fixed sources, each of which alternates between active and inactive periods. The active periods have lengths τ as above, and the inactive periods are exponentially distributed with parameter λ_0.

Section 2 considers scaling limits of the above models that give deterministic "fluid approximations" and corresponding central limit theorems. The proofs of these results extend ideas developed in Kurtz (1983), Solomon (1987), and Garcia (1995). Section 3 considers models in which the activation times occur at a constant rate (that is, form a Poisson process). Extending the definition of the process to the time interval $(-\infty, \infty)$, the input process has stationary increments. Under "heavy traffic" scaling, we obtain Gaussian limit processes having stationary increments. In Section 4, we show that fractional Brownian motion can be obtained as a limit under the heavy traffic scaling. Section 5 suggests two variations of the model indicating how a type of control can be introduced and how models with a fixed number of regenerative sources can be developed. Finally, Section 6 is an appendix containing several technical lemmas.

Iglehart (1973) considers functional limit theorems for models of the form (1.1) in which N is a renewal process. His results have substantial overlap with the results of Section 3. In particular, his proof of relative compactness is similar to that used here. Models of this form have also been considered by Klüppelberg and Mikosch (1995a,b) in the context of risk analysis, where X is interpreted as the cumulative payout of an insurance claim. They assume that N is a Poisson process. Their results also have substantial overlap with the results of Section 3, although the approach is quite different.

Our motivation comes from the analysis that has been carried out on network traffic data (see Willinger, Taqqu, Leland, and Wilson (1995)) which indicates that the traffic exhibits "long-term dependence" that is not consistent with simpler compound-Poisson models of workload input. In this context, Willinger, Taqqu, Sherman, and Wilson (1995) obtain a limit theorem for an on/off source model similar to the results of Section 3. Their model is not covered by the class of models considered in detail here (except in the case of exponential off periods as in Example 1.2). Their model is included in a class of regenerative source models that can be studied using techniques similar to those developed here. These models are described briefly in Section 5, but the details of the convergence results will be given elsewhere.

2 Law of large numbers and central limit theorem

A variety of limit theorems can be proved for solutions of systems of the above form based on the law of large numbers and the central limit theorem for the underlying Poisson random measure ξ. For example, suppose that $\lambda_n(v, t) = n\lambda(n^{-1}v, t)$ (where $v \in D_{\mathbb{R}^3}[0, \infty)$) and that ξ_n is a Poisson random measure with mean measure $nm \times \nu$. Let N_n be the activation process, U_n the work input process, and L_n the active source process, and define

$$X_n(t) = \frac{1}{n}N_n(t), \qquad Y_n(t) = \frac{1}{n}U_n(t), \qquad Z_n(t) = \frac{1}{n}L_n(t) .$$

Then we can write

$$X_n(t) = \frac{1}{n}\xi_n(B_n(t))$$

$$Y_n(t) = \int_{B_n(t)} u(r \wedge \gamma_t^n(s))\frac{1}{n}\xi_n(ds \times du \times dr) \qquad (2.1)$$

$$Z_n(t) = X_n(t) - \frac{1}{n}\xi_n(A_n(t))$$

where

$$B_n(t) = \left\{(s, u, r) : s \leq \int_0^t \lambda(X_n, Y_n, Z_n, z)dz\right\}$$

$$A_n(t) = \left\{ (s, u, r) : s \leq \int_0^{t-r} \lambda(X_n, Y_n, Z_n, z) dz \right\}$$

$$\int_0^{\beta_n(s)} \lambda(X_n, Y_n, Z_n, z) dz = s$$

and

$$\gamma_t^n(s) = t - \beta_n(s).$$

Using the fact that $\frac{1}{n}\xi_n(A) \to m \times \nu(A)$ in probability for each Borel set A and uniformity results of Stute (1976), under appropriate uniqueness conditions, it is not difficult to show that (X_n, Y_n, Z_n) converges to the solution of the system

$$\begin{aligned} X(t) &= \int_0^t \lambda(X, Y, Z, s) ds \\ Y(t) &= \int_0^t \mu(t-s)\lambda(X, Y, Z, s) ds \\ Z(t) &= \int_0^t \lambda(X, Y, Z, s)\nu_\tau(t-s, \infty) ds \end{aligned} \qquad (2.2)$$

where $\mu(t) = \int_{D_\mathbb{R}[0,\infty) \times [0,\infty)} u(r \wedge t)\nu(du \times dr) = E[X_i(\tau_i \wedge t)]$ and ν_τ is the distribution of τ_i, so that $\nu_\tau(t, \infty) = P\{\tau_i > t\} = \nu(D_\mathbb{R}[0,\infty) \times (t,\infty))$. In particular, we have the following theorem.

Theorem 2.1 *Suppose that there exists a finite, nondecreasing function K such that*

$$\begin{aligned} |\lambda(x_1, y_1, z_1, t) &- \lambda(x_2, y_2, z_2, t)| \\ &\leq K(t) \sup_{s \leq t}(|x_1(s) - x_2(s)| + |y_1(s) - y_2(s)| + |z_1(s) - z_2(s)|). \end{aligned}$$

Then there exists a unique solution to (2.2) and

$$\sup_{s \leq t}(|X_n(s) - X(s)| + |Y_n(s) - Y(s)| + |Z_n(s) - Z(s)|) \to 0$$

in probability.

Proof Observing, for example, that $m \times \nu(B_n(t)) = \int_0^t \lambda(X_n, Y_n, Z_n, s) ds$, we can rewrite (2.1) as

$$\begin{aligned} X_n(t) &= \frac{1}{n}\tilde{\xi}_n(B_n(t)) + \int_0^t \lambda(X_n, Y_n, Z_n, s) ds \\ Y_n(t) &= \int_{B_n(t)} u(r \wedge \gamma_t^n(s))\frac{1}{n}\tilde{\xi}_n(ds \times du \times dr) \\ &\quad + \int_0^t \mu(t-s)\lambda(X_n, Y_n, Z_n, s) ds \end{aligned} \qquad (2.3)$$

$$Z_n(t) = \frac{1}{n}\tilde{\xi}_n(B_n(t) - A_n(t)) + \int_0^t \nu_\tau(t - s, \infty)\lambda(X_n, Y_n, Z_n, s)ds,$$

where $\tilde{\xi}_n(B) = \xi_n(B) - m \times \nu(B)$. The terms involving $\tilde{\xi}_n$ go to zero uniformly in bounded time intervals, and the theorem follows by the Lipschitz assumption and Gronwall's inequality. □

To derive the corresponding central limit theorem, define $\tilde{X}_n(t) = \sqrt{n}(X_n(t) - X(t))$, $\tilde{Y}_n(t) = \sqrt{n}(Y_n(t) - Y(t))$, $\tilde{Z}_n(t) = \sqrt{n}(Z_n(t) - Z(t))$, and

$$\Xi_n(A) = \frac{1}{\sqrt{n}}(\xi(A) - nm \times \nu(A)).$$

Note that $\Xi_n(A) \Rightarrow \Xi(A)$ where Ξ is Gaussian white noise with $E[\Xi(A)\Xi(B)] = m \times \nu(A \cap B)$. Then

$$\tilde{X}_n(t) = \Xi_n(B_n(t)) + \int_0^t \sqrt{n}(\lambda(X_n, Y_n, Z_n, s) - \lambda(X, Y, Z, s))ds$$

$$\tilde{Y}_n(t) = \int_{B_n(t)} u(r \wedge \gamma_t^n(s))\Xi_n(ds \times du \times dr) \qquad (2.4)$$

$$+ \int_0^t \mu(t - s)\sqrt{n}(\lambda(X_n, Y_n, Z_n, s) - \lambda(X, Y, Z, s))ds$$

$$\tilde{Z}_n(t) = \Xi_n(B_n(t) - A_n(t))$$

$$+ \int_0^t \nu_\tau(t - s, \infty)\sqrt{n}(\lambda(X_n, Y_n, Z_n, s) - \lambda(X, Y, Z, s))ds.$$

Assume that there exist K_1 and K_2 such that $\lambda(v, t) \leq K_1 + K_2 \sup_{s \leq t}|v(s)|$ and that λ is differentiable in an appropriate sense. For example, we can assume that $D\lambda : D_{\mathbb{R}^6}[0, \infty) \to D_{\mathbb{R}}[0, \infty)$ satisfies

$$|\lambda(v + \epsilon\tilde{v}, t) - \lambda(v, t) - \epsilon D\lambda(v, \tilde{v}, t)| \leq \epsilon^2 K_3 \sup_{s \leq t}|\tilde{v}(s)|^2. \qquad (2.5)$$

In particular, if $\lambda(v, t) = h(v(t))$, then $D\lambda(v, \tilde{v}, t) = \nabla h(v(t)) \cdot \tilde{v}(t)$.

Theorem 2.2 *Suppose that for each $T > 0$, $\int u(r \wedge T)^2\nu(du \times dr) < \infty$ and there exists $C > 0$ such that for $0 \leq h \leq 1$,*

$$\sup_{t \leq T} \int (u(r \wedge (t + h)) - u(r \wedge t))^2(1 + u(r \wedge t)^2)\nu(du \times dr) \leq Ch$$

and

$$\sup_{h \leq t \leq T} \int (u(r \wedge (t + h)) - u(r \wedge t))^2(u(r \wedge t) - u(r \wedge (t - h)))^2\nu(du \times dr) \leq Ch^2.$$

Suppose that λ is bounded and satisfies (2.5) and that $D\lambda$ satisfies the Lipschitz condition

$$|D\lambda(v_1,\tilde{v}_1,t) - D\lambda(v_2,\tilde{v}_2,t)| \leq K(t)\sup_{s\leq t}(|v_1(s) - v_2(s)| + |\tilde{v}_1(s) - \tilde{v}_2(s)|).$$

Then $(\tilde{X}_n,\tilde{Y}_n,\tilde{Z}_n) \Rightarrow (\tilde{X},\tilde{Y},\tilde{Z})$ *where*

$$\tilde{X}(t) = \Xi(B(t)) + \int_0^t D\lambda(X,Y,Z,\tilde{X},\tilde{Y},\tilde{Z},z)dz \qquad (2.6)$$

$$\tilde{Y}(t) = \int_{B(t)} u(r \wedge \gamma_t(s))\Xi(ds \times du \times dr)$$

$$+ \int_0^t \mu(t-z)D\lambda(X,Y,Z,\tilde{X},\tilde{Y},\tilde{Z},z)dz$$

$$\tilde{Z}(t) = \Xi(B(t) - A(t)) + \int_0^t \nu(t-z,\infty)D\lambda(X,Y,Z,\tilde{X},\tilde{Y},\tilde{Z},z)dz.$$

Proof The proof is similar to the arguments in Kurtz (1983) and Solomon (1987). Note that (2.4) can be viewed as a system of integral equations driven by the random processes $R_n^1(t) = \Xi_n(B_n(t))$, $R_n^2(t) = \int_{B_n(t)} u(r \wedge \gamma_t^n(s))\Xi_n(ds \times du \times dr)$, and $R_n^3(t) = \Xi_n(B_n(t) - A_n(t))$. By the Lipschitz assumptions on $D\lambda$ and Gronwall's inequality, the result will follow by the continuous mapping theorem, provided we verify the functional convergence of (R_n^1, R_n^2, R_n^3).

Let $E = D_{\mathbb{R}}[0,\infty) \times [0,\infty)$, $\Lambda_n(t) = \int_0^t \lambda(X_n,Y_n,Z_n,s)ds$, and $\Lambda(t) = \int_0^t \lambda(X,Y,Z,s)ds$. If we take $V(s,u,r) = I_{[0,\Lambda_n(t))}(s) - I_{[0,\Lambda(t))}(s)$, then

$$\Xi_n(B_n(t)) - \Xi_n(B(t)) = \int_{[0,t]\times E} V(s-,u,r)\Xi_n(ds \times du \times dr)$$

and by (6.3),

$$E[(\Xi_n(B_n(t)) - \Xi_n(B(t)))^2] = E[m \times \nu(B_n(t)\triangle B(t))] = E[|\Lambda_n(t) - \Lambda(t)|].$$

We also have

$$E\left[\left(\int_{B_n(t)} u(r \wedge \gamma_t^n(s))\Xi_n(ds \times du \times dr) - \int_{B(t)} u(r \wedge \gamma_t(s))\Xi_n(ds \times du \times dr)\right)^2\right]$$

$$= E\left[\int \left(I_{B_n(t)}(s,u,r)u(r \wedge \gamma_t^n(s)) - I_{B(t)}(s,u,r)u(r \wedge \gamma_t(s))\right)^2\right.$$

$$\otimes m \times \nu(ds \times du \times dr)\bigg]$$

$$\leq \int_E u^2(r \wedge t)\nu(du \times dr)E[|\Lambda_n(t) - \Lambda(t)|]$$

$$+ E\left[\int_{[0,\infty]\times E} I_{[0,\Lambda(t)]}(s)(u(r \wedge \gamma_t^n(s)) - u(r \wedge \gamma_t(s)))^2 m \times \nu(ds \times du \times dr)\right]$$

with a similar inequality holding for $\Xi_n(B_n(t) - A_n(t)) - \Xi_n(B(t) - A(t))$. The right sides of these inequalities go to zero by the law of large numbers, Theorem

2.1, and the convergence of the finite dimensional distributions for (R_n^1, R_n^2, R_n^3) follows, that is replace $B_n(t)$, $A_n(t)$ and γ_t^n by their deterministic limits and apply the ordinary central limit theorem. (See Theorem 6.1.)

Since the limits are continuous, to obtain relative compactness of the sequence $\{(R_n^1, R_n^2, R_n^3)\}$, it is sufficient to check relative compactness of each component. We employ the results of Chenčov (1956) (see Ethier and Kurtz (1986), Theorem 3.8.8) summarized in Theorem 6.2. We restrict our attention to

$$
\begin{aligned}
R_n^2(t) &= \int_{B_n(t)} u(r \wedge \gamma_t^n(s)) \Xi_n(ds \times du \times dr) \\
&= \int_{[0,t] \times E} I_{[0, \Lambda_n(t)]}(s) u(r \wedge \gamma_t^n(s)) \Xi_n(ds \times du \times dr).
\end{aligned}
$$

The estimates in the other cases are similar. We apply Corollary 6.8 with $U(t) = R_n^2(t+h) - R_n^2(t)$, $V(t) = R_n^2(t) - R_n^2(t-h)$, and $C(X,Y,M) = c\sqrt{h}$, for appropriate c. Note that $M = \Xi_n$ and $\eta_k = n^{1-k/2}\nu$, so

$$
\begin{aligned}
\int_0^t &\int_E \left(I_{[0, \Lambda_n(t+h)]}(s) u(r \wedge \gamma_{t+h}^n(s)) - I_{[0, \Lambda_n(t)]}(s) u(r \wedge \gamma_t^n(s)) \right)^2 \nu(du \times dr) ds \\
&\leq \int_0^t \int_E I_{(\Lambda_n(t), \Lambda_n(t+h)]}(s) u^2(r \wedge \gamma_{t+h}^n(s)) \nu(du \times dr) ds \\
&\quad + \int_0^t \int_E I_{[0, \Lambda_n(t)]}(u(r \wedge \gamma_t^n(s)) - u(r \wedge \gamma_{t+h}^n(s)))^2 \nu(du \times dr) ds \\
&\leq \int u^2(r \wedge t)\nu(du \times dr)(\Lambda_n(t+h) - \Lambda_n(t)) + tCh \\
&\leq \left(\int u^2(r \wedge t)\nu(du \times dr)\|\lambda\|_\infty + tC \right) h
\end{aligned}
$$

gives (6.6) for $i = 2$ and $j = 0$. The calculation for $i = 0$ and $j = 2$ is the same. For $i = j = 2$, we have

$$
\begin{aligned}
\int_0^t &\int_E \left(I_{[0, \Lambda_n(t+h)]}(s) u(r \wedge \gamma_{t+h}^n(s)) - I_{[0, \Lambda_n(t)]}(s) u(r \wedge \gamma_t^n(s)) \right)^2 \\
&\otimes \left(I_{[0, \Lambda_n(t)]}(s) u(r \wedge \gamma_t^n(s)) - I_{[0, \Lambda_n(t-h)]}(s) u(r \wedge \gamma_{t-h}^n(s)) \right)^2 \nu(du \times dr) ds \\
&\leq \int_0^t \int_E I_{(\Lambda_n(t-h), \Lambda_n(t)]}(s)(u(r \wedge \gamma_{t+h}^n(s)) - u(r \wedge \gamma_t^n(s)))^2 \\
&\qquad\qquad \otimes u^2(r \wedge \gamma_t^n(s))\nu(du \times dr) ds \\
&\quad + \int_0^t \int_E I_{[0, \Lambda_n(t-h)]}(s)(u(r \wedge \gamma_{t+h}^n(s)) - u(r \wedge \gamma_t^n(s)))^2 \\
&\qquad\qquad \otimes (u(r \wedge \gamma_t^n(s)) - u(r \wedge \gamma_{t-h}^n(s)))^2 \nu(du \times dr) ds \\
&\leq (\Lambda_n(t) - \Lambda_n(t-h))Ch + tCh^2 \\
&\leq (C\|\lambda\|_\infty + tC) h^2.
\end{aligned}
$$

The estimates for $i = 1$ and $j = 2$ and for $i = 2$ and $j = 1$ follow as in Remark 6.9, and Corollary 6.8 gives the inequalities

$$E[(R_n^2(h) - R_n^2(0))^2] \leq c_0 h$$

and

$$E[(R_n^2(t + h) - R_n^2(t))^2 (R_n^2(t) - R_n^2(t - h))^2] \leq c_0 h^2$$

for some $c_0 > 0$ independent of n. Similar estimates hold for R_n^1 and R_n^3, and Lemma 6.2 gives the relative compactness of $\{R_n^k\}$. As noted, the asymptotic continuity then ensures the relative compactness of $\{(R_n^1, R_n^2, R_n^3)\}$, and convergence of $\{(\tilde{X}_n, \tilde{Y}_n, \tilde{Z}_n)\}$ follows by the continuous mapping theorem and the uniqueness of the solution of the limiting system (2.6). $\qquad\square$

3 Models with stationary increments

We now assume that the arrival rate λ is a constant and that the input process has been running for ever. Let ξ be the Poisson random measure on $(-\infty, \infty) \times D_\mathbb{R}[0, \infty) \times [0, \infty)$ with mean measure $\lambda m \times \nu$. For $t_1 < t_2$, let $U(t_1, t_2)$ be the work input into the system during the time interval $(t_1, t_2]$. Then, letting $E = D_\mathbb{R}[0, \infty) \times [0, \infty)$, we can write

$$
\begin{aligned}
U(t_1, t_2) &= \int_{(-\infty, t_1] \times E} (u(r \wedge (t_2 - s)) - u(r \wedge (t_1 - s))) \, \xi(ds \times du \times dr) \\
&\quad + \int_{(t_1, t_2] \times E} u(r \wedge (t_2 - s)) \xi(ds \times du \times dr)
\end{aligned}
$$

where the first term gives the work input into the system in the time interval $(t_1, t_2]$ by sources that activated at or before time t_1 and the second term is work input by sources that activate during the time interval $(t_1, t_2]$. Note that U is additive in the sense that $U(t_1, t_3) = U(t_1, t_2) + U(t_2, t_3)$ for $t_1 < t_2 < t_3$, and the amount of work input into the system has stationary increments in the sense that the distribution of $U(t_1 + t, t_2 + t)$ does not depend on t. It follows that $E[U(t_1, t_2)] = \alpha(t_2 - t_1)$ where $\alpha = \lambda \int_E u(r) \nu(du \times dr)$. We can simplify notation if we define $u(s) = 0$ for $s < 0$, and we have

$$U(t_1, t_2) - \alpha(t_2 - t_1) = \int_{(-\infty, t_2] \times E} (u(r \wedge (t_2 - s)) - u(r \wedge (t_1 - s))) \, \tilde{\xi}(ds \times du \times dr)$$

where $\tilde{\xi}(A) = \xi(A) - \lambda m \times \nu(A)$.

In a simple service model with a single server working at "rate" 1 and employing U as the work input, queued work would satisfy

$$Q(t) = Q(0) + U(0, t) - t + \Lambda(t) \tag{3.1}$$

where Λ measures the idle time of the server, and assuming the server works at maximal rate whenever there is work to be done,

$$\Lambda(t) = 0 \vee \sup_{s \leq t}(t - U(0,s) - Q(0)).$$

See, for example, Harrison (1985), Section 2.2. A heavy traffic limit for such a model involves rescaling the work measurements and the arrival and service rates under the assumption that the workload arrival and service rates are asymptotically the same. Letting λ and ν (and hence α) depend on n, (3.1) becomes

$$\frac{Q_n(t)}{\sigma_n} = \frac{Q_n(0)}{\sigma_n} + \frac{1}{\sigma_n}(U_n(0,t) - \alpha_n t) + \frac{\alpha_n - \beta_n}{\sigma_n}t + \frac{1}{\sigma_n}\Lambda_n(t),$$

and $\sigma_n^{-1}Q_n$ converges in distribution provided $\sigma_n^{-1}(\alpha_n - \beta_n)$ converges and $\sigma_n^{-1}(U_n(0,t) - \alpha_n t)$ converges in distribution.

Consequently, we consider the behavior of

$$\begin{aligned}
V_n(t_1, t_2) &= \sigma_n^{-1}(U_n(t_1, t_2) - \alpha_n(t_2 - t_1)) \\
&= \int_{(-\infty, t_2] \times E} (u(r \wedge (t_2 - s)) - u(r \wedge (t_1 - s)))\, \Xi_n(ds \times du \times dr)
\end{aligned}$$

where $\Xi_n(A) = \sigma_n^{-1}(\xi_n(A) - \lambda_n m \times \nu_n(A))$ and σ_n and λ_n tend to infinity. Note that $Var(\Xi_n(A)) = \sigma_n^{-2}\lambda_n m \times \nu_n(A)$. We assume that $\nu_n(D_{\mathbb{R}}[0,\infty) \times \{0\}) = 0$ and that

$$\lim_{n \to \infty} \sigma_n^{-2}\lambda_n \nu_n = \zeta \tag{3.2}$$

in a "somewhat vague" topology on $\mathcal{M}(D_{\mathbb{R}}[0,\infty) \times (0,\infty))$, that is

$$\sigma_n^{-2}\lambda_n \int_{D_{\mathbb{R}}[0,\infty) \times (0,\infty)} h(u,r)\nu_n(du \times dr) \to \int_{D_{\mathbb{R}}[0,\infty) \times (0,\infty)} h(u,r)\zeta(du \times dr) \tag{3.3}$$

for all bounded continuous functions on $D_{\mathbb{R}}[0,\infty) \times (0,\infty)$ for which there exists an $r_0 > 0$ such that $h(u,r) = 0$ for all $r < r_0$. In addition, we assume that the measures on E defined by

$$\gamma_n(B) = \sigma_n^{-2}\lambda_n \int_B u(r)^2 \wedge 1\nu_n(du \times dr)$$

converge weakly to the measure

$$\gamma(B) = \int_B u(r)^2 \wedge 1\zeta(du \times dr).$$

Formally, these assumptions imply that V_n converges in distribution to V defined by

$$V(t_1, t_2) = \int_{(-\infty, t_2] \times E} (u(r \wedge (t_2 - s)) - u(r \wedge (t_1 - s)))\, W(ds \times du \times dr) \tag{3.4}$$

where W is Gaussian white noise on $(-\infty, \infty) \times E$ corresponding to $m \times \zeta$. The variance of $V(t_1, t_2)$ is

$$E[V(t_1, t_2)^2] = \int_{(-\infty, t_2] \times E} \left(u(r \wedge (t_2 - s)) - u(r \wedge (t_1 - s)) \right)^2 ds\zeta(du \times dr)$$

(3.5)

again recalling our convention that $u(s) = 0$ for $s < 0$.

Theorem 3.1 *Let φ be nonnegative and convex on $[0, \infty)$ with $\varphi(0) = 0$ and $\lim_{x \to \infty} \varphi(x)/x = \infty$. Suppose that for each $T > 0$,*

$$\sup_n \lambda_n \int_E \varphi(\sigma_n^{-2} u(r \wedge T)^2) \nu_n(du \times dr) < \infty, \qquad (3.6)$$

$$\lim_{t \to \infty} \sup_n \sigma_n^{-2} \lambda_n \int_{D_{\mathbb{R}}[0,\infty) \times [t,\infty)} \int_t^r (u(r \wedge (T + s)) - u(r \wedge s))^2 ds\nu_n(du \times dr) = 0,$$

(3.7)

and there exists $C > 0$ such that for $0 \le h \le 1$,

$$\sup_n \int_E \int_0^r (u(r \wedge (h + s)) - u(r \wedge s))^2 ds\sigma_n^{-2} \lambda_n \nu_n(du \times dr) \le Ch \qquad (3.8)$$

and

$$\sup_n \int_E \int_0^r (u(r \wedge (s + h)) - u(r \wedge s))^2$$
$$\otimes (u(r \wedge s) - u(r \wedge (s - h)))^2 ds\sigma_n^{-4} \lambda_n \nu_n(du \times dr)$$
$$\le Ch^2. \qquad (3.9)$$

Suppose that the mapping $(s, u, r) \to u(r \wedge s)$ from $[0, \infty) \times D_{\mathbb{R}}[0, \infty) \times [0, \infty)$ is continuous a.e. $m \times \zeta$. Then $V_n(0, \cdot) \Rightarrow V(0, \cdot)$ in the Skorohod topology on $D_{\mathbb{R}}[0, \infty)$.

Remark 3.2 *For fixed t_1 and t_2, convergence in distribution of $V_n(t_1, t_2)$ is essentially a central limit theorem for a "triangular array". Conditions (3.6) and (3.7) give the "uniform integrability" that such results require. (See Theorem 6.1.) In particular, if $\lambda_n = n$, $\sigma_n = \sqrt{n}$, and $\nu_n \equiv \nu$, then $\int_E u(r)^2 \nu(du \times dr) < \infty$ implies (3.6) and (3.7) for appropriate choice of φ. If, in addition, the input process has finite second moments and stationary, independent increments, the remaining conditions follow.*

Proof Let $t_1 \le t_2$. Note that $E[V_n(0, t_1) V_n(0, t_2)] = \int_E \int_0^r (u(r \wedge (t_2 + s)) - u(r \wedge s))(u(r \wedge (t_1 + s)) - u(r \wedge s)) ds\sigma_n^{-2} \lambda_n \nu_n(du \times dr) \to \int_E \int_0^r (u(r \wedge (t_2 + s)) - u(r \wedge s))(u(r \wedge (t_1 + s)) - u(r \wedge s)) ds\zeta(du \times dr)$ where the convergence follows from the convergence assumption (3.2) and the convergence of γ_n to γ, the a.e. continuity assumption on the mapping $(s, u, r) \to u(r \wedge s)$, and the uniform integrability assumptions (3.6) and (3.7). The convergence of the finite dimensional distributions of $V_n(0, \cdot)$ then follows from Theorem 6.1. The relative compactness of $\{V_n(0, \cdot)\}$ follows from Theorem 6.2 by essentially the same argument as in Theorem 2.2. $\qquad \square$

4 A representation of fractional Brownian motion

Fractional Brownian motion is a mean-zero Gaussian process Z with stationary increments satisfying $E[(Z(t_2) - Z(t_1))^2] = c(t_2 - t_1)^\alpha$ for some $c > 0$ and $1 < \alpha < 2$. (Of course, if $\alpha = 1$, Z is an ordinary Brownian motion.) If we assume the source in the previous section broadcasts at rate 1, that is $X(s) \equiv s$, $s \geq 0$, and if we let W be Gaussian white noise on $(-\infty, \infty) \times [0, \infty)$ corresponding to

$$ds\,\zeta(dr) = \lambda \frac{1}{r^\beta} ds\,dr,$$

where $2 < \beta < 3$, then (3.4) becomes

$$
V(t_1, t_2) = \int_{(-\infty, t_1] \times [0, \infty)} (r \wedge (t_2 - s) - r \wedge (t_1 - s))\, W(ds \times dr)
$$
$$
+ \int_{(t_1, t_2] \times [0, \infty)} (r \wedge (t_2 - s))W(ds \times dr) \qquad (4.1)
$$

and (3.5) gives

$$
\begin{aligned}
E[V(t_1, t_2)^2] &= \int_{-\infty}^{t_1} \int_0^\infty (r \wedge (t_2 - s) - r \wedge (t_1 - s))^2\, \lambda r^{-\beta}\,dr\,ds \\
&\quad + \int_{t_1}^{t_2} \int_0^\infty (r \wedge (t_2 - s))^2 \lambda r^{-\beta}\,dr\,ds \\
&= \int_0^\infty \int_0^\infty (r \wedge (t_2 - t_1 + u) - r \wedge u)^2\, \lambda r^{-\beta}\,dr\,du \\
&\quad + \int_0^{t_2 - t_1} \int_0^\infty (r \wedge u)^2 \lambda r^{-\beta}\,dr\,du \\
&= \int_0^\infty \int_u^{t_2 - t_1 + u} (r - u)^2\, \lambda r^{-\beta}\,dr\,du \\
&\quad + \int_0^\infty \int_{t_2 - t_1 + u}^\infty (t_2 - t_1)^2 \lambda r^{-\beta}\,dr\,du \\
&\quad + \int_0^{t_2 - t_1} \int_0^u \lambda r^{2 - \beta}\,dr\,du + \int_0^{t_2 - t_1} \int_u^\infty u^2 \lambda r^{-\beta}\,dr\,du \\
&= \left(\frac{\lambda}{(\beta - 1)(4 - \beta)} + \frac{\lambda}{(\beta - 1)(\beta - 2)} + \frac{\lambda}{(3 - \beta)(4 - \beta)} \right. \\
&\quad \left. + \frac{\lambda}{(\beta - 1)(4 - \beta)} \right) (t_2 - t_1)^{4 - \beta}
\end{aligned}
$$

and hence V is fractional Brownian motion with $\alpha = 4 - \beta$.

To see that ζ can arise as a limit as in (3.2), let $\lambda_n = n^\alpha$, with $\alpha > \beta - 1$, $\nu_n(dr) = (\beta - 1)n(nr + 1)^{-\beta}dr$, and $\sigma_n = n^{\frac{\alpha - \beta + 1}{2}}$. It is easy to check that (3.2) holds and that $\gamma_n \Rightarrow \gamma$. For (3.6), take $\varphi(x) = x^2$. The integral on the left of (3.7) is bounded by

$$\int_t^\infty T^2(\beta - 1)r^{-\beta+1}dr$$

which is convergent since $\beta > 2$. Since $r \wedge (h + s) - r \wedge s \leq r \wedge h$, the left side of (3.8) is bounded by

$$(\beta - 1)\int_0^h r^{-\beta+3}dr + h^2(\beta - 1)\int_h^\infty r^{-\beta+1}dr = \left(\frac{\beta - 1}{4 - \beta} + \frac{\beta - 1}{\beta - 2}\right)h^{4-\beta},$$

which, since $4 - \beta > 1$, implies the relative compactness in $C_{\mathbb{R}}[0, \infty)$ by the Kolmogorov criterion without checking (3.9).

For a more complex example, let ν be the distribution of (X, τ) where X has stationary, ergodic increments (for example, X can be a compound Poisson process) and τ is independent of X and satisfies

$$\lim_{r \to \infty} r^{\beta-1}P\{\tau > r\} = c,$$

that is τ is in the domain of attraction of a stable law of index $\beta - 1$. Assume that $\sup_{t \geq 1} E[\varphi(t^{-2}(X(t))^2)] < \infty$, where φ is as in Theorem 3.1, and that for some $C > 0$, $Var(X(t)) < Ct$ for $0 \leq t \leq 1$ and

$$E[(X(t_3) - X(t_2))^2(X(t_2) - X(t_1))^2] \leq C(t_3 - t_2)(t_2 - t_1).$$

Let ν_n be the distribution of $(\frac{1}{n}X(n\cdot), \frac{1}{n}\tau)$. Note that this transformation corresponds to rescaling time and workload measurements and that $\frac{1}{n}X(nt) \to mt$ a.s. and in L^2. Then with h as in (3.3) and $\sigma_n^{-2}\lambda_n = n^{\beta-1}$,

$$\begin{aligned}\lim_{n \to \infty} \sigma_n^{-2}\lambda_n \int h(u, r)\nu_n(du \times dr) &= \lim_{n \to \infty} \sigma_n^{-2}\lambda_n E[h(\tfrac{1}{n}X(n\cdot), \tfrac{1}{n}\tau)]\\ &= \int_0^\infty h(m\cdot, r)\frac{c(\beta - 1)}{r^\beta}dr,\end{aligned}$$

and the estimates in Theorem 3.1 are easy to check.

5   Variations of the model

5.1   Admission control

The estimates used in proving relative compactness in the convergence theorems, Theorems 2.2 and 3.1, depend on the integrand being "predictable" with respect to the filtration generated by the Poisson random measures. Similar estimates will also hold for more general predictable integrands. For example, we can associate with each source another variable Z_i, say, with values in $[0, 1]$. Now let ξ be the Poisson random measure on $[0, \infty) \times D_{\mathbb{R}}[0, \infty) \times [0, \infty) \times [0, 1]$ with mean measure $\lambda m \times \hat{\nu}$. Let U^R satisfy

$$U^R(t_1, t_2) = \int_{(-\infty, t_2] \times E \times [0,1]} (u(r \wedge (t_2 - s)) - u(r \wedge (t_1 - s)))$$
$$\otimes I_{[0, R(s)]}(z)\xi(ds \times du \times dr \times dz)$$

where $R(t) \in [0,1]$ depends on the process up to time t (for example, on the number of active sources and/or the workload backlog). Note that if the ith (attempted) activation occurs at time S_i, but $R(S_i) < Z_i$, then the source contributes no work to the system. The introduction of R allows one to model an admission control policy. The value of Z_i can be thought of as specifying the priority level of the source. If $R(s) = 1$, any source that attempts to activate at time s is allowed to input work into the system. If $R(s) = 0$, any source that attempts to activate at time s is rejected. Otherwise, whether or not a source is accepted depends on its "priority level" Z_i.

For simplicity, let $\hat{\nu}_n = \nu_n \times m$. As before, letting $R_n = 1 - \frac{1}{\beta_n}\tilde{R}_n$ and setting $\Xi_n(A) = \sigma_n^{-1}(\xi_n(A) - \lambda_n m \times \hat{\nu}_n(A))$, we have

$$
\begin{aligned}
V_n^R(t_1, t_2) &= \sigma_n^{-1}(U_n^R(t_1, t_2) - \alpha_n(t_2 - t_1)) \\
&= \int_{(-\infty, t_2] \times E} (u(r \wedge (t_2 - s)) - u(r \wedge (t_1 - s))) \\
&\qquad \otimes I_{[0, R_n(s)]}(z)\Xi_n(ds \times du \times dr \times dz) \\
&\quad - \int_{(-\infty, t_2] \times E} (u(r \wedge (t_2 - s)) - u(r \wedge (t_1 - s))) \\
&\qquad \otimes \tilde{R}_n(s)\frac{\lambda_n}{\beta_n \alpha_n}\nu_n(du \times dr).
\end{aligned}
$$

For example, under the conditions of Remark 3.2, if $\beta_n = \sqrt{n}$ and $\tilde{R}_n$ is a continuous function of the normalized number of active sources, a limit theorem analogous to Theorem 3.1 will hold.

5.2   Regenerative sources

As noted previously, the models considered above do not include many models of on/off sources. Suppose that there is a fixed number n of potential sources and that each of the n sources inputs work over a period of time of length distributed as τ and that during that period the cumulative work is distributed as X. We assume that at the end of each such time period, the source immediately begins a new period with the same statistics and that the behavior in different periods is independent (hence the regenerative terminology). An on/off source that inputs work at a constant rate c during the on period satisfies

$$
X(t) = c(t \wedge \tau^{(1)}), \quad \tau = \tau^{(1)} + \tau^{(2)}
$$

where $\tau^{(1)}$ is the length of the on (active) portion of the cycle and $\tau^{(2)}$ is the length of the off (inactive) portion of the cycle.

Let $\{(X_i, \tau_i)\}$ be i.i.d. with the same distribution as (X, τ). Define $\xi_n([0, z] \times A) = \sum_{i=1}^{[nz]} \delta_{(X_i, \tau_i)}$. For $k = 1, \ldots, n$, let τ_k^0 denote the first cycle completion (regeneration) time after time 0 for the kth source, and let $X_k^0(t)$, $0 \le t \le \tau_k^0$, denote the cumulative work input by the kth source in the time interval $[0, t]$.

Let

$$U_n(t) = \sum_{k=1}^{n} X_k^0(t \wedge \tau_k^0) + \int_0^t X_{N_n(s)}(\tau_{N(s)} \wedge (t - s)) dN_n(s)$$

and

$$N_n(t) = \sum_{k=1}^{n} I_{[\tau_k^0, \infty)}(t) + \int_0^t I_{[\tau_{N_n(s)}, \infty)}(t - s) dN_n(s)$$

where $N_n(t)$ counts the total number of cycle completions in the time interval $[0, t]$ (each source may have completed more than one cycle) and $U_n(t)$ denotes the total work input during the time interval $[0, t]$. U_n and N_n can be represented in terms of ξ_n and results analogous to Theorems 2.2 and 3.1 can be proved using the central limit theorem for ξ_n. These results will be discussed elsewhere.

6 Appendix

The central limit theorem for integrals against Poisson random measures plays an important role in the limit theorems considered in this paper. If $\{\xi_n\}$ is a sequence of Poisson random measures on U with mean measures η^n, $\tilde{\xi}_n(A) = \xi_n(A) - \eta^n(A)$ and f^n is a sequence of functions on U, then the characteristic function for

$$Z^n = \int_U f^n(u)\tilde{\xi}_n(du)$$

is

$$E[e^{i\theta Z^n}] = \exp\left\{ \int_U (e^{i\theta f^n(u)} - 1 - i\theta f^n(u))\eta^n(du) \right\}$$

and the following version of the Lindeberg–Feller central limit theorem holds.

Theorem 6.1 *For $k = 1, \ldots, d$, $n = 1, 2, \ldots$, let $f_k^n \in L^2(\eta^n)$ and*

$$Z_k^n = \int_U f_k^n(u)\tilde{\xi}_n(du).$$

Suppose

$$\lim_{n \to \infty} \int_U f_k^n(u)f_l^n(u)\eta^n(du) = \sigma_{kl}^2$$

and for each $\epsilon > 0$ and k,

$$\lim_{n \to \infty} \int_U I_{\{|f_n(u)| > \epsilon\}}(f_k^n(u))^2\eta^n(du) = 0.$$

Then $Z^n \Rightarrow Z$, where Z is normal with mean zero and covariance matrix $\sigma^2 = ((\sigma_{kl}^2))$.

Because the processes of interest in Sections 2 and 3 involve stochastic convolutions, it is difficult or impossible to employ standard martingale estimates in order to verify relative compactness in the proofs of convergence in distribution. Consequently, we exploit the classical results of Chenčov (1956) (see Ethier and Kurtz (1986), Theorem 3.8.8) which involve estimating moments of products of increments of the processes. The following simplified version of those results is sufficient for our purposes.

Theorem 6.2 *Let $\{X_n\}$ be a sequence of cadlag $\mathbb{R}^d$-valued processes. Suppose that for each $t \geq 0$, $\{X_n(t)\}$ is relatively compact (in the sense of convergence in distribution), that for each $\epsilon > 0$,*

$$\lim_{t \to 0} \sup_n P\{|X_n(t) - X_n(0)| > \epsilon\} = 0, \tag{6.1}$$

and that for each $T > 0$ there exists a $C_T > 0$ and $\theta > 1$ such that for $0 \leq h \leq 1$ and $h \leq t \leq T$

$$E[(X_n(t+h) - X_n(t))^2 (X_n(t) - X_n(t-h))^2] \leq C_T h^\theta.$$

Then $\{X_n\}$ is relatively compact in the sense of convergence in distribution in the Skorohod topology.

Remark 6.3 *Note that to verify the relative compactness of $\{X_n(t)\}$, it is sufficient to verify $\sup_n E[|X_n(t)|^a] < \infty$ for some $a > 0$, and to verify (6.1), it is sufficient to show that $\lim_{t \to 0} \sup_n E[|X_n(t) - X_n(0)|^a] = 0$.*

The necessary moment estimates can be obtained using results on integrals against orthogonal martingale random measures. Let E be a complete, separable metric space with Borel sets $\mathcal{B}(E)$, and let $\mathcal{U} \subset \mathcal{B}(E)$ be closed under finite unions and satisfy $A \in \mathcal{U}$ and $B \in \mathcal{B}(E)$ implies $A \cap B \in \mathcal{U}$. M is an $\{\mathcal{F}_t\}$-orthogonal martingale random measure on E if for each $A \in \mathcal{U}$, $M(A, \cdot)$ is an $\{\mathcal{F}_t\}$-martingale and for disjoint $A_1, \ldots, A_m \in \mathcal{U}$, $M(A_1, \cdot), \ldots, M(A_m, \cdot)$ are orthogonal martingales (that is, the product of any two is a martingale) and $\sum_{i=1}^m M(A_i, t) = M(\cup_i^m A_i, t)$ a.s. If $M(A, \cdot)$ is square integrable for each $A \in \mathcal{U}$, then there exists a a random measure $\tilde{\eta}$ on $E \times [0, \infty)$ such that the Meyer process for $M(A, \cdot)$ and $M(B, \cdot)$ is given by $\langle M(A, \cdot), M(B, \cdot) \rangle_t = \tilde{\eta}(A \cap B \times [0, t])$. For simplicity, we assume that

$$\tilde{\eta}(A \times [0, t]) = \int_0^t \eta(s, A) ds,$$

where η is an $\mathcal{M}(E)$-valued process adapted to $\{\mathcal{F}_t\}$. Note that if

$$M(A, t) = \frac{N(A \times [0, t]) - \nu(A)t}{\sigma}$$

where N is a Poisson random measure on $E \times [0, \infty)$ with mean measure $\nu \times m$, then M is an orthogonal martingale random measure with $\mathcal{U} = \{A \in \mathcal{B}(E) : \nu(A) < \infty\}$ and $\tilde{\eta} = \sigma^{-2}\nu \times m$, that is $\eta(s, A) = \sigma^{-2}\nu(A)$, and if W is Gaussian white noise on $E \times [0, \infty)$ with $Var(W(C)) = \nu \times m(C)$, then $M(A, t) = W(A \times [0, t])$ is an orthogonal martingale random measure with $\mathcal{U} = \{A \in \mathcal{B}(E) : \nu(A) < \infty\}$ and $\tilde{\eta} = \nu \times m$. For details on integration against orthogonal martingale random measures see Walsh (1986) or Kurtz and Protter (1995).

Lemma 6.4 *Let η be a deterministic measure on E and $\mathcal{U} = \{A \in \mathcal{B}(E) : \eta(A) < \infty\}$, and let M be an orthogonal martingale random measure on E*

adapted to $\{\mathcal{F}_t\}$ *such that* $M(A,\cdot)$ *is square integrable for each* $A \in \mathcal{U}$ *and* $\langle M(A,\cdot), M(B,\cdot)\rangle_t = \eta(A \cap B)t$, $A, B \in \mathcal{U}$. *For* $l > 2$, *let* $[M(A,\cdot)]_t^l = \sum_{s \leq t}(M(A,s) - M(A,s-))^l$ *(that is, the sum is over all discontinuities) and suppose that for* $2 < l \leq l_0$, *there exist measures* η_l *such that*

$$[M(A,\cdot)]_t^l - \eta_l(A)t$$

is a martingale, and define $\eta_2 \equiv \eta$. *(Recall that for any* A *with* $\eta(A) < \infty$, $[M(A,\cdot)]_t - \eta(A)t$ *is a martingale.)*

Fix m, $2 \leq m \leq l_0$, *and let* X *and* Y *be cadlag,* $L^2(\eta)$-*valued processes such that*

$$E\left[\left(\int_0^t \int_E (|X(u,s)|^k + |Y(u,s)|^k)\eta_k(du)ds\right)^{\frac{m}{k}}\right] < \infty \qquad (6.2)$$

for $2 \leq k \leq m$. *Define*

$$U(t) = \int_{E \times [0,t]} X(u,s-)M(du \times ds), \qquad V(t) = \int_{E \times [0,t]} Y(u,s-)M(du \times ds).$$

Then for $k + l \leq m$

$$E[U^k(t)V^l(t)] = \sum_{\substack{i \leq k, j \leq l \\ 2 \leq i+j}} \binom{k}{i}\binom{l}{j} E\left[\int_{E \times [0,t]} U^{k-i}(s-)V^{l-j}(s-)X^i(u,s-)\right.$$

$$\left. \otimes Y^j(u,s-)\eta_{i+j}(du)ds\right].$$

In particular,

$$E[U(t)V(t)] = E\left[\int_{E \times [0,t]} X(u,s-)Y(u,s-)\eta(du)ds\right]. \qquad (6.3)$$

Remark 6.5 *If* M *is a centered Poisson process, then* $\eta_k = \eta$ *for all* $k \geq 2$. *If* $M(A,\cdot)$ *is continuous for all* $A \in \mathcal{U}$, *then* $\eta_k = 0$ *for* $k > 2$.

Proof U *and* V *are martingales, and Itô's formula implies*

$$U^k(t)V^l(t) = \int_0^t kU^{k-1}(s-)V^l(s-)dU(s) + \int_0^t lU^k(s-)V^{l-1}(s-)dV(s)$$

$$+ \frac{1}{2}\int_0^t k(k-1)U^{k-2}(s-)V^l(s-)d[U]_s$$

$$+ \int_0^t klU^{k-1}(s-)V^{l-1}(s-)d[U,V]_s$$

$$+ \int_0^t l(l-1)U^k(s-)V^{l-2}(s-)d[V]_s \qquad (6.4)$$

$$+ \sum_{\substack{i \leq k, j \leq l \\ 2 < i+j}} \binom{k}{i}\binom{l}{j}\sum_{s \leq t} U^{k-i}(s-)V^{l-j}(s-)(\Delta U(s))^i(\Delta V(s))^j,$$

where $\Delta U(s) = U(s) - U(s-)$. For $2 < i + j \leq m$, the orthogonality and the moment assumptions imply

$$\sum_{s \leq t} (\Delta U(s))^i (\Delta V(s))^j - \int_{E \times [0,t]} X^i(u, s-) Y^j(u, s-) \eta_{i+j}(du) ds$$

is a martingale as are

$$[U]_t - \int_{E \times [0,t]} X^2(u, s-) \eta(du) ds, \qquad [U, V]_t - \int_{E \times [0,t]} X(u, s-) Y(u, s-) \eta(du) ds,$$

and the analogous centering of $[V]_t$. The desired identity then follows by taking expectations of both sides of (6.4). For the moment estimates needed to justify this argument, see Section 2 of Kurtz and Protter (1995). $\qquad \square$

Corollary 6.6 *Let k be even, and let*

$$C(X, M) = \max_{2 \leq i \leq k} E\Big[\Big(\int_0^t | \int_E X^i(s-, u) \eta_i(du)| ds \Big)^{\frac{k}{i}} \Big],$$

and let K be the (unique) positive solution of

$$K = \sum_{i=2}^{k} \binom{k}{i} K^{\frac{k-i}{k}}.$$

Then

$$E[U^k(t)] \leq K C(X, M). \tag{6.5}$$

Remark 6.7 *For a similar estimate on $E[|U(t)|^k]$ for k odd, see Section 2 of Kurtz and Protter (1995).*

Proof Since $|U(t)|^j$ is a submartingale for $j \geq 1$,

$$E[U^k(t)] \leq \sum_{2 \leq i \leq k} \binom{k}{i} E\Big[|U^{k-i}(t)| \int_0^t | \int_E X^i(s, u) \eta_i(du)| ds \Big]$$

$$\leq \sum_{2 \leq i \leq k} \binom{k}{i} E[U^k(t)]^{\frac{k-i}{k}} E\Big[\Big(\int_0^t | \int_E X^i(s, u) \eta_i(du)| ds \Big)^{\frac{k}{i}} \Big]^{\frac{i}{k}}$$

$$\leq \sum_{2 \leq i \leq k} \binom{k}{i} E[U^k(t)]^{\frac{k-i}{k}} C(X, M)^{\frac{i}{k}}.$$

Dividing both sides of this inequality by $C(X, M)$ we see that (6.5) must hold. $\square$

Corollary 6.8 *Suppose in addition to (6.2), for $i + j \geq 2$, $i, j \leq 2$,*

$$\int_0^t | \int_E X^i(s-, u) Y^j(s-, u) \eta_{i+j}(du)| ds \leq C^{i+j}(X, Y, M) \quad \text{a.s.} \tag{6.6}$$

Then $E[\|U(t)\|], E[\|V(t)\|] \leq C(X,Y,M)$, $E[U^2(t)], E[V^2(t)] \leq C^2(X,Y,M)$ *and*

$$E[U^2(t)V^2(t)] \leq 11C^4(X,Y,M).$$

Remark 6.9 *Note that*

$$\int_0^t |\int_E X^2(s-,u)Y^1(s-,u)\eta_3(du)|ds$$

$$\leq \int_0^t \sqrt{\int_E X^2(s-,u)\eta_3(du) \int_E X^2(s-u)Y^2(s-,u)\eta_3(du)}ds$$

$$\leq \sqrt{\int_0^t \int_E X^2(s-,u)\eta_3(du)ds \int_0^t \int_E X^2(s-u)Y^2(s-,u)\eta_3(du)ds}$$

so, for example, in the Poisson case, the estimates for $(i,j) = (2,0), (0,2)$ *and* $(2,2)$ *imply the estimates for* $(2,1)$ *and* $(1,2)$.

Proof We have

$$E[U^2(t)] = E\Big[\int_{E\times[0,t]} X^2(s-,u)\eta_2(du)ds\Big] \leq C^2(X,Y,M),$$

and similarly for $E[V^2(t)]$, which in turn implies the inequalities for $E[\|U(t)\|]$ and $E[\|V(t)\|]$. Using the fact that $|U(t)|, |V(t)|, U^2(t)$ and $V^2(t)$ are submartingales and that $|U(s)V(s)| \leq \frac{1}{2}(U^2(s) + V^2(s))$, we have

$$E[U^2(t)V^2(t)]$$

$$= \sum_{\substack{i\leq 2, j\leq 2 \\ 2\leq i+j}} \binom{2}{i}\binom{2}{j} E\Big[\int_{E\times[0,t]} U^{2-i}(s-)V^{2-j}(s-)$$

$$\otimes X^i(s-,u)Y^j(s-,u)\eta_{i+j}(du)ds\Big].$$

$$\leq \sum_{j=0}^2 \binom{2}{j} E\Big[|V^{2-j}(t)| \int_0^t |\int_E X^2(s-,u)Y^j(s-,u)\eta_{2+j}(du)|ds\Big]$$

$$+ \sum_{i=0}^1 \binom{2}{i} E\Big[|U^{2-i}(t)| \int_0^t |\int_E X^i(s-,u)Y^2(s-,u)\eta_{i+2}(du)|ds\Big]$$

$$+ 2E\Big[(U^2(t) + V^2(t)) \int_0^t |\int_E X^1(s-,u)Y^1(s-,u)\eta_2(du)|ds\Big]$$

$$\leq 11C^4(X,Y,M).$$

$\square$

Corollary 6.10 *If* $XY = 0$ *and*

$$\max\Big\{\int_0^t |\int_E X^2(s,u)\eta(du)|ds, \int_0^t |\int_E Y^2(s,u)\eta(du)|ds\Big\} \leq \tilde{C}(X,Y,M)$$

then

$$E[U^2(t)V^2(t)] \leq 2\tilde{C}^2(X,Y,M).$$

Proof Note that $E[U^2(t)], E[V^2(t)] \leq \tilde{C}(X,Y,M)$. By Lemma 6.4 and the argument used in the proof above

$$
\begin{aligned}
E[U^2(t)V^2(t)] &= E\Big[\int_{E\times[0,t]} U^2(s-)Y^2(s-,u)\eta(du)ds\Big] \\
&\quad + E\Big[\int_{E\times[0,t]} V^2(s-)X^2(s-,u)\eta(du)ds\Big] \\
&\leq E[U^2(t)]\tilde{C}(X,Y,M) + E[V^2(t)]\tilde{C}(X,Y,M)
\end{aligned}
$$

which gives the desired result. □

Bibliography

1. Chenčov, N. N. (1956). Weak convergence of stochastic processes whose trajectories have no discontinuities of the second kind and the heuristic approach to the Kolmogorov-Smirnov tests. *Theor. Probab. Appl.*, **1**, 140–149.

2. Cutland, Nigel J., Kopp, P. Ekkehard, and Willinger, Walter (1993). Stock price returns and the Joseph effect: a fractional version of the Black-Scholes model. Preprint.

3. Ethier, Stewart N. and Kurtz, Thomas G. (1986). *Markov Processes: Characterization and Convergence*. Wiley, New York.

4. Foley, R. D. (1982). The non-homogeneous $M/G/\infty$ queue. *OPSEARCH*, **19**, 40–48.

5. Garcia, Nancy Lopes (1995). Birth and death processes as projections of higher dimensional Poisson processes. *Adv. Appl. Probab.* **27**, 911–930.

6. Harrison, J. Michael (1985). *Brownian Motion and Stochastic Flow Systems*. Wiley, New York.

7. Iglehart, Donald L. (1973). Weak convergence of compound stochastic process, I. *Stochastic Process. Appl.*, **1**, 11–31.

8. Klüppelberg, C. and Mikosch, T. (1995a). Explosive Poisson shot noise processes with application to risk reserves. *Bernoulli* **1**, 125–147.

9. Klüppelberg, C. and Mikosch, T. (1995b). Delay in claim settlement and ruin probability approximations. *Scand. Actuarial J.* 154–168.

10. Kurtz, Thomas G. (1983). Gaussian approximations for Markov chains and counting processes. *Bulletin of the International Statistical Institute, Proceedings of the 44th Session, Invited Paper Vol. 1*, 361–376.

11. Kurtz, Thomas G. and Protter, Philip (1995). Weak convergence of stochastic integrals and differential equations: infinite dimensional case. *CIME School in Probability.* (to appear).

12. Solomon, Wiremu (1987). Representation and approximation of large popu-

lation age distributions using Poisson random measures. *Stochastic Process. Appl.*, **26**, 237–255.

13. Stute, W. (1976). On a generalization of the Glivenko–Cantelli theorem. *Z. Wahrsch. Verw. Geb.*, **35**, 167–175.

14. Walsh, John (1986). An introduction to stochastic partial differential equations. *Lecture Notes in Mathematics*, 1180, 265–439.

15. Willinger, W., Taqqu, M., Leland, W. E., and Wilson, D. V. (1995). Self-similarity in high-speed packet traffic: analysis and modeling of Ethernet traffic measurements. *Stat. Sci.*, **10**, 67–85.

16. Willinger, W., Taqqu, M., Sherman, R., and Wilson, D. V. (1995). Self-similarity through high-variability: statistical analysis of Ethernet LAN traffic at the source level. (Extended version.) Preprint.

8

Notes on effective bandwidths

Frank Kelly

University of Cambridge

Abstract

This paper presents a personal view of work to date on effective bandwidths, emphasizing the unifying role of the concept: as a summary of the statistical characteristics of sources over different time and space scales; in bounds, limits and approximations for various models of multiplexing under quality of service constraints; and as the basis for simple and robust tariffing and connection acceptance control mechanisms for poorly characterized traffic. The framework assumes only stationarity of sources, and illustrative examples include periodic streams, fractional Brownian input, policed and shaped sources, and deterministic multiplexing.

1 Introduction

Within a broadband network, the usage of a network resource may not be well assessed by a simple count of the number of bits carried. For example, to provide an acceptable performance to bursty sources with tight delay and loss requirements it may be necessary to keep the average utilization of a link below 10%, while for constant rate sources or sources able to accommodate substantial delays it may be possible to push the average utilization well above 90%.

This paper attempts a unified perspective on *effective bandwidth*, a concept that has been developed by several authors over recent years to provide a measure of resource usage which adequately represents the trade-off between sources of different types, taking proper account of their varying statistical characteristics and quality of service requirements. The concept has attracted much attention and some criticism, but in the author's view there is emerging an elegant and powerful mathematical theory with important technological applications. It seems appropriate to describe the paper as a personal view, since there is not yet a generally accepted definition of an effective bandwidth, and since other frameworks for the interpretation of the material are certainly possible.

This paper is based on talks given to the INFORMS Telecommunications Conference held in Boca Raton, Florida, in March 1995, and to the Royal Statistical Society Research Workshop in Stochastic Networks held at Heriot-Watt University, Edinburgh, in August 1995; I am grateful to participants at these meetings for many comments. The figures for the paper were prepared by Damon Wischik, while supported by the Commission of the European Communities under ACTS project AC039. Pointers to related information resources will be available for a period on http://www.statslab.cam.ac.uk/~frank/ .

In Section 2 we present our definition of the effective bandwidth of a source, describe some of its simpler properties, and present a variety of examples. The effective bandwidth of a source depends upon two free parameters, representing a space and time scaling respectively, and, as Gibbens (1996) demonstrates, this dependence provides a convenient tool for the description and analysis of real sources. The appropriate choice of space and time scale will depend upon characteristics of the resource such as its capacity, buffer size, traffic mix and scheduling policy.

In Section 3 we compare and contrast several multiplexing models, and describe how the effective bandwidth provides a measure associated with a source such that a resource can deliver a performance guarantee expressed in terms of loss or delay by limiting the sources served so that their effective bandwidths sum to less than a threshold. Under different models this result may be expressed as a conservative global bound, or as an asymptotic local limit, or as an approximation capable of successive refinements; but the ubiquity of the single functional form, described in Section 2, is striking.

The effective bandwidth of a source depends sensitively upon the statistical properties of the source, yet these properties may not be known with certainty, either to the user responsible for the source or to the network. It is sometimes thought that this limits the applicability of the concept. On the contrary, the concept is central to any understanding of just how well described a source needs to be, and to the discussion, in Section 4, of tariffing and connection acceptance control mechanisms for sources that may be poorly characterized.

Whitt (1993) and de Veciana and Walrand (1995) provide valuable reviews of earlier work on effective bandwidths. The term itself was first used by Gibbens and Hunt (1991) and Kelly (1991) in their investigation of linear acceptance regions for certain buffered resources, although the essential concept for unbuffered resources had been described earlier in the seminal paper of Hui (1988), and a closely related notion was described by Guérin *et al.* (1991).

2 Effective bandwidths

In this section we define the effective bandwidth associated with a stationary source, describe some of its simpler properties, and illustrate the definition with several contrasting examples.

2.1 Definition

Let $X[0,t]$ be the amount of work that arrives from a source in the interval $[0,t]$. Assume that $X[0,t]$ has stationary increments. Define the *effective bandwidth* of the source to be

$$\alpha(s,t) = \frac{1}{st} \log \mathbb{E}\big[e^{sX[0,t]}\big] \qquad 0 < s,t < \infty. \tag{2.1}$$

2.2 Properties

(i) If $X[0,t]$ has independent increments, then $\alpha(s,t)$ does not depend upon t.

(ii) If there exists a random variable X such that $X[0, t] = Xt$ for $t > 0$, then $\alpha(s, t) = \alpha(st, 1)$, and so $\alpha(s, t)$ depends on s, t only through the product st. Otherwise $\alpha(s/t, t)$ is strictly decreasing in t.

(iii) If $X[0, t] = \sum_i X_i[0, t]$, where $(X_i[0, t])_i$ are independent, then

$$\alpha(s, t) = \sum_i \alpha_i(s, t). \tag{2.2}$$

(iv) For any fixed value of t, $\alpha(s, t)$ is increasing in s, and lies between the mean and peak of the arrival rate measured over an interval of length t: that is,

$$\frac{\mathbb{E}X[0, t]}{t} \leq \alpha(s, t) \leq \frac{\bar{X}[0, t]}{t} \tag{2.3}$$

where $\bar{X}[0, t]$ is the (possibly infinite) essential supremum

$$\bar{X}[0, t] = \sup\Big\{x : P\big\{X[0, t] > x\big\} > 0\Big\}.$$

The form of $\alpha(s, t)$ near $s = 0$ is determined by the mean, variance and higher moments of $X[0, t]$, while the form of $\alpha(s, t)$ near $s = \infty$ is primarily influenced by the distribution of $X[0, t]$ near its maximum: if $\alpha(s, t)$ is finite for some $s > 0$ then for given t

$$\alpha(s, t) = \frac{1}{t}\mathbb{E}X[0, t] + \frac{s}{2t}\operatorname{Var}X[0, t] + o(s) \quad \text{as } s \to 0 \tag{2.4}$$

while if $\alpha(s, t)$ is bounded above as $s \to \infty$ then for given t

$$\alpha(s, t) = \frac{\bar{X}[0, t]}{t} + \frac{1}{st}\log P\big\{X[0, t] = \bar{X}[0, t]\big\} + o\left(\frac{1}{s}\right) \quad \text{as } s \to \infty. \tag{2.5}$$

Write $\alpha(0, t)$ and $\alpha(\infty, t)$ for the lower and upper bounds respectively of the range (2.3); note that the mean rate $\alpha(0, t)$ does not depend on t, since $X[0, t]$ has stationary increments.

The definition (2.1) may be motivated in several ways. The logarithmic moment generating function is naturally associated with the additive property (iii), while the scalings with t and s beget properties (i) and (iv) respectively. Properties (i)–(iv) are straightforward consequences of convexity and of results on moment generating functions – see Chang (1994) for several relevant observations, as well as a discussion of the case, not considered here, when the increments of $X[0, t]$ may be non-stationary. Although $(\alpha(s, t), 0 < s, t < \infty)$ does not in general determine the distribution of $(X[0, t], 0 < t < \infty)$, it follows from the analyticity of the moment generating function (Billingsley 1986, Exercise 26.7) that if $\alpha(s, t)$ is finite for $s = \varepsilon > 0$ then $(\alpha(s, t), 0 < s < \varepsilon)$ determines the distribution of $X[0, t]$ and, further, $\alpha(s, t)$ is infinitely differentiable with respect to s on the interior of the interval on which $\alpha(s, t)$ is finite.

Courcoubetis *et al.* (1995) and Duffield *et al.* (1995) have explored the estimation of $\alpha(s, t)$ for large values of t, and its relation to the tail behaviour of

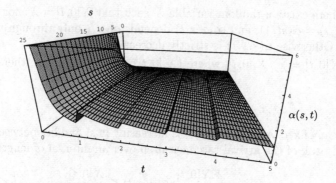

FIG. 1. Effective bandwidth of a periodic source. The source produces a single
   unit of workload at the end of every unit interval, but the phase of the source
   is random. Note the growth of the effective bandwidth over intervals shorter
   than the period of the source.

queues. In contrast, Sriram and Whitt (1986) have emphasized the importance
of identifying the relevant time scale for queueing phenomena: from relation (2.4)

$$\alpha(s,t) = \alpha(0,t)\left(1 + \frac{s}{2}I[0,t] + o(s)\right) \quad \text{as } s \to 0$$

where $I[0,t] = \operatorname{Var} X[0,t]/\mathbb{E}X[0,t]$ is their index of dispersion for counts.

2.3   Examples

2.3.1   *Periodic sources*

For a source which produces b units of workload at times $\{Ud+nd, n = 0, 1, \ldots\}$,
where U is uniformly distributed on the interval $[0,1]$,

$$\alpha(s,t) = \frac{b}{t}\left\lfloor\frac{t}{d}\right\rfloor + \frac{1}{st}\log\left[1 + \left(\frac{t}{d} - \left\lfloor\frac{t}{d}\right\rfloor\right)(e^{bs} - 1)\right]. \tag{2.6}$$

Observe that

$$\lim_{t\to 0}\alpha(s,t) = \frac{e^{bs} - 1}{ds};$$

the growth of the effective bandwidth as t decreases is apparent in Fig. 1, which
plots the function (2.6) with parameters $b = d = 1$. The model has been used to
describe the packet streams arising from constant rate information sources: for
a review see Roberts (1992, Section 6). We shall consider the model further in
Sections 3.5 and 3.6.1.

2.3.2   *Fluid sources*

Consider a stationary fluid source described by a two-state Markov chain. The
transition rate from state 2 to state 1 is λ and the transition rate from state 1
to state 2 is μ. While the Markov chain is in state 1 workload is produced at a
constant rate h; while it is in state 2 no workload is produced. Then

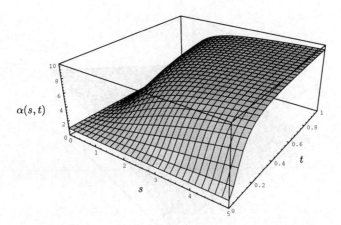

FIG. 2. Effective bandwidth of an on–off fluid source, with parameters $\lambda = 1, \mu = 9, h = 10$. The effective bandwidth $\alpha(s,t)$ approaches the mean rate $\lambda h/(\lambda + \mu)$ as either s or t approaches zero.

$$\alpha(s,t) = \frac{1}{st} \log \left\{ \left(\frac{\lambda}{\lambda + \mu}, \frac{\mu}{\lambda + \mu} \right) \exp \left[\begin{pmatrix} -\mu + hs & \mu \\ \lambda & -\lambda \end{pmatrix} t \right] \begin{pmatrix} 1 \\ 1 \end{pmatrix} \right\},$$

and

$$\lim_{t \to \infty} \alpha(s,t) = \frac{1}{2s} \left(hs - \mu - \lambda + \left((hs - \mu + \lambda)^2 + 4\lambda\mu \right)^{\frac{1}{2}} \right),$$

a central expression of Gibbens and Hunt (1991) and Guérin *et al.* (1991), and there obtained from the seminal work on stochastic fluid models of Anick *et al.* (1982). The function $\alpha(s,t)$ is illustrated in Fig. 2.

More generally, consider a stationary source described by a finite Markov chain with stationary distribution π and q-matrix Q, where workload is produced at rate h_i while the chain is in state i. Then from the backward equations for the Markov chain one can deduce (Kesidis *et al.* 1993, p.427) that

$$\alpha(s,t) = \frac{1}{st} \log\{\pi \exp[(Q + \mathbf{h}s)t]\mathbf{1}\} \tag{2.7}$$

where $\mathbf{h} = \text{diag}(h_i)_i$, and

$$\lim_{t \to \infty} \alpha(s,t) = \frac{1}{s} \phi(s)$$

where $\phi(s)$ is the largest real eigenvalue of the matrix $Q + \mathbf{h}s$ (Elwalid and Mitra 1993). If $h_1 > h_i$, $i \neq 1$, then relation (2.5) becomes

$$\alpha(s,t) = h_1 - \frac{1}{s} \left(\mu_1 - \frac{1}{t} \log \pi_1 \right) + o\left(\frac{1}{s} \right) \qquad \text{as } s \to \infty$$

where μ_1 is the transition rate out of the state with peak rate. Chang and Thomas (1995, p.1097) discuss this expansion in the case $t = \infty$: for a fluid

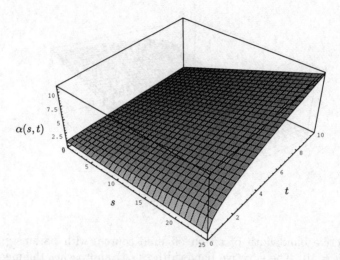

FIG. 3. Effective bandwidth of a Gaussian source. This example has Hurst parameter $H = 0.75$: long range order is indicated by the continued growth of the effective bandwidth with large t.

source, the relevant limits in s and t commute.

2.3.3 *Gaussian sources*

Suppose that

$$X[0, t] = \lambda t + Z(t)$$

where $Z(t)$ is normally distributed with zero mean; as usual, the facility of calculation under Gaussian assumptions outweighs any problem of interpretation for negative increments. Then

$$\alpha(s, t) = \lambda + \frac{s}{2t} \operatorname{Var} Z(t),$$

and so $\alpha(s, t)$ is determined, for all s and t, by the first two terms of the expansion (2.4).

The case $\operatorname{Var} Z(t) = \sigma^2 t$ commonly arises from heavy traffic models (Harrison 1985). The more general case $\operatorname{Var} Z(t) = \sigma^2 t^{2H}$ arises when the process Z is fractional Brownian motion, with Hurst parameter $H \in (0, 1)$. Then

$$\alpha(s, t) = \lambda + \frac{\sigma^2 s}{2} t^{2H-1}, \tag{2.8}$$

and the behaviour of $\alpha(s, t)$ as $t \to \infty$ depends upon whether $H < \frac{1}{2}$, $H = \frac{1}{2}$ or $H > \frac{1}{2}$. Respectively $\lim_{t \to \infty} \alpha(s, t)$ is finite and does not depend upon s, as in example (i); or the limit depends upon s, as in example (ii); or $\alpha(s, t)$ grows as a fractional power of t. The third case exhibits long range order (Norros 1994; Willinger 1995), and has been proposed as a model for Ethernet traffic

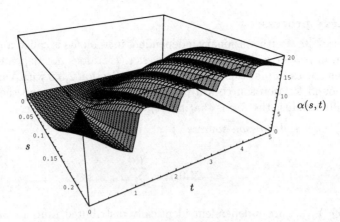

FIG. 4. Effective bandwidth of an on–off periodic source. Note the increase of the effective bandwidth as t either decreases below the period of the source, or increases towards the interval over which the source remains 'on' or 'off'.

data (Willinger *et al.* 1995). Figure 3 illustrates the form (2.8) with parameters $H = 0.75$, $\lambda = 1$ and $\sigma^2 = 0.25$.

Courcoubetis and Weber (1995) and Weber (1994) discuss the approximation of the effective bandwidth of an arbitrary stationary source by a Gaussian source with general autocovariance structure, using the first two terms of the expansion (2.4).

We shall consider Gaussian models further in Sections 3.1.3, 3.2.2, 3.3.1, 3.5, 3.6.1 and 3.6.2.

2.3.4 *General on–off sources*

Suppose next that a source alternates between long periods in an 'on' state, where it behaves as a source with effective bandwidth $\alpha_1(s,t)$, and long periods in an 'off' state, where it produces no workload. Let p be the proportion of time spent in the 'on' state. Then for values of t small compared with periods spent in an 'on' or 'off' state,

$$\alpha(s,t) = \frac{1}{st} \log\Big[1 + p\Big(\exp\big(st\alpha_1(s,t)\big) - 1\Big)\Big]. \tag{2.9}$$

Figure 4 illustrates this function when $p = 0.05$ and $\alpha_1(s,t)$ is given by expression (2.6) with $b = 20, d = 1$.

The above example shares similarities with examples 2.3.1, 2.3.3 or 2.3.2, over short, intermediate or long time scales respectively. Observe that the construction can be endlessly repeated: the 'on' period of a fluid source may at a finer time scale appear as a periodic source, whose bursts may themselves have structure on a still finer time scale, and so on. By variations of this hierarchical construction it is possible to define sources whose effective bandwidth $\alpha(s,t)$ may resemble any or all of Figs 1, 2, 3 and 4, depending on the range of s and t values plotted.

2.4   Lévy processes

A process $X[0, t]$ with stationary independent increments is called a *Lévy process*; as noted in property (i), for such a process $\alpha(s, t)$ does not depend upon t. We have seen one example: the Gaussian source of Section 2.3.3 with Var $Z(t) = \sigma^2 t$. A compound Poisson source will provide another example, and these two cases essentially exhaust the forms that $\alpha(s, t) = \alpha(s)$, say, may take.

2.4.1   *Compound Poisson sources*

If

$$X[0, t] = \sum_{n=1}^{N(t)} Y_n$$

where $Y_1, Y_2, \ldots$ are independent identically distributed random variables with distribution function F, and $N(t)$ is an independent Poisson process of rate ν, then

$$\alpha(s) = \frac{1}{s} \int (e^{sx} - 1) \nu dF(x).$$

For example, if $Y_1, Y_2, \ldots$ are exponentially distributed with parameter μ, then

$$\alpha(s) = \frac{\nu}{\mu - s} \quad \text{for } s < \mu. \tag{2.10}$$

2.4.2   *Infinitely divisible sources*

If $X[0, t]$ has stationary independent increments, then $X[0, 1]$ is infinitely divisible. Hence, by the Lévy–Khinchin representation of any infinitely divisible random variable as the limit of a mixture of compound Poisson random variables (Feller 1971, Chapter XVII),

$$\alpha(s) = \lambda + \frac{\sigma^2 s}{2} + \frac{1}{s} \int_{-\infty}^{+\infty} (e^{sx} - 1) d\nu(x) \tag{2.11}$$

is the most general form possible for $\alpha(s)$, where $\nu(\cdot)$ is a measure on $(-\infty, \infty)$.

If, in addition, the increments of $X[0, t]$ are non-negative, then the most general form possible for $\alpha(s)$ is

$$\alpha(s) = \lambda + \frac{1}{s} \int_0^\infty (e^{sx} - 1) d\nu(x)$$

where $\nu(\cdot)$ is a measure on $(0, \infty)$: it follows that $\alpha(s)$ is convex, and indeed that all derivatives are positive. For example, if $\lambda = 0$ and $d\nu(x) = x^{-1} e^{-x} dx$, then

$$\alpha(s) = -\frac{1}{s} \log(1 - s)$$

and $X[0, t]$ is a gamma process, with increments distributed as gamma random variables (Kingman 1993, Chapter 8,9). If $\int_0^\infty d\nu(x) = \infty$, as for the gamma process, then jumps of $X[0, t]$ are everywhere dense. Jumps of height greater than δ form a Poisson process of rate $\int_\delta^\infty d\nu(x)$.

2.5   Policing and shaping

Say that a stationary source is *policed by parameters* (ρ, β) if

$$\bar{X}[0,t] \leq \rho t + \beta \qquad 0 < t < \infty, \tag{2.12}$$

or, equivalently,

$$\alpha(\infty, t) \leq \rho + \frac{\beta}{t} \qquad 0 < t < \infty.$$

The parameters ρ and $(\beta - 1)/\rho$ are the *peak cell rate* and *cell delay variation tolerance* of ITU Recommendation I371 (1994). For example, the periodic source described in Section 2.3.1 is policed by parameters (ρ, β) provided $\rho \geq \frac{b}{d}$ and $\beta \geq b$. A fluid source, as described in Section 2.3.2, is policed by parameters (ρ, β) provided $\rho \geq \max_i \{h_i\}$.

Sources which do not satisfy constraint (2.12) may be *shaped* to do so, by either delaying or discarding some of the arriving workload. In general, shaping will alter the effective bandwidth of a source for larger values of s, and on short, intermediate and long time scales, as we next illustrate.

Suppose that a source is shaped to conform with parameters (ρ, β) by passage through a device that delays or discards some of the workload. Let $X_{sh}[0,t], t > 0$, describe the stationary departure stream from the device, the *shaped* process, and let $\alpha_{sh}(s,t)$ be its effective bandwidth. Since $0 \leq X_{sh}[0,t] \leq \rho t + \beta$, a simple upper bound is

$$\alpha_{sh}(s,t) \leq \frac{1}{st} \log \left[1 + \frac{t\alpha_{sh}(0,t)}{\rho t + \beta} \left(e^{s(\rho t + \beta)} - 1 \right) \right]$$

where $\alpha_{sh}(0,t)$ is the mean rate of the shaped process. Example 2.3.1 illustrates that this bound may become tight as t approaches zero.

To explore the impact of shaping at intermediate time scales we describe a simple example. Consider a Lévy process, shaped by passage through a queue with service rate C. The queue size is a Markov process: assume further that it may be represented by a stationary Markov chain with q-matrix Q, for example the q-matrix of an M/M/1 queue for the arrival process leading to expression (2.10). Then the shaped process can be described as a fluid source, possibly with infinite state space. Indeed the shaped process is just an alternating renewal process, taking the level C for a busy period and the level 0 for an exponentially distributed idle period. If $\alpha_{sh}(s,t)$ is the effective bandwidth of the departure stream from the queue, then $\alpha_{sh}(s,t)$ may be calculated from expression (2.7), where h_i takes values 0 or C.

De Veciana *et al.* (1994) have extensively explored the impact of shaping on the limiting form of the effective bandwidth as $t \to \infty$. An example of their results is that a Gaussian source, with effective bandwidth $\alpha(s,t) = \lambda + \sigma^2 s/2$, shaped by passage through a queue with service rate C, has

$$\alpha_{sh}(s,\infty) = \lambda + \frac{\sigma^2 s}{2} \qquad s < \frac{C - \lambda}{\sigma^2} \tag{2.13}$$

$$= \quad C - \frac{(C-\lambda)^2}{2\sigma^2 s} \quad \text{otherwise.} \tag{2.14}$$

More generally, de Veciana *et al.* (1994) consider a wide class of arrival processes for which $\alpha(s,\infty) < \infty$ for some s, and show that $\alpha_{sh}(s,\infty) = \alpha(s,\infty)$ for values of s less than a critical level, while above this level the impact of the peak rate C is felt and $\alpha_{sh}(s,\infty) = C - \kappa/s$, where the constant κ may be calculated. See de Veciana and Walrand (1995) for further discussion of shaping.

3   Multiplexing models

In this section we suppose the arrival process is

$$X[0,t] = \sum_{j=1}^{J} \sum_{i=1}^{n_j} X_{ji}[0,t] \tag{3.1}$$

where $(X_{ji}[0,t])_{ji}$ are independent processes with stationary increments whose distributions may depend upon j but not upon i, and that there is a resource that has to cope with the aggregate arriving stream of work. We interpret n_j as the number of sources of type j, and shall write $\alpha_j(s,t)$ for the effective bandwidth of a source of type j. Thus

$$\alpha(s,t) = \sum_{j=1}^{J} n_j \alpha_j(s,t). \tag{3.2}$$

We shall explore several multiplexing models, and shall be interested in the relationship between constraints of the form

$$\sum_{j=1}^{J} n_j \alpha_j(s^*,t^*) \le C^* \tag{3.3}$$

for one or several choices of (s^*,t^*,C^*) and the acceptance region, defined as the set of vectors $(n_1,n_2,\ldots,n_J)$ for which a given performance, described in terms of queueing delay or buffer overflow, can be guaranteed.

   In Section 3.1 we describe the result of Hui (1988, 1990) which establishes inequality (3.3) as a conservative bound on the non-linear acceptance region for a bufferless model. In Section 3.2, based on Kelly (1991), we see relation (3.3) emerge as the linear limiting form of, and as a conservative bound on, the acceptance region for a buffered model with Lévy input. A linear limiting form was established for more general input processes, including the fluid sources studied in detail by Gibbens and Hunt (1991) and Elwalid and Mitra (1993), by Kesidis *et al.* (1993): we review this result in Section 3.3, together with its recent generalization by Duffield and O'Connell (1996). In the models of Sections 3.1 and 3.3 time scales essentially degenerate: the time scale t^* appearing in (3.3) approaches zero or infinity. In Sections 3.4 and 3.5 we describe two tractable models illustrating phenomena when time scales do not degenerate. In Section 3.6 we discuss the important recent results of Botvitch and Duffield (1995),

Simonian and Guibert (1995) and Courcoubetis and Weber (1996) on an asymptotic regime where the form (3.3) emerges, for finite values of t^*, as a tangent to the limiting acceptance region. In Section 3.7 we briefly discuss priority models, which provide further important examples where several constraints of the form (3.3) may be needed to approximate the acceptance region.

3.1 Bufferless models

We look first at a simple model where

$$X = \sum_{j=1}^{J} \sum_{i=1}^{n_j} X_{ji}$$

and X_{ji} are independent random variables with scaled logarithmic moment generating functions

$$\alpha_j(s) = \frac{1}{s} \log \mathbb{E}[e^{sX_{ji}}]. \tag{3.4}$$

We might suppose that X_{ji} is the instantaneous arrival rate of work from a source of type j at a bufferless resource of capacity C, corresponding to the choice $\alpha_j(s) = \lim_{t \to 0} \alpha_j(s/t, t)$. Alternatively we might suppose that $X_{ji}[0, t] = X_{ji}t$, so that $\alpha_j(s/t, t) = \alpha_j(s)$ for all values of t.

Chernoff's bound gives

$$\log P\{X \geq C\} \leq \log \mathbb{E}[e^{s(X-C)}] = s(\alpha(s) - C) \tag{3.5}$$

where $\alpha(s) = \sum_j n_j \alpha_j(s)$. Thus the constraint $\log P\{X \geq C\} \leq -\gamma$ will certainly be satisfied if the vector $n = (n_1, n_2, \ldots, n_J)$ lies within the set

$$A = \left\{ n : \inf_s \left[s \left(\sum_{j=1}^{J} n_j \alpha_j(s) - C \right) \right] \leq -\gamma \right\} \tag{3.6}$$

where throughout $n \geq 0$. The region A has a convex complement in $\mathbb{R}_+^J$, since this complement is defined as the intersection of $\mathbb{R}_+^J$ with a family of half-spaces. The half-space touching at a point n^* on the boundary of the region A is

$$\sum_{j=1}^{J} n_j \alpha_j(s^*) \leq C - \frac{\gamma}{s^*}, \tag{3.7}$$

where s^* attains the infimum appearing in relation (3.6) with n replaced by n^*. Thus condition (3.7) is a conservative global bound, of the form (3.3), on the acceptance region: if n satisfies this condition then the performance guarantee $\log P\{X \geq C\} \leq -\gamma$ is assured.

Let $A(\gamma, C)$ be the subset of $\mathbb{R}_+^J$ such that $n \in A(\gamma, C)$ implies $\log P\{X \geq C\} \leq -\gamma$. Chernoff's theorem (Billingsley 1986) gives that

$$\lim_{N \to \infty} \frac{1}{N} \log P \left\{ \sum_{j=1}^{J} \sum_{i=1}^{n_j N} X_{ji} \geq CN \right\} = \inf_s \left[s \left(\sum_{j=1}^{J} n_j \alpha_j(s) - C \right) \right]. \tag{3.8}$$

Except in the trivial circumstances where the infimum is zero or minus infinity or where a source type has zero mean rate, the infimum (3.8) is strictly increasing in each component of n. It follows that

$$\lim_{N \to \infty} \frac{A(\gamma N, CN)}{N} = A, \tag{3.9}$$

and in this sense the approximation leading to the region A becomes more accurate as the number of sources increases, and the tail probability decreases.

The convergence statement (3.9) requires a comment on topology. Throughout we shall use the Hausdorff distance (Császár 1978) to define a pseudo-metric over the set of subsets of $\mathbb{R}_+^J$. This induces a topology over the quotient space formed by identifying subsets of $\mathbb{R}_+^J$ which share the same closure. Our notation will use a subset of $\mathbb{R}_+^J$ to represent its equivalence class; the intersection operator, as will appear in for example relation (3.25), is defined on equivalence classes in the natural way using *closed* subsets of $\mathbb{R}_+^J$ as representatives. Thus limits such as (3.9) are uninformative about the limiting behaviour of performance measures at points on the boundary of the set A.

3.1.1 *Stream-based measures*

The discussion leading to relation (3.9) concerned a resource-based congestion measure, the probability of resource overload, rather than a stream-based congestion measure, such as the proportion of work from an arriving stream that is lost. To convert from resource-based measures to stream-based measures requires two steps: relating the expected size of overloads to tail probabilities of overloads; and dividing by stream rates.

From Chernoff's bound the expected rate of load loss

$$
\begin{aligned}
\mathbb{E}(X - C)^+ &= \int_0^\infty P\{X \geq C + x\}dx \\
&\leq \int_0^\infty \exp\Big[s\big(\alpha(s) - (C + x)\big)\Big]dx \\
&= \frac{1}{s} \exp\Big[s\big(\alpha(s) - C\big)\Big].
\end{aligned}
$$

We deduce that

$$\mathbb{E}(X - C)^+ \leq \frac{1}{s^*} \exp\Big[s^*\big(\alpha(s^*) - C\big)\Big] \tag{3.10}$$

where s^* attains the infimum in (3.8). Thus if condition (3.7) is satisfied, and hence $P\{X > C\} \leq e^{-\gamma}$ is assured, then also assured is that $\mathbb{E}(X - C)^+ \leq e^{-\gamma}/s^*$. Note that the proportion of load lost is just $\mathbb{E}(X - C)^+/\mathbb{E}X$. Let $A_{st}(\gamma, C)$ be the subset of $\mathbb{R}_+^J$, such that $n \in A_{st}(\gamma, C)$ implies that the proportion of work lost is not greater than $e^{-\gamma}$. Then a further consequence of Chernoff's theorem is that

$$\lim_{N \to \infty} \frac{A_{st}(\gamma N, CN)}{N} = A.$$

3.1.2 *Improved approximations*

The inequalities (3.5) and (3.10) provide bounds on probability of resource overload or the proportion of work lost: closely related *tilted approximations* may be developed by various techniques reviewed by Reid (1988, Section 6.3), Bucklew (1990, Chapter VII), and Jensen (1995). For example, the estimates

$$P\{X \geq C\} \sim \frac{1}{s^*(2\pi\sigma^2(s^*))^{\frac{1}{2}}} e^{s^*(\alpha(s^*)-C)} \qquad (3.11)$$

and

$$\mathbb{E}(X - C)^+ \sim \frac{1}{s^{*2}(2\pi\sigma^2(s^*))^{\frac{1}{2}}} e^{s^*(\alpha(s^*)-C)}, \qquad (3.12)$$

where $\sigma^2(s) = \frac{\partial^2}{\partial s^2}(s\alpha(s))$, have been discussed by Hui (1988), Roberts (1992, p. 154), and Hsu and Walrand (1995): the prefactor of the exponential term considerably improves accuracy. Note that the prefactor depends upon whether the measure of interest is resource-based or stream-based.

3.1.3 *Approximate linearity*

How well approximated is the region (3.6) by the linearly constrained region (3.7)? Some insight may be obtained from the Gaussian case, where explicit calculations are easy to perform. Suppose that

$$\alpha_j(s) = \lambda_j + \frac{s\sigma_j^2}{2},$$

corresponding to a normally distributed load with mean λ_j and variance σ_j^2. Then the region (3.6) becomes

$$\sum_j n_j\lambda_j + \left(2\gamma \sum_j n_j\sigma_j^2\right)^{\frac{1}{2}} \leq C. \qquad (3.13)$$

The tangent plane at a point n^* on the boundary of the region (3.13) is of the form (3.7) with

$$s^* = \frac{C - \sum_j n_j^*\lambda_j}{\sum_j n_j^*\sigma_j^2}$$

and hence

$$\alpha_j(s^*) = \lambda_j + \frac{\gamma\sigma_j^2}{C(1 - \delta^*)} \qquad (3.14)$$

where $\delta^* = \sum_j n_j^*\lambda_j/C$, the traffic intensity. Thus the coefficients (3.14) will be relatively insensitive to the traffic mix n^*, provided $(1 - \delta^*)^{-1}$ does not vary too greatly with n^*, or, equivalently, provided the traffic intensity is not too close to 1 on the boundary of the acceptance region.

Let $C^* = C - \gamma/s^*$, the *effective capacity* appearing on the right hand side of inequality (3.7). Then

$$C^* = C - \gamma \frac{\sum_j n_j^* \sigma_j^2}{C(1 - \delta^*)}$$

$$= C - \gamma \frac{\text{variance of load}}{\text{mean free capacity}} .$$

We shall refer again to this simple model in Sections 3.3.1 and 3.6.2.

3.2   M/G/1 models

Next suppose that each of the processes $X_{ji}[0, t]$ has independent increments, as discussed in Section 2.5, and write $\alpha_j(s) = \alpha_j(s, t)$, $\alpha(s) = \alpha(s, t)$. Let Q be distributed as the stationary workload in a queue with a server of capacity C and an infinite buffer, fed by the arrival stream $X[0, t]$. (More formally we could define the queue size at time τ as

$$Q(\tau) = (X[0, \tau] - C\tau) - \inf_{0 < t < \tau} \{X[0, t] - Ct\},$$

and let $\tau \to \infty$ – see Harrison 1985, p. 19, or Asmussen 1987, Chapter III, 7–8.) Then the Pollaczek–Khinchin formula (see, for example, Asmussen 1987, p. 206, or Kella and Whitt 1992) is simply

$$\mathbb{E}[e^{sQ}] = \frac{C - \alpha(0)}{C - \alpha(s)} . \tag{3.15}$$

Cramér's estimate (Feller 1971) describes the tail behaviour of the distribution for Q. Suppose there exists a finite constant κ such that $\alpha(\kappa) = C$, and suppose that κ is in the interior of the interval on which $\alpha(s)$ is finite, so that $\alpha'(\kappa)$ is necessarily finite. Then Cramér's estimate is

$$P\{Q \geq b\} \sim \frac{C - \alpha(0)}{\kappa \alpha'(\kappa)} e^{-\kappa b} \quad \text{as } b \to \infty. \tag{3.16}$$

Let $A(\gamma, b)$ be the subset of $\mathbb{R}_+^J$ such that $n \in A(\gamma, b)$ implies $\log P\{Q \geq b\} \leq -\gamma$. Then a consequence of Cramér's estimate is that

$$\lim_{N \to \infty} A(\gamma N, bN) = A, \tag{3.17}$$

where

$$A = \left\{ n : \sum_j n_j \alpha_j\left(\frac{\gamma}{b}\right) \leq C \right\}, \tag{3.18}$$

again a region defined by a constraint of the form (3.3). Kelly (1991) notes that $A \subset A(\gamma, b)$, and so the linearly constrained region A is a conservative global bound, as well as an asymptotic limit.

3.2.1   *Finite buffers*

The above discussion concerned the proportion of time the buffer occupancy exceeded a level b, in a queue with an infinite buffer. Next we consider what happens if there is a finite buffer of size b, and any excess workload over this level is lost. Note that we can construct a sample path of this process from the sample path of an M/G/1 queue with infinite buffer: just remove the time intervals when the workload is above b. That this construction works is a consequence of the simple rule for overflow, and the assumption that the arrival process has independent increments. The stationary distribution for the workload in an M/G/1 queue with finite buffer b is thus obtained from that for the infinite buffer case by conditioning on the event that the workload does not exceed b. From Cramér's estimate (3.16) it can be deduced that the proportion of workload lost with a finite buffer of size b, $L(b)$, satisfies

$$L(b) \sim \frac{C(C - \alpha(0))^2}{\kappa \alpha'(\kappa)\alpha(0)} e^{-\kappa b} \quad \text{as } b \to \infty.$$

It follows that if $A_{\text{prop}}(\gamma, b)$ is the subset of $\mathbb{R}_+^J$ such that $n \in A_{\text{prop}}(\gamma, b)$ implies $\log L(b) \leq -\gamma$ then

$$\lim_{N \to \infty} A_{\text{prop}}(\gamma N, bN) = A.$$

3.2.2 *Brownian input*

Suppose that

$$X_{ji}[0, t] = \lambda_j t + \sigma_j Z(t)$$

where $Z(t)$ is a standard Brownian motion. Then superpositions can also be expressed in terms of a Brownian motion, Z^1, as

$$X[0, t] = \left(\sum_j n_j \lambda_j \right) t + \left(\sum_j n_j \sigma_j^2 \right)^{\frac{1}{2}} Z^1(t),$$

and hence, from basic results on reflected Brownian motion (Harrison 1985),

$$P\{Q \geq b\} = \exp \left\{ \frac{-2b(C - \sum_j n_j \lambda_j)}{\sum_j n_j \sigma_j^2} \right\}.$$

Thus the constraint $\log P\{Q \geq b\} \leq -\gamma$ becomes *precisely* the condition

$$\sum_j n_j \left(\lambda_j + \sigma_j^2 \frac{\gamma}{2b} \right) \leq C, \tag{3.19}$$

which is just the canonical constraint (3.3) with $s^* = \gamma/b$.

3.2.3 *Mean delays*

From the Pollaczek–Khinchin formula (3.15) it follows that $\mathbb{E}Q = \alpha'(0)/(C - \alpha(0))$, and hence that a constraint of the form $\mathbb{E}Q \leq L$ is satisfied if and only if

$$\sum_{j=1}^J n_j \left[\alpha_j(0) + \frac{\alpha_j'(0)}{L} \right] \leq C.$$

This provides a linear acceptance region which accords with the previous example in the case of Brownian input, but which is not, in general, of the canonical form (3.3). In Kelly (1991) this and other possible definitions of an effective bandwidth were considered, with emphasis on the linearity of the acceptance region under a variety of performance criteria. In this paper we explore a different perspective, one which emphasizes the unifying role of the definition (2.1) under a variety of multiplexing models.

3.3   Buffer asymptotic models

Tail probabilities decay exponentially in models more general than the M/G/1 queue. Suppose that Q is distributed as the stationary workload in a queue with a server of capacity C and an infinite buffer, fed by an arrival stream $X[0, t]$ with stationary and ergodic increments. Thus we weaken the M/G/1 assumption of independent increments to an assumption of ergodic increments. Suppose that

$$\lim_{t \to \infty} \alpha(s, t) = \alpha(s) \tag{3.20}$$

and that there exists a finite constant κ such that $\alpha(\kappa) = C$, and $\alpha'(\kappa)$ is finite. Then

$$\lim_{b \to \infty} \frac{1}{b} \log P\{Q \ge b\} = -\kappa \tag{3.21}$$

(Kesidis *et al.* 1993; Chang 1994; Glynn and Whitt 1994). Thus the relations (3.17), (3.18) hold in this more general context.

The examples of Section 2.3 show that even if the limit (3.20) exists, convergence to the limit may be arbitrarily slow; further, for finite values of t, $\alpha(s, t)$ may be much smaller or larger than the limit $\alpha(s)$. The examples of Choudhury *et al.* (1994) can be interpreted as further illustrations of this phenomenon. The usefulness of the limit (3.21) thus depends on the rate of convergence to this limit, and whether convergence has essentially occurred on the time scales of interest. Interestingly, the GI/G/1 or M/G/1 models provide a natural choice of time scale. For example, suppose the limit (3.20) is approximately of the form (2.11) appropriate for an M/G/1 model. Under this model the time taken to empty a full buffer is of order $t_1 = b/(C - \alpha(0))$, while the time taken to fill an empty buffer is of order $t_2 = b/(\kappa \alpha'(\kappa))$, as b increases (see Tse *et al.* 1995 for a valuable discussion of regenerative structure in this model). For the asymptotic (3.21) to be appropriate, $\alpha(\kappa, t)$ should have essentially converged to its limit $\alpha(\kappa)$ by time scales t in the region of t_1, t_2; and $\alpha(s, t)$ evaluated in this region should be used in estimates such as (3.16).

The limit (3.20) may not, however, be capable of representation in the form (2.11) appropriate for an M/G/1 model, or even in the form (3.4) for *any* random variable. The form (3.4) is differentiable, while, for example, the function (2.13), (2.14) obtained from a shaped process has a discontinuous derivative at the critical point $s = (C - \lambda)/\sigma^2$, where there is a transition away from a regenerative regime.

Duffield and O'Connell (1996) have extended buffer asymptotics to examples where the limit (3.20) does not exist, but where, with a suitable rescaling, a large deviation principle may still be applied. We illustrate their result with a simple example.

3.3.1 *Fractional Brownian input*

Suppose that $\alpha(s,t)$ is given by expression (2.8), corresponding to fractional Brownian motion with Hurst parameter H. Then Duffield and O'Connell (1996) show that

$$\lim_{b \to \infty} \frac{\log P\{Q \geq b\}}{b^{2(1-H)}} = -\frac{1}{2\sigma^2} \left(\frac{C - \lambda}{H}\right)^{2H} (1 - H)^{-2(1-H)} \qquad (3.22)$$

agreeing with an earlier bound of Norros (1994).

Next suppose that

$$\alpha_{ji}(s,t) = \lambda_j + \frac{\sigma_j^2}{2} st^{2H-1} \qquad (3.23)$$

so that $\alpha(s,t)$, given by relation (3.2), corresponds to a superposition of fractional Brownian sources, all sharing the same Hurst parameter H. Then from the result (3.22) it is possible to deduce (Duffield *et al.* 1994) that the condition $P\{Q \geq b\} \leq e^{-\gamma}$ becomes (asymptotically, as $\gamma, b \to \infty$ with $\gamma/b^{2(1-H)}$ held constant) the condition

$$2\gamma \sum_{j=1}^{J} n_j \sigma_j^2 \leq b^{2(1-H)} \left(\frac{C - \sum_{j=1}^{J} n_j \lambda_j}{H}\right)^{2H} (1 - H)^{-2(1-H)}. \qquad (3.24)$$

Observe that for $H = 1/2$ this is just the (exact) condition (3.19), while as $H \to 1$ it approaches the condition (3.13). The major effect of long range order is thus on the scaling relationship between γ, b and C, as discussed by Norros (1994), rather than on the geometrical form of the acceptance region A. We shall return to this point later, in Section 3.6.2, where we shall also discuss the connection between inequality (3.24) and the form (3.3).

3.4 Deterministic multiplexing

Suppose the arriving work is dealt with by a server of capacity C with a finite buffer of capacity b, initially empty. Under what condition is the capacity of the buffer *never* exceeded? The condition is (Cruz 1991) that $n \in A$ where

$$A = \bigcap_{0 < t < \infty} A_t, \qquad (3.25)$$

an intersection of linearly constrained regions

$$A_t = \left\{ n : \sum_j n_j \alpha_j(\infty, t) \leq C + \frac{b}{t} \right\}. \qquad (3.26)$$

3.4.1 *Policed sources*

Recall that in Section 2.5 we discussed policed sources. If

$$\alpha_j(\infty, t) = \rho_j + \frac{\beta_j}{t}$$

corresponding to a source policed by parameters (ρ_j, β_j), then

$$A = A_0 \cap A_\infty$$

where

$$A_0 = \left\{ n : \sum_j n_j \beta_j \le b \right\}, \quad A_\infty = \left\{ n : \sum_j n_j \rho_j \le C \right\}.$$

Note that if β_j/ρ_j does not vary with the source type j, then the boundaries of the regions A_0 and A_∞ are parallel.

3.4.2 *Multiple policers*

If

$$\alpha_j(\infty, t) = \min_{k \in K_j} \left\{ \rho_{jk} + \frac{\beta_{jk}}{t} \right\}$$

corresponding to a source policed by a finite set K_j of parameter choices, then A can be written as an intersection of a finite collection of sets A_t. For example, if $K_j = \{1, 2, \ldots, K\}$, if $(\beta_{jk}, k = 1, 2, \ldots, K)$ is an increasing sequence for each $j = 1, 2, \ldots, J$, and if the ratios

$$t_k = \frac{\beta_{jk+1} - \beta_{jk}}{\rho_{jk} - \rho_{jk+1}} \tag{3.27}$$

do not vary with the source type j and are increasing in k, then

$$A = \bigcap_{k=0}^{K} A_{t_k},$$

where $t_0 = 0$ and $t_K = \infty$.

3.5 Brownian bridge models

When several independent periodic sources, of type (2.6), are superimposed, the resulting process can be approximated by a Brownian bridge (for a recent review see Hajek 1994). This motivates study of the source

$$X_{ji}[0, t] = \lambda_j t + \sigma_j Z_0(t - \lfloor t \rfloor)$$

where $Z_0(t), 0 \le t \le 1$, is a standard Brownian bridge. Then

$$\alpha_j(s, t) = \lambda_j + \frac{\sigma_j^2 s}{2t} (t - \lfloor t \rfloor)(1 + \lfloor t \rfloor - t). \tag{3.28}$$

For example, a periodic source, with period 1 and burst size β_j, might be approximated by $\lambda_j = \beta_j, \sigma_j = \beta_j$: this example is of some interest as a conservative description of sources policed by parameters (ρ_j, β_j), where the ratio β_j/ρ_j does not vary with source type, and where setting the ratio to 1 simply fixes the time unit. Superpositions can also be expressed in terms of a Brownian bridge, Z_0^1, as

$$X[0,t] = \sum_j n_j \lambda_j t + \left(\sum_j n_j \sigma_j^2\right)^{\frac{1}{2}} Z_0^1(t - \lfloor t \rfloor).$$

The condition for the queue to be stable is just

$$\sum_j n_j \lambda_j < C. \tag{3.29}$$

Given this, the stationary probability $P\{Q \geq b\}$ is

$$P\left\{\max_{0 \leq t \leq 1}\{X[0,t] - Ct\} \geq b\right\} = \exp\left\{\frac{-2b}{\sum_j n_j \sigma_j^2}(b + C - \sum n_j \lambda_j)\right\}$$

(Hajek 1994, p. 150), and this probability is less than $e^{-\gamma}$ if and only if

$$\sum_j n_j \left(\lambda_j + \sigma_j^2 \frac{\gamma}{2b}\right) < b + C. \tag{3.30}$$

This constraint does not, in general, imply the condition (3.29).

Thus *two* linear constraints (3.29) and (3.30) are equivalent to the condition that $\log P\{Q \geq b\} < -\gamma$. Constraint (3.29) is of the canonical form (3.3) with $t^* = \infty$. Constraint (3.30) may be thrown into the form (3.3), with for example the choice $(s^*, t^*) = (2\gamma/b, 1/2)$. This example will be explored further in Section 3.6.1.

3.6 Buffer and source asymptotics

In Sections 3.1 and 3.3 we described asymptotic results when the number of sources or the buffer size, respectively, increased. Recently Botvich and Duffield (1995), Simonian and Guibert (1995) and Courcoubetis and Weber (1996) have obtained important results when the number of sources *and* the buffer size increase together, the regime considered in a key early paper of Weiss (1986).

Again suppose that the arrival process is given by the superposition (3.1), where the increments of $X_{ji}[0,t]$ are stationary. Let $L(C, b, n)$ be the proportion of workload lost, through overflow of a buffer of size $b > 0$, when the server has rate C and as usual $n = (n_1, n_2, \ldots, n_J)$. Then the above authors establish that

$$\lim_{N \to \infty} \frac{1}{N} \log L(CN, bN, nN) = \sup_t \inf_s \left[st \sum_j n_j \alpha_j(s,t) - s(b + Ct)\right]. \tag{3.31}$$

Let $A(\gamma, C, b)$ be the subset of $\mathbb{R}_+^J$ such that $n \in A(\gamma, C, b)$ implies $\log L(C, b, n) \leq -\gamma$. As in Section 3.2, the limit (3.31) is strictly increasing in each component of n, and hence

$$\lim_{N \to \infty} \frac{A(\gamma N, CN, bN)}{N} = A$$

where

$$A = \bigcap_{0 < t < \infty} A_t \tag{3.32}$$

with

$$A_t = \left\{ n : \inf_s \left[st \sum_j n_j \alpha_j(s,t) - s(b + Ct) \right] \leq -\gamma \right\}, \qquad (3.33)$$

a region with convex complement in $\mathbb{R}_+^J$. Moreover, if the boundary of the region A is differentiable at a point n^*, then the tangent plane is

$$\sum_j n_j \alpha_j(s^*, t^*) = C + \frac{b}{t^*} - \frac{\gamma}{s^* t^*} \qquad (3.34)$$

where (s^*, t^*) is an extremizing pair in relation (3.31) with n replaced by n^*. Thus a constraint of the canonical form (3.3) emerges as an asymptotic local limit, local in variations of the traffic mix n.

It is interesting to compare the regions (3.32) and (3.33) with corresponding regions obtained in earlier sections. Consider the model of Section 3.1, where several formal comparisons are possible. If $b = 0$, then A_t is increasing in t, by the final remark of property (ii), and so $A = A_0$. We recover the region (3.6), with the interpretation $\alpha_j(s) = \lim_{t \to 0} \alpha_j(s/t, t)$. Or, if $b > 0$ and $\alpha_j(s,t)$ depends on s, t only through the product st, then A_t is decreasing in t and so $A = A_\infty$. Again we recover the region (3.6) with the interpretation $\alpha_j(s) = \alpha_j(s/t, t)$, $t > 0$. If $\alpha_j(s,t)$ is independent of t, as discussed in Section 3.2, then the envelope of the regions A_t is the linear boundary of the region (3.18).

The results of Section 3.1 concern a regime where the time taken to fill a buffer is much *shorter* than the time periods over which sources fluctuate, while the results of Section 3.3 concern a regime where it is much *longer*. In both cases the limit results concern the behaviour of $\alpha(s,t)$ for t near zero or infinity. The great advantage of the limiting regime described in this section is that it allows the shape of the effective bandwidth $\alpha(s,t)$ to identify the relevant time scale implicitly, and, in general, the region (3.32) will depend upon $\alpha(s,t)$ evaluated at finite values of t. A simple illustration of this is provided by the limit as $\gamma \to \infty$, when the region (3.33) shrinks to the region (3.26) of Section 3.4; thus in example 3.4.2 there is a single linear constraint for each of the time constants (3.27). A more subtle illustration is provided by examples 3.6.1 and 3.6.2 below.

Several further examples are discussed in detail by Botvitch and Duffield (1995), Simonian and Guibert (1995) and Courcoubetis and Weber (1996). Simonian and Guibert (1995) also describe bounds and estimates that parallel the tilted approximations (3.11), (3.12).

3.6.1   *A Brownian bridge model*

Suppose that $\alpha_{ji}(s,t)$ is given by expression (3.28). Then the set (3.32) becomes

$$A = \bigcap_{0 < t < 1} A_t \cap A_\infty = A_{(0,1)} \cap A_\infty$$

where $A_{(0,1)}$ and A_∞ are simply the regions (3.30) and (3.29) respectively; recall that for the Brownian bridge model of Section 3.5 the acceptance region A is exact for finite values of γ, C and b.

3.6.2 *Fractional Brownian input*

If $\alpha_{ji}(s,t)$ is given by expression (3.23), then the tangent plane (3.34) uses the space and time scales

$$s^* = \frac{b + t^*(C - \sum_j n_j^* \lambda_j)}{t^{*2H} \sum_j n_j^* \sigma_j^2}, \quad t^* = \left(2\gamma \sum_j n_j^* \sigma_j^2\right)^{\frac{1}{2H}} \left(\frac{b}{1-H}\right)^{\frac{1}{H}},$$

and the acceptance region (3.32) becomes

$$H\left(\frac{1-H}{b}\right)^{\frac{1}{H}-1} \left(2\gamma \sum_j n_j \sigma_j^2\right)^{\frac{1}{2H}} + \sum_j n_j \lambda_j \leq C. \tag{3.35}$$

This is just condition (3.24), although the limiting regime is different. Note that the region (3.35) is convex or concave (has convex complement) according as $H \leq \frac{1}{2}$ or $H \geq \frac{1}{2}$. Regions that are neither convex nor concave can be constructed by allowing the Hurst parameter H to vary with source type.

If $H = \frac{1}{2}$ the condition (3.35) is just the linear constraint (3.19). Even the most extreme values of H produce rather well-behaved acceptance regions: as $H \to 1$ the inequality (3.35) approaches the condition (3.13), and as $H \to 0$ it approaches the conditions

$$\sum_j n_j \lambda_j \leq C, \quad 2\gamma \sum_j n_j \sigma_j^2 \leq b^2,$$

a limiting acceptance region with a similar geometrical form to that found in example 3.4.1.

3.7 Priorities

Multiple time and space scales may also arise for certain priority mechanisms. Suppose a single resource gives strict priority to sources $j \in J_1$, which have a strict delay requirement, but also serves sources $j \in J_2$, which have a much less stringent delay requirement. Then two constraints of the form

$$\sum_{j \in J_1} \alpha_j(s_1, t_1) \leq C_1, \quad \sum_{j \in J_1 \cup J_2} \alpha_j(s_2, t_2) \leq C_2 \tag{3.36}$$

may be needed to ensure that both sets of requirements are met (for several examples, see Bean 1994, Elwalid and Mitra 1995, de Veciana and Walrand 1995). If the less stringent delay requirement becomes very weak, corresponding to a *very* large buffer and almost *no* sensitivity to delay, then s_2 will approach zero, and $\alpha_j(s_2, t_2)$ will approach $\mathbb{E}X[0,t]/t$, the mean load produced by source j. The second constraint of (3.36) then becomes the simple constraint that the mean loads of all sources should not exceed the capacity of the resource.

With several priority classes the key point remains that each priority class may have its own characteristic space and time scale: under strict priority a source is unaffected by lower priority sources, but will be affected by the behaviour of higher priority sources on its characteristic space and time scale.

Kulkarni *et al.* (1995) study an alternative priority mechanism, where first-in-first-out scheduling is used and arriving work of low priority is rejected if the workload is above a threshold.

In Section 3 we have reviewed a variety of results, emphasizing their interpretation in terms of effective bandwidths. Of course other perspectives are possible. In particular, Shwartz and Weiss (1995) explore several more detailed aspects of buffer behaviour for on–off fluid sources of the type defined in Section 2.3.2, using this model to illustrate the considerable power of large deviation theory.

4   Tariffs and connection acceptance

The effective bandwidth of a source depends sensitively upon its statistical characteristics. The source, however, may have difficulty providing such information. Uncertain characterization of sources raises challenging practical and theoretical issues for the design of tariffing and connection acceptance control mechanisms. Suppose, for example, that mechanisms are based on attempts to measure the effective bandwidth of a connection, perhaps by estimating expression (2.1) using an empirical averaging to replace the expectation operator. Is this satisfactory? Suppose a user requests a connection policed by a high peak rate, but then happens to transmit very little traffic over the connection. Then an *a posteriori* estimate of quantity (2.1) will be near zero, even though an *a priori* expectation may be much larger, as assessed by either the user or the network. If tariffing and connection acceptance control are primarily concerned with expectations of *future* quality of service, and if sources may be non-ergodic over the relevant time scales, then the distinction matters.

In this section we describe an approach to tariffing and connection acceptance control mechanisms that can make effective and robust use of both prior declarations and empirical averages. The key idea is the use of prior declarations to choose a linear function that bounds the effective bandwidth (as illustrated in Fig. 5); tariffs and connection acceptance can then be based upon the relatively simple measurements needed to evaluate this function.

Although an *individual* source may be poorly characterized, certain features of the *aggregate* load on a resource may be known. In this section we assume that the key constraints (3.3), and the critical space and time scales appearing in these constraints, have been identified.

4.1   Charging mechanisms

Let

$$Z = \mathbb{E}e^{sX[\tau, \tau+t]}, \tag{4.1}$$

and rewrite expression (2.1) as

$$\alpha(Z) = \frac{1}{st} \log Z, \tag{4.2}$$

where the notation now emphasizes the dependence of the effective bandwidth on the summary Z of the statistical characteristics of the source.

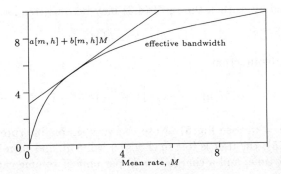

FIG. 5. Implicit pricing of an effective bandwidth. The effective bandwidth is shown as a function of the mean rate, M. The user is free to choose the declaration m, and is then charged an amount $a[m, h]$ per unit time, and an amount $b[m, h]$ per unit volume. The values of $a[m, h]$ and $b[m, h]$ are determined from the tangent at the point $M = m$.

Suppose that, before the call's admission, the network requires the user to announce a value z, and then charges for the call an amount $f(z; Z)$ per unit time, where Z is estimated by an empirical averaging. We suppose that the user is risk-neutral and attempts to select z so as to minimize the expected cost per unit time: call a minimizing choice of z, $\hat{z}$ say, an *optimal* declaration for the user. What properties would the network like the optimal declaration $\hat{z}$ to have? Well, first of all the network would like to be able to deduce from $\hat{z}$ the user's *a priori* expectation (4.1). A second desirable property would be that the expected cost per unit time under the optimal declaration $\hat{z}$ be proportional to the effective bandwidth (4.2) of the call. In Kelly (1994a) it is shown that these two requirements essentially characterize the tariff $f(z; Z)$ as

$$f(z; Z) = a(z) + b(z)Z, \qquad (4.3)$$

defined as the tangent to the curve $\alpha(Z)$ at the point $Z = z$.

4.1.1 *On–off sources*

Consider the very simple case of an on–off source which produces workload at a constant rate h while in an 'on' state, and produces no workload while in an 'off' state. Suppose the periods spent in 'on' and 'off' states are large, so that the effective bandwidth is given by expression (2.9) with $\alpha_1(s, t) = h$ and $p = M/h$. Here M and h are respectively the mean and peak of the source. If h is fixed and known (it may, for example, be policed) then

$$Z = 1 + \frac{M}{h} \left(e^{sth} - 1 \right). \qquad (4.4)$$

(Kelly 1994b provides a numerical illustration of the choice, discussed theoretically in Section 3.1, of the parameter st.) If we let z be defined by expression (4.4) with M replaced by m then the tariff (4.3) may be rewritten as

$$a[m, h] + b[m, h]M, \qquad (4.5)$$

the tangent to the function

$$\alpha[M, h] = \frac{1}{st} \log \left[1 + \frac{M}{h} \left(e^{sth} - 1 \right) \right]$$

at the point $M = m$ (see Fig. 5). Note the very simple interpretation possible for the tariff (4.5): the user is free to choose a value m, and then incurs a charge $a[m, h]$ per unit time, and a charge $b[m, h]$ per unit of volume carried.

4.1.2 Priorities

Next we consider an example where it may be important to tariff several constraints of the form (3.3) simultaneously. Consider the model of Section 3.7, where there are several priority classes. Let $Z_k, \alpha_k(Z_k), z_k, f_k, a_k, b_k$ be defined as in relations (4.1)–(4.3), but with (s, t) replaced by (s_k, t_k). Then a tariff for priority class j of

$$f^{(j)}\big((z_k)_k; (Z_k)_k\big) = \sum_{k \geq j} c_k \big[a_k(z_k) + b_k(z_k)Z_k \big]$$

has the required incentive properties, where c_k is a weight, or shadow price, attached to the kth constraint from the collection (3.36).

4.2 Connection acceptance control

We now describe how the coefficients defined in Section 4.1 can be used as the basis of a simple and effective connection acceptance control.

Suppose that a resource has accepted connections $1, 2, \ldots, I$, and write (a_i, b_i) for the coefficients $(a(z_i), b(z_i))$ chosen by the user responsible for connection i at the time that the connection was accepted. Suppose also that the resource measures the load $X_i[\tau, \tau + t]$ produced by connection i over a period of length t, and let $Y_i = \exp(sX_i[\tau, \tau + t])$. Define the *effective load* on the resource to be

$$\sum_{i=1}^{I} (a_i + b_i Y_i).$$

Then a connection acceptance control may be defined as follows. A new request for a connection should be accepted or rejected according as the most recently calculated effective load is below or above a threshold value, with the proviso that if a request is rejected then later requests are also rejected until an existing connection terminates.

4.2.1 On–off sources

Consider again the simple case of on–off sources described in Section 4.1.1. Let h_i be the fixed and known peak of connection i, write (a_i, b_i) for the coefficients $(a[m_i, h_i], b[m_i, h_i])$ chosen by the user, and let the measured load from connection i be $M_i = X_i[\tau, \tau + t]/t$. Then the effective load on the resource becomes

$$\sum_{i=1}^{I} (a_i + b_i M_i),$$

to be compared with a threshold value.

An advantage of the on–off model, both for tariffing and connection acceptance control, is that it bounds other more complex source models. The reader who is surprised that schemes using only simple load measurements can guarantee strict quality of service requirements should see Gibbens *et al.* (1995), where issues of robustness and performance are investigated in some detail.

Of course on–off sources may, on a finer time scale, have more detailed structure, as in example 2.3.4. This may give rise to additional constraints of the form (3.3). There are a range of responses possible, ranging in complexity and conservatism. The models of Sections 2.3.1 or 3.5 might be appropriate as a conservative bound when source i is policed by parameters (ρ_i, β_i), where $\beta_i/\rho_i = 1$; this approach is described for a single source type in Gibbens *et al.* (1995, Section 6). A less conservative approach would use the same space and time scales, around $(s, t) = (2\gamma/b, 1/2)$, to assess the aggregate fine time scale load (4.1). Work in progress concerns how such connection acceptance controls might be implemented.

Bibliography

1. Anick, D., Mitra, D. and Sondhi, M.M. (1982). Stochastic theory of a data-handling system with multiple sources. *Bell Syst. Tech. J.*, **61**, 1871–1894.

2. Asmussen, S. (1987). *Applied Probability and Queues*. Wiley, Chichester.

3. Bean, N. (1994). Effective bandwidths with different quality of service requirements. In *IFIP Transactions, Integrated Broadband Communication Networks and Services* (ed. V.B. Iverson). Elsevier, Amsterdam, 241–252.

4. Billingsley, P. (1986). *Probability and Measure* (2nd edn). Wiley, New York.

5. Botvich, D.D. and Duffield, N. (1995). Large deviations, the shape of the loss curve, and economies of scale in large multiplexers. *Queueing Syst.*, **20**, 293–320.

6. Bucklew, J.A. (1990). *Large Deviation Techniques in Decision, Simulation and Estimation*. Wiley, New York.

7. Chang, C.-S. (1994). Stability, queue length, and delay of deterministic and stochastic queueing networks. *IEEE Trans. Autom. Control*, **39**, 913–931.

8. Chang, C.-S. and Thomas, J.A. (1995). Effective bandwidth in high-speed digital networks. *IEEE J. Sel. Areas Commun.*, **13**, 1091–1100.

9. Choudhury, G., Lucantoni, D. and Whitt, W. (1994). On the effectiveness of effective bandwidths for admission control in ATM networks. In Labetoulle and Roberts (1994), 411–420.

10. Courcoubetis, C., Kesidis, G., Ridder, A., Walrand, J. and Weber, R.

(1995). Admission control and routing in ATM networks using inferences from measured buffer occupancy. *IEEE Trans. Commun.*, **43**, 1778-1784.

11. Courcoubetis, C. and Weber, R. (1995). Effective bandwidths for stationary sources. *Probab. Eng. Inf. Sci.*, **9**, 285–296.

12. Courcoubetis, C. and Weber, R. (1996). Buffer overflow asymptotics for a switch handling many traffic sources. *J. Appl. Probab.*, **33**.

13. Cruz, R.L. (1991). A calculus for network delay. *IEEE Trans. Inf. Theor.*, **37**, 114–141.

14. Császár, A. (1978). *General Topology*. Adam Hilger, Bristol.

15. de Veciana, G., Courcoubetis, C. and Walrand, J. (1994). Decoupling bandwidths for networks: a decomposition approach to resource management for networks. In *Proc. IEEE INFOCOM*, Vol. 2, 466–474.

16. de Veciana, G. and Walrand, J. (1995). Effective bandwidths: call admission, traffic policing and filtering for ATM networks. *Queueing Syst.*, **20**, 37–59.

17. Duffield, N.G., Lewis, J.T., O'Connell, N., Russell, R. and Toomey, F. (1994). Predicting quality of service for traffic with long-range fluctuations. Preprint.

18. Duffield, N.G., Lewis, J.T., O'Connell, N., Russell, R. and Toomey, F. (1995). Entropy of ATM traffic streams: a tool for estimating QoS parameters. *IEEE J. Sel. Areas Commun.*, **13**, 981–990.

19. Duffield, N.G. and O'Connell, N. (1996). Large deviations and overflow probabilities for the general single-server queue, with applications. *Math. Proc. Cambridge Philos. Soc.*

20. Elwalid, A.I. and Mitra, D. (1993). Effective bandwidth of general Markovian traffic sources and admission control of high speed networks. *IEEE/ACM Trans. Networking*, **1**, 329–343.

21. Elwalid, A.I. and Mitra, D. (1995). Analysis, approximations and admission control of a multi-service multiplexing system with priorities. In *Proc. IEEE INFOCOM*, 463–472.

22. Feller, W. (1971). *An Introduction to Probability Theory and Its Applications, Volume II* (2nd edn). Wiley, New York.

23. Gibbens, R.J. (1996). Traffic characterisation and effective bandwidths for broadband network traces. This volume.

24. Gibbens, R.J. and Hunt, P.J. (1991). Effective bandwidths for the multitype UAS channel. *Queueing Syst.*, **9**, 17–28.

25. Gibbens, R.J., Kelly, F.P. and Key, P.B. (1995). A decision-theoretic approach to call admission control in ATM networks. *IEEE J. Sel. Areas Commun.*, **13**, 1101–1114.

26. Glynn, P.W. and Whitt, W. (1994). Logarithmic asymptotics for steady-state tail probabilities in a single-server queue. *J. Appl. Probab.*, **31A**, 131–156.

27. Guérin, R., Ahmadi, H. and Naghshineh, M. (1991). Equivalent capacity and its application to bandwidth allocation in high-speed networks. *IEEE J. Sel. Areas Commun.*, **9**, 968–981.

28. Hajek, B. (1994). A queue with periodic arrivals and constant service rate. In Kelly (1994c), 147–157.

29. Harrison, J.M. (1985). *Brownian Motion and Stochastic Flow Systems*. Wiley, New York.

30. Hsu, I. and Walrand, J. (1995). Admission control for ATM networks. In Kelly and Williams (1995), 413–429.

31. Hui, J.Y. (1988). Resource allocation for broadband networks. *IEEE J. Sel. Areas Commun.*, **6**, 1598–1608.

32. Hui, J.Y. (1990). *Switching and Traffic Theory for Integrated Broadband Networks*. Kluwer, Boston.

33. ITU Recommendation I371 (1994). Traffic control and congestion control in B–ISDN. Geneva.

34. Jensen, J.L. (1995). *Saddlepoint Approximations*. Oxford University Press.

35. Kella, O. and Whitt, W. (1992). A tandem fluid network with Lévy input. In *Queueing and Related Models* (ed. U.N. Bhat and I.V. Basawa). Oxford University Press, 112–128.

36. Kelly, F.P. (1991). Effective bandwidths at multi-class queues. *Queueing Syst.*, **9**, 5–16.

37. Kelly, F.P. (1994a). On tariffs, policing and admission control of multiservice networks. *Oper. Res. Lett.*, **15**, 1–9.

38. Kelly, F.P. (1994b). Tariffs and effective bandwidths in multiservice networks. In Labetoulle and Roberts (1994), 401–410.

39. Kelly, F.P. (ed.) (1994c). *Probability, Statistics and Optimisation: a Tribute to Peter Whittle*. Wiley, Chichester.

40. Kelly, F.P. and Williams, R.J. (eds) (1995). *Stochastic Networks*. Springer Verlag, New York.

41. Kesidis, G., Walrand, J. and Chang, C.-S. (1993). Effective bandwidths for multiclass Markov fluids and other ATM sources. *IEEE/ACM Trans. Networking*, **1**, 424-428.

42. Kingman, J.F.C. (1993). *Poisson Processes*. Clarendon Press, Oxford.

43. Kulkarni, V.G., Gün, L. and Chimento, P.F. (1995). Effective bandwidth vectors for multiclass traffic multiplexed in a partitioned buffer. *IEEE J. Sel. Areas Commun.*, **13**, 1039–1047.

44. Labetoulle, J. and Roberts, J.W. (eds) (1994). *The Fundamental Role of Teletraffic in the Evolution of Telecommunication Networks*. Elsevier, Amsterdam.

45. Norros, I. (1994). A storage model with self-similar input. *Queueing Syst.*, **16**, 387–396.

46. Reid, N. (1988). Saddlepoint methods and statistical inference. *Stat. Sci.*, **3**, 213–238.

47. Roberts, J.W. (ed.) (1992). *Performance Evaluation and Design of Multiservice Networks.* Office for Official Publications of the European Communities, Luxembourg.

48. Shwartz, A. and Weiss, A. (1995). *Large Deviations for Performance Analysis: Queues, Communication and Computing.* Chapman and Hall, London.

49. Simonian, A. and Guibert, J. (1995). Large deviations approximation for fluid sources fed by a large number of on/off sources. *IEEE J. Sel. Areas Commun.*, **13**, 1017–1027.

50. Sriram, K. and Whitt, W. (1986). Characterizing superposition arrival processes in packet multiplexers for voice and data. *IEEE J. Sel. Areas Commun.*, **4**, 833–846.

51. Tse, D.N.C., Gallager, R.E. and Tsitsiklis, J.N. (1995). Statistical multiplexing of multiple time-scale Markov streams. *IEEE J. Sel. Areas in Commun.*, **13**, 1028–1038.

52. Weber, R. (1994). Large deviation and fluid approximations in control of stochastic systems. In Kelly (1994c), 159–171.

53. Weiss, A. (1986). A new technique for analyzing large traffic systems. *Adv. Appl. Probab.*, **18**, 506–532.

54. Whitt, W. (1993). Tail probabilities with statistical multiplexing and effective bandwidths in multi-class queues. *Telecommun. Syst.*, **2**, 71–107.

55. Willinger, W. (1995). Traffic modelling for high-speed networks: theory versus practice. In Kelly and Williams (1995), 395–409.

56. Willinger, W., Taqqu, M.S., Leland, W.E. and Wilson, D.V. (1995). Self-similarity in high-speed packet traffic: analysis and modelling of ethernet traffic measurements. *Stat. Sci.*, **10**, 67–85.

9

Traffic characterisation and effective bandwidths for broadband network traces

R. J. Gibbens

University of Cambridge

Abstract

The trend of current developments in telecommunication networks is towards systems which will allow a number of widely disparate traffic streams to share common resources through statistical sharing. How efficiently this can be managed depends critically upon the statistical characteristics of the traffic. Our aim in this paper is to investigate the application of a simple and robust graphical descriptor of a given traffic stream, designed to emphasise those aspects of the data that are important for the statistical sharing of common resources. Two examples considered in detail are an MPEG-1 video encoding of the *Star Wars* movie and traces of ethernet traffic taken at Bellcore.

1 Introduction

There has been considerable progress in recent years in the development of queueing models which predict the performance of resources dealing with random traffic, and these models have given important insights into just which characteristics of traffic are most important. Traffic characterisation has an important role to play in the quest to understand how to design and control networks able to cope with the combination of bursty multi-media traffic and guaranteed quality of service. The recent collection of articles in Gallager *et al.* (1995) gives a comprehensive overview and introduction of this broad field.

Kelly (1995, 1996) has considered the notion of *effective bandwidth* in the context of stochastic models for the statistical sharing of resources. The effective bandwidth is given by the statistical descriptor

$$\alpha(s,t) = \frac{1}{st} \log \mathbb{E}\left[e^{sX[\tau, \tau + t]}\right] \tag{1.1}$$

where s is the *space scale* and t is the *time scale*. For example, s might be measured in units of bytes^{-1} or cells^{-1} and t might be measured in seconds. Here, $X[\tau, \tau + t]$ is the workload arriving at a resource in the period $[\tau, \tau + t]$ of

The author wishes to acknowledge the support of the Royal Society through a University Research Fellowship and of the EPSRC (Grant GR/J31896) for computing equipment.

length t and the expectation is taken over the distribution of such random time periods.

The appropriate time and space scales are determined not by the traffic source alone but rather together with the characteristics of the resource such as the capacity or buffer lengths and the required quality of service. Kelly (1996) derives the form of $\alpha(s,t)$ for several of the most common stochastic models of traffic sources, including Bernoulli bufferless models, periodic, fluid and fractional Brownian motion input sources. Furthermore, models for the statistical sharing of such sources at resources lead to admissible regions given explicitly in terms of $\alpha(s,t)$ for a determined choice of the space and time scales, s and t, respectively.

This paper does not develop these models further but instead investigates the form of the effective bandwidth surface, $\alpha(s,t)$, estimated from data on real traffic sources. Data on broadband traffic is now much more easily measured and readily available (see, for example, the Internet Traffic Archive). Two traces which have received considerable attention in the literature are the measurements taken from ethernet segments at Bellcore (see Willinger (1995a, 1995b, 1995c)) and the various video encodings produced of the *Star Wars* movie (see Garrett and Willinger (1994) and Rose (1995)) in motion JPEG and MPEG-1 formats.

In a trace of real traffic measurements each packet is assumed to have a corresponding record giving the packet's time of arrival and size. We shall represent this information by the collection $\mathcal{X} = \{(t_i, x_i); i = 1, \ldots, N\}$ for a trace consisting of N packets with arrival times and sizes given by t_i and x_i respectively. The two following important special cases often arise in practice.

(1) The packets are all of a fixed size. For example, $x_i = 53$ bytes would correspond to the size of an ATM cell.

(2) The interarrival times are constant. Suppose that $t_{i+1} - t_i = \delta$ for each $i = 1, \ldots, N-1$ representing, for example, the packets encoding each successive video frame of a movie produced at a constant frequency.

Given the trace, $\mathcal{X}$, we have

$$X[\tau, \tau + t] = \sum_{i=1}^{N} x_i I(\tau \leq t_i \leq \tau + t), \qquad 0 \leq \tau \leq t_N - t \qquad (1.2)$$

as the amount of data arriving during the interval $[\tau, \tau + t]$. Hence, the effective bandwidth is given by

$$\alpha(s,t) = \frac{1}{st} \log \frac{1}{t_N - t} \int_0^{t_N - t} e^{sX[\tau, \tau + t]} \, d\tau. \qquad (1.3)$$

The integrand in this expression is a piece-wise constant function with jumps at the time epochs when a packet either enters or leaves the recording interval $[\tau, \tau + t]$. Accordingly, we can relatively easily compute this integral expression without truncation error, leaving just round-off errors due to finite precision arithmetic.

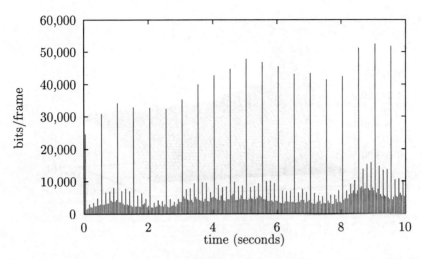

FIG. 1. *Star Wars* movie dataset showing periodic structure arising from the GOP structure within MPEG-1 encoded video streams. Only the first 10 seconds are shown and the large frames every 0.5 seconds are the I-frames.

2   *Star Wars* **movie dataset**

In this section we apply the calculation of the effective bandwidth surface, $\alpha(s,t)$, to an MPEG-1-encoded trace of the *Star Wars* movie. In the MPEG-1 encoding for this trace the individual packets correspond to the data in video frames produced at a constant frequency of 24 frames per second. The trace records the number of bits per frame in a sequence of 40,000 consecutive frames. Figure 1 shows the bits per frame for the first 10 seconds (240 frames) of the trace. A strong periodic component can be seen in the trace which arises from the MPEG-1 encoding format. In this format groups of 12 frames are encoded into a fixed pattern (known as the *Group of Pictures* (GOP)) of I-frames, P-frames and B-frames according to varying amounts of motion compensation between frames. For this trace, I-frames carry the most data and occur every 12 frames (0.5 seconds) as can easily be seen in the figure. For further discussion of the MPEG-1 format and its statistical properties see Rose (1995).

Figure 2 shows the effective bandwidth surface, $\alpha(s,t)$, computed according to expression (1.3) using the entire trace. We can see a periodic ripple effect in the surface at constant values of t. This is not unexpected given the periodic nature of the MPEG-1 encoding and the results in Kelly (1996) for a strict periodic source model. For illustration, suppose that a source instantaneously produces b units of data at the random times given by $\{Ud + id, \ i = 0, 1, \ldots\}$ where the random variable U is uniformly distributed on $[0, 1]$ (as shown in Fig. 3). Then Kelly (1996) gives the closed form expression for the effective bandwidth as

$$\alpha_P(s,t) = \frac{b}{t} \left\lfloor \frac{t}{d} \right\rfloor + \frac{1}{st} \log \left(1 + \left(\frac{t}{d} - \left\lfloor \frac{t}{d} \right\rfloor \right) \left(e^{bs} - 1 \right) \right). \qquad (2.1)$$

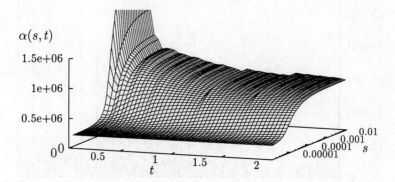

FIG. 2. This figure shows the effective bandwidth surface $\alpha(s,t)$ computed from the *Star Wars* MPEG-1 video trace. The space scale s is measured in bits^{-1} (using a logarithmic (base 10) axis), the time scale is measured in seconds and $\alpha(s,t)$ is measured in bits per second. Notice the ripple effect and the high cliff-like region for small t. As $s \to 0$, the effective bandwidth approaches the mean rate whereas when s increases the effective bandwidth approaches the peak rate measured over intervals of time t.

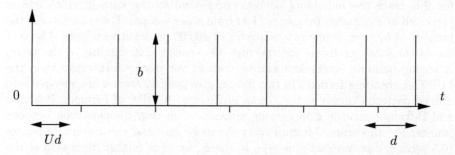

FIG. 3. This figure shows the periodic source model with parameters d (seconds), b (bits) and where $U \sim [0,1]$.

Thus, the superposition of a periodic source together with a *constant* bit-rate source of rate λ is

$$\alpha(s,t) = \lambda + \alpha_P(s,t). \tag{2.2}$$

Figure 4 shows this surface for the parameters $\lambda = 143,516$ bits per second, $b = 40,000$ bits and $d = 0.5$ seconds (this has the same mean rate as the trace). A comparison with Fig. 2 shows a close fit between the two surfaces when s is small. The ripple effects are clearly seen in the surface and are caused by the decreased uncertainty about the amount of data in a time interval whose length, t, is nearly

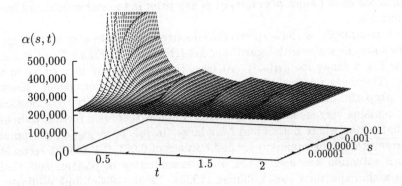

FIG. 4. This figure shows the effective bandwidth computed for the superposition of a constant bit-rate source and a periodic source with parameters $d = 0.5$ seconds, $b = 40,000$ bits and with a constant bit-rate component of $\lambda = 143,516$ bits per second. The space scale s is measured in bits^{-1}
(using a logarithmic (base 10) axis), the time scale is measured in seconds
and $\alpha(s,t)$ is measured in bits per second.

a multiple of d. Similarly, there is a great deal of uncertainty (or burstiness)
about the quantity of data arriving during particularly short intervals of time
and the effective bandwidth surfaces reflect this by rising rapidly to form a cliff
as t nears zero.

This choice of parameters attempts to model the periodic nature of the I-
frames (which occur every 0.5 seconds) but ignores the variation in the size of
the individual I-frames. Such variation becomes increasingly important when
the relevant space scale s increases and the performance becomes more highly
dependent on the peak bit rate of the source over intervals of time t. This
additional variation accounts for the level shift in the surface as s increases.

3 Bellcore ethernet dataset

A second example is presented by the traces of ethernet packets taken on ethernet
LANs at Bellcore in 1989. We consider a single trace comprising one million
packets measured on an internal ethernet segment in August 1989. In this case
we have packets of varying sizes x_i and of irregularly varying arrival times t_i
with $i = 1, \ldots, N$ and $N = 10^6$.

The calculation of $\alpha(s,t)$ for this trace over a grid of 50 by 50 points was
found to take approximately 100 minutes to compute on a current high performance workstation. For exploratory work, we have considered an approximate
numerical algorithm based on an *ad hoc* sampling approach presented in Section 5. This approach has the advantage that we can relatively quickly examine

the basic features of the surface and by varying the degree of sampling used trade off truncation error (which manifests itself as a jitter in the surface) against speed of calculation.

For this example, we have shown the effective bandwidths over a wider range of space and time scales with logarithmic axes (to base 10). Figure 5 shows $\alpha(s,t)$ whereas Fig. 6 shows $\log_{10} \alpha(s,t)$. At least three regions are discernible in the surface. There is a relatively flat region where s and t are both small. There is a very steep cliff-like region in common with the MPEG-1 surface of Section 2 when t remains very small but s increases. Finally, there is a further comparatively flat region where both s and t are large. In this latter region we can also observe several minor ripples at around t values of 0.001, 0.1 and 0.5 seconds.

Much attention has been given to an examination of possible self-similar aspects within this data (see Willinger (1995a, 1995b, 1995c) and Willinger *et al.* (1996) for a bibliographic survey of work in this area) such as would be present if the traffic followed a fractional Brownian motion model described as follows (see also Norros (1994)). Suppose now we model the data arriving in a time interval of length t by

$$X[\tau, \tau + t] = \lambda t + Z(t) \tag{3.1}$$

where $Z(t)$ is fractional Brownian motion with $\mathrm{Var}[Z(t)] = \sigma^2 t^{2H}$ for $H \in (0,1)$, the Hurst parameter, so that then (Kelly (1996))

$$\alpha(s,t) = \lambda + \frac{\sigma^2 t^{2H-1}}{2} s. \tag{3.2}$$

Figure 7 shows the effective bandwidth surface, $\alpha(s,t)$, of a fractional Brownian motion model with the parameters $\lambda = 138{,}185$ bytes per second, $\sigma = 89{,}668$ bytes per second and $H = 0.81$ obtained by fitting a crude linear regression to the log variance–time plot of the data. Figure 8 shows $\log_{10} \alpha(s,t)$ instead. These surfaces appear to follow those computed from the trace in the first flat region and in the transition to higher values of the effective bandwidth as s and t both increase. However, the surface for the fractional Brownian motion model carries on increasing and does not flatten off in the manner shown by the surface for the trace. There are several potential explanations for why the effective bandwidth for the trace levels off. The effect is most naturally explained in this case by the finite number of packets in the trace but a second explanation, which would occur however long the trace, is if the traffic was shaped by inherent limits imposed by the LAN due to, for example, hardware constraints at the various bottleneck components of the network or host interfaces attached to the network. Finally, a further significant difference between the trace and the fractional Brownian motion model is that there is definitely no cliff-like region for the model surface with small t and increasing s.

4   Conclusions

In this short paper we have investigated the effective bandwidth surface, $\alpha(s,t)$, computed for two traces of real broadband traffic. A direct computational

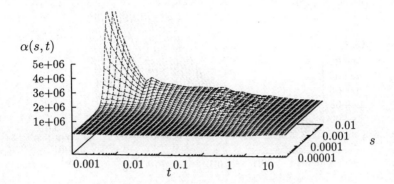

FIG. 5. This figure shows the effective bandwidth surface, $\alpha(s, t)$, for the Bellcore ethernet trace. The space and time scales are logarithmic (to base 10) with units of s in bytes^{-1}, t in seconds and $\alpha(s, t)$ in bytes per second.

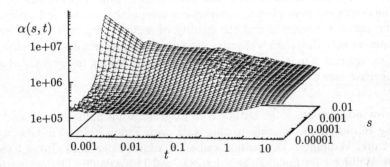

FIG. 6. This figure shows the effective bandwidth surface, $\alpha(s, t)$, for the Bellcore ethernet trace. Each axis has a logarithmic scale (to base 10) with units of s in bytes^{-1}, t in seconds and $\alpha(s, t)$ in bytes per second.

method is available. For exploratory work, a sampling-based approach is also considered.

The effective bandwidth surface provides both a graphical and a quantitative summary of the performance implications of real broadband traffic sources in resources with statistical sharing.

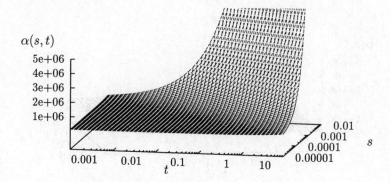

FIG. 7. This figure shows the effective bandwidth surface, $\alpha(s,t)$, for the fractional Brownian motion model with parameter values of $\lambda = 138{,}185$ bytes per second, $\sigma = 89{,}668$ bytes per second and Hurst parameter $H = 0.81$. The space and time scales are logarithmic (to base 10) with units of s in bytes^{-1}, t in seconds and $\alpha(s,t)$ in bytes per second.

Effective bandwidths, $\alpha(s,t)$, have many important applications to issues such as statistical multiplexing, connection acceptance control and tariffs. The specific network resources and the quality of service guarantees considered in conjunction with the traffic's effective bandwidth surface will determine the space and time regions of most critical importance in establishing the overall behaviour and performance of the system.

Acknowledgements The author is most grateful for many helpful and stimulating conversations relating to this work with Frank Kelly, Yih-Choung Teh and Walter Willinger. The author wishes to acknowledge both Oliver Rose and Walter Willinger for making their datasets widely available. The author is also grateful for the helpful suggestions of the referee.

5 Appendix

Consider forming the quantity

$$a = \sum_{q=1}^{Q} e^{sX[\tau_q, \tau_q + t]} \tag{5.1}$$

where $(\tau_q, q = 1, 2, \ldots, Q)$ forms a realisation of a Poisson process on $[0, t_N - t]$ of rate λ. Then Q is a Poisson random variable of mean $(t_N - t)\lambda$, and by varying λ we may vary the expected sample size. An approximately unbiased estimate of $\alpha(s,t)$ is

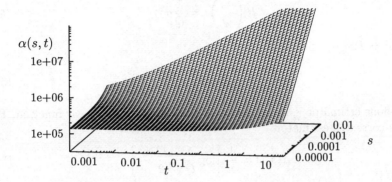

FIG. 8. This figure shows the effective bandwidth surface, $\alpha(s,t)$, for the fractional Brownian motion model with parameter values of $\lambda = 138{,}185$ bytes per second, $\sigma = 89{,}668$ bytes per second and Hurst parameter $H = 0.81$. Each axis has a logarithmic scale (to base 10) with units of s in bytes^{-1}, t in seconds and $\alpha(s,t)$ in bytes per second.

$$\widehat{\alpha}(s,t) = \frac{1}{st}\left[\log\frac{a}{Q} + \frac{Q\sigma^2}{2a^2}\right], \tag{5.2}$$

where σ is defined below, and this estimate has an approximate standard error $\sigma\sqrt{Q}/ast$.

Divide the sample of Q time periods into $r = 1, 2, \ldots, R$ sub-samples with each sub-sample being selected with equal probability. Calculate

$$a_r = \sum_{q=1}^{Q_r} e^{sX[\tau_q,\tau_q+t]} \tag{5.3}$$

for $r = 1, 2, \ldots, R$, where $Q = \sum_r Q_r$. Let $a = \sum_r a_r$. What is the sampling variability of estimate a? Let

$$\sigma_a^2 = \frac{1}{R-1}\sum_{r=1}^{R}\left(a_r - \frac{a}{R}\right)^2. \tag{5.4}$$

Then a has approximate variance $R\sigma_a^2$ and

$$\frac{1}{Q}\sum_{q=1}^{Q} e^{sX[\tau_q,\tau_q+t]} = \frac{1}{Q}\sum_{r=1}^{R} a_r = \frac{a}{Q} \tag{5.5}$$

has mean

$$\mathbb{E}[e^{sX[\tau,\tau+t]}] \tag{5.6}$$

and approximate variance

$$R\left(\frac{\sigma_a}{Q}\right)^2 = \frac{1}{Q}\sigma^2 \tag{5.7}$$

where $\sigma^2 = R\sigma_a^2/Q$. Let

$$\alpha(s,t) = \frac{1}{st}\log \mathbb{E}\left[e^{zX[\tau,\tau+t]}\right]. \tag{5.8}$$

The obvious estimator $\frac{1}{st}\log\frac{a}{Q}$ is biased, since $\log$ is a non-linear function. But the bias-corrected estimator

$$\widehat{\alpha}(s,t) = \frac{1}{st}\left[\log\frac{a}{Q} + \frac{Q}{2}\frac{\sigma^2}{a^2}\right] \tag{5.9}$$

has approximate expectation $\alpha(s,t)$ and approximate variance $\left(\frac{1}{st}\right)^2\frac{Q\sigma^2}{a^2}$.

In order to see this consider the following. Let ε have mean zero and (small) variance η^2. Then

$$\mathbb{E}[f(x+\varepsilon)] = \mathbb{E}\left[f(x) + \varepsilon f'(x) + \frac{\varepsilon^2}{2}f''(x) + \ldots\right] \approx f(x) + \frac{f''(x)}{2}\eta^2 \tag{5.10}$$

and the variance of $f(x+\varepsilon)$ is approximately $(f'(x))^2\eta^2$ (see, for example, Cox and Hinkley (1974)). But

$$\frac{1}{st}\log\left(\frac{a}{Q}\right) = \frac{1}{st}\left[\log\left(\mathbb{E}\left(\frac{a}{Q}\right)\right) + \log\left(1 - \frac{\mathbb{E}\left(\frac{a}{Q}\right) - \frac{a}{Q}}{\mathbb{E}\left(\frac{a}{Q}\right)}\right)\right] \tag{5.11}$$

$$= \alpha(s,t) + \frac{1}{st}\log(1-Y) \tag{5.12}$$

where

$$Y = \frac{\mathbb{E}\left(\frac{a}{Q}\right) - \frac{a}{Q}}{\mathbb{E}\left(\frac{a}{Q}\right)}. \tag{5.13}$$

Finally note that

$$\mathbb{E}(Y) = 0, \qquad \mathrm{Var}(Y) = \frac{\sigma^2/Q}{a^2/Q^2} = \frac{Q\sigma^2}{a^2}. \tag{5.14}$$

Bibliography

1. Cox, D. R. and Hinkley, D. V. (1974). *Theoretical Statistics*. Chapman and Hall, London.
2. Gallager, R. G., Hajek, B., Kelly, F. P., Mitra, D. and Varaiya, P. P. (1995). *IEEE Journal on Selected Areas in Communications*, special issue on the Advances in the Fundamentals of Networking, part I: Bridging fundamental theory and networking. **13**, No 6.

3. Garrett, M. W. and Willinger, W. (1994). Analysis, modelling and generation of self-similar VBR video traffic. *Proceedings of the ACM Sigcomm '94*, 269–280.

4. Internet Traffic Archive, World Wide Web resource available at `http://town.hall.org/Archives/pub/ITA/`.

5. Kelly, F. P. (1995). Modelling communication networks, present and future. *The Clifford Patterson Lecture*. Royal Society public lecture.

6. Kelly, F. P. (1996). Notes on effective bandwidths. *This volume*.

7. Norros, I. (1994). A storage model with self-similar input. *Queueing Systems*, **16**, 387–396.

8. Rose, O. (1995). Statistical properties of MPEG video traffic and their impact on traffic modeling in ATM systems. University of Wuerzburg. Institute of Computer Science Research Report Series. Report No 101.

9. Willinger, W. (1995a). Traffic modelling for high-speed networks: theory and practice. In *Stochastic Networks* (eds Kelly, F. P. and Williams, R. J.). Springer-Verlag, New York, 395–409.

10. Willinger, W. (1995b). Self-similarity in high-speed packet traffic: analysis and modelling of ethernet traffic measurements. *Statistical Science*, **10**, No 1, 67–85.

11. Willinger, W. (1995c). Self-similarity through high-variability: statistical analysis of ethernet LAN traffic at the source level. *Proceedings of the ACM Sigcomm '95*, 100–113.

12. Willinger, W., Taqqu, M.S. and Erramilli, A. (1996). A bibliographical guide to self-similar traffic and performance modeling for modern high-speed networks. *This volume*.

10

Nonparametric estimation for quantities of interest in queues

Susan M. Pitts

University of Cambridge

Abstract

Various techniques from nonparametric statistics may be applied to stochastic models. The theoretical results are summarized here, and their application is illustrated with reference to the $GI/GI/1$ queue.

1 Introduction

Suppose that a particular queueing situation is modelled by an $M(\lambda)/M(\mu)/1$ queue. A common procedure in practice is to derive estimators for the parameters λ and μ from observations of the arrival and service times. Depending on the purpose of the investigation, a possible next step is statistical inference for such quantities as the traffic intensity or the mean stationary waiting time.

Suppose that we move to a nonparametric context, where the exponential assumption on the interarrival and service times is relaxed, leading to a consideration of the $GI/GI/1$ queue. Now we construct nonparametric estimators of the interarrival and service time distribution functions and proceed to (nonparametric) estimation of various other quantities of interest for this queueing model.

Our aim is to show how established techniques from nonparametric statistics may be applied in situations like this. These are general techniques, which can be (and have been) applied to various other stochastic models. Here we illustrate the methods as applied to the $GI/GI/1$ queue. First we give a precise description of the model, and introduce notation.

Suppose that customers arrive in a renewal process to a single server FIFO queue, and that successive service times are independent identically distributed positive random variables $S_1, S_2, \ldots$ defined on some probability space $(\Omega, \mathcal{A}, P)$. Let $T_1, T_2, \ldots$ denote the interarrival times. Assume that they are independent and identically distributed positive random variables, also defined on $(\Omega, \mathcal{A}, P)$, independent of the service times. Suppose that the first customer arrives at time $t = 0$ to an empty queue. Let F_S and F_T be the distribution functions of S_1 and T_1 respectively, and for convenience, assume further that F_S and F_T

are continuous. We also assume throughout that $E(S_1) < \infty$, $E(T_1) < \infty$ and $E(S_1)/E(T_1) < 1$. This means that the queue is stable and that successive customer waiting times (i.e. the times from the moment of a particular customer's arrival until the moment when service for that customer begins) converge in distribution to a proper random variable W called the stationary waiting time. It is a standard result that the distribution of W is the same as that of the maximum of a random walk with independent identically distributed steps $X_1, X_2, \ldots$, where X_1 has the same distribution as $S_1 - T_1$. For results about the $GI/GI/1$ queue, see Asmussen (1987), Chapter VIII. Write F and F_W for the distribution functions of X_1 and W respectively. We assume that $F(0) < 1$.

Another standard result states that if there exists $R > 0$ such that $M(R) = 1$, where $M(r)$ is the moment generating function $E(e^{rX_1})$ of X_1, then

$$P(W > u) \sim ae^{-Ru} \text{ as } u \to \infty,$$

where a is a positive constant (depending on F), see Grandell (1991). Notice that the condition $M(R) = 1$ is essentially a requirement that S_1 is not heavy tailed.

We can think of a quantity of interest like R or F_W as the output of a functional applied to F (or to (F_S, F_T)), so that for example $R = \Xi(F)$ where Ξ is a map from an appropriate function space, which will be defined precisely in Section 2, to $\mathbf{R}$. Typically there is no easy explicit expression for the output except in certain special cases, for instance if S_1 and T_1 are both exponentially distributed.

The statistical problem under consideration here is that of estimating an output quantity when data are available on the input quantity. Thus, given a random sample $\{X_1, \ldots, X_n\}$ from F, a particular estimator of the stationary waiting time tail decay parameter R will be proposed, and statistical properties of this estimator will be established. An obvious way to do this is to use the data to construct an estimator $\tilde{F}$ of F and then to estimate $R = \Xi(F)$ by the plug-in estimator $\Xi(\tilde{F})$. A nonparametric estimator of F is the *empirical distribution function* $\hat{F}_n$ based on $\{X_1, \ldots, X_n\}$, defined by

$$\hat{F}_n(x, \omega) = \frac{1}{n} \sum_{i=1}^{n} 1_{[X_i(\omega), \infty)}(x) \text{ for } x \in \mathbf{R} \text{ and } \omega \in \Omega,$$

where 1_A is the indicator function of the set A. With $\tilde{F} = \hat{F}_n$, we obtain a nonparametric estimator $\hat{R}_n = \Xi(\hat{F}_n)$ of R.

The above estimator can be placed within a framework often encountered in statistics where we consider a functional Φ and estimate $\Phi(F)$ by $\Phi(\hat{F}_n)$. Suppose that Φ takes elements of $(B_1, \| \cdot \|_1)$ to $(B_2, \| \cdot \|_2)$, where $(B_i, \| \cdot \|_i)$ is a normed space, $i = 1, 2$. Roughly, proofs about properties of $\Phi(\hat{F}_n)$ as an estimator of $\Phi(F)$ can then be broken down into two parts: establishing the appropriate statistical property for $\hat{F}_n$ as an estimator of F, and then establishing an appropriate property for Φ that will ensure that the statistical property carries over from the input to the output.

As an easy example, it is clear that

$$\|\hat{F}_n - F\|_1 \to 0 \text{ as } n \to \infty \text{ almost surely } \text{ and } \Phi \text{ continuous}$$
$$\Rightarrow \quad \|\Phi(\hat{F}_n) - \Phi(F)\|_2 \to 0 \text{ as } n \to \infty \text{ almost surely.} \tag{1.1}$$

When B_1 is the space of right-continuous real-valued functions with left-hand limits on $[-\infty, \infty]$ and $\| \cdot \|_1$ is the supremum norm, then the strong consistency result for $\hat{F}_n$ is the Glivenko–Cantelli Theorem (see, for example, Chow and Teicher (1988), 8.2). (1.1) is a general statement about plug-in estimators. In order to apply it to a particular case, for example to $\hat{R}_n$, we must choose the spaces $(B_i, \| \cdot \|_i)$ so that $\hat{F}_n$ is strongly consistent for F in B_1 and so that Ξ is $\| \cdot \|_1$-$\| \cdot \|_2$ continuous. There is a tension between making $\| \cdot \|_1$ sufficiently strong so that convergence of $\hat{F}_n$ in B_1 implies convergence of $\Phi(\hat{F}_n)$ in B_2, and making $\| \cdot \|_1$ such that a Glivenko–Cantelli-like result holds in B_1. In addition to the above two requirements, a further aesthetic aim is to choose spaces such that the conditions that a distribution function must satisfy in order to be in $(B_1, \| \cdot \|_1)$ are related as far as possible to natural probabilistic conditions such as existence of certain moments. Although consistency is not discussed further here, similar considerations about the choice of spaces apply to asymptotic normality described in Section 3.

In the remaining sections we describe a particular setting and choice of B_i in which application of results from nonparametric statistics to various quantities of interest in stochastic models has proved successful. The framework described is not the most sophisticated or general, but is one which is easily accessible and rich enough for many applications. In the remaining sections we illustrate the general methods with reference to nonparametric estimators of R and of F_W, and refer to these as Examples I and II respectively. In Section 2 we describe the spaces forming the domains and codomains for the relevant functionals, and in Section 3 we illustrate how asymptotic normality may be proved for such plug-in estimators and demonstrate these procedures for Examples I and II. In Section 4 we discuss bootstrap confidence intervals for the unknown quantities.

The results for the two examples are not new and appear in Pitts *et al.* (1996) and Pitts (1994a), and the reader is referred to these papers for technical details. In the first paper, the above methods are applied to the adjustment coefficient in the Sparre Andersen model of risk theory. Results in this risk model can be translated into results about the $GI/GI/1$ queue. The correspondence between risk theory and queueing theory means that later references include many in the insurance literature, see for example Hipp (1994). The problem of estimating R in Example I arises in the estimation of quality of service parameters for ATM traffic. Duffield *et al.* (1995) adopt a similar empirical approach to ours, for stationary input streams, rather than the renewal input streams of the $GI/GI/1$ queue.

It should be noted that alternative approaches to estimation of such quantities of interest include parametric estimation, see, for example, Basawa and Prabhu (1988). Another approach is to observe the output quantities themselves and to construct estimators from these observations (Heyde 1988).

2 Spaces

The input quantities in our examples are distribution functions, and a natural space for these is D_∞, the set of all right-continuous, real-valued functions on $[-\infty, \infty]$ with left-hand limits and left continuous at infinity. For f in D_∞, let $\|f\|_\infty = \sup_t |f(t)|$, so that $(D_\infty, \|\cdot\|_\infty)$ is a (nonseparable) Banach space. A real-valued right-continuous function f on $\mathbf{R}$ with left-hand limits, and with finite limits at $\pm\infty$, may be extended in an obvious way to an element of D_∞.

It turns out that weighted versions of D_∞ are appropriate spaces for our purposes. In general let $\phi : \mathbf{R} \to \mathbf{R}^+$ be a continuous function, and let $D(\phi)$ be the set of all f such that ϕf is in D_∞, and for f in $D(\phi)$ put $\|f\|_{D(\phi)} = \|\phi f\|_\infty$. Write $C(\phi)$ for the subset of $D(\phi)$ consisting of continuous functions.

Example Consider Example I. For $r > 0$, let $\phi_r(x) = \exp((rx)^+)$, where $y^+ = \max\{y, 0\}$, and write D_r for $D(\phi_r)$, $\|\cdot\|_r$ for $\|\cdot\|_{D(\phi_r)}$, and C_r for $C(\phi_r)$. Since the coefficient $R > 0$ is defined by $E(e^{RX_1}) = 1$, we see that the functional Ξ operates on distribution functions whose moment generating functions exist at least in $[0, r]$ for some $r > 0$. Since

$$E(e^{rX_1}) < \infty \Rightarrow P(X_1 > x) = o(e^{-rx}) \text{ as } x \to \infty,$$

(Widder 1946, VI.3), we see that

$$M(r) < \infty \Rightarrow 1 - F \in D_r.$$

This leads us to revise our view of Ξ so that it operates on $1 - F$ and takes D_r to $\mathbf{R}$. From Mammitzsch (1986), F is such that $R > 0$ exists if and only if $\int x F(dx) < 0$, $F(0) < 1$, and, if $b = \sup\{r \in \mathbf{R} : M(r) < \infty\}$ is in $\mathbf{R}$ with $M(b) < \infty$, then in addition we also have $M(b) \geq 1$.

Example Another possible choice for ϕ is, for α and β in $\mathbf{R}$,

$$\phi_{\alpha\beta}(x) = \begin{cases} (1 + |x|)^\alpha & \text{if } x < 0 \\ (1 + x)^\beta & \text{if } x \geq 0. \end{cases}$$

Now write $D_{\alpha\beta}$ for $D(\phi_{\alpha\beta})$, $\|\cdot\|_{\alpha\beta}$ for $\|\cdot\|_{D(\phi_{\alpha\beta})}$, and $C_{\alpha\beta}$ for $C(\phi_{\alpha\beta})$. Since, for $\alpha > 0$,

$$\int |x|^\alpha F(dx) < \infty \Rightarrow 1_{[0,\infty)} - F \in D_{\alpha\alpha},$$

we see that these spaces relate to moment conditions on F. For Example II, we require moment conditions on the input distribution functions F_S and F_T. For $\alpha > 1$ and $\beta > 1$, let Ψ be the map taking $(1_{[0,\infty)} - F_S, 1_{[0,\infty)} - F_T)$ in $D_{0\alpha} \times D_{0\beta}$ to F_W in D_∞, for those F_S, F_T satisfying the stability condition $\int x F_S(dx) < \int x F_T(dx)$.

Our nonparametric estimators are $\hat{R}_n = \Xi(1 - \hat{F}_n)$, and $\hat{F}_{W,n} = \Psi(1_{[0,\infty)} - \hat{F}_{S,n}, 1_{[0,\infty)} - \hat{F}_{T,n})$, where $\hat{F}_{S,n}$ and $\hat{F}_{T,n}$ are the empirical distribution functions based on $\{S_1, \ldots, S_n\}$ and $\{T_1, \ldots, T_n\}$ respectively. Suitable modifications must be made in defining the estimators when the input empirical distribution functions do not satisfy the relevant conditions. For instance, in Example I, if the sample mean is nonnegative, or if $\hat{F}_n(0) = 1$, we define $\hat{R}_n$ to be zero, and recognize that, by the strong law of large numbers, with probability one, eventually the necessary conditions will be satisfied.

3 Asymptotic normality

In this section we discuss what is meant by asymptotic normality for random elements of the D-spaces, and indicate the property of Φ that ensures that asymptotic normality of the input estimator carries over to asymptotic normality of the output estimator. This is the delta method, which roughly says that asymptotic normality of $\hat{F}_n$ in B_1 (in a sense made precise below) together with differentiability of Φ (again in a sense to be made precise below) implies asymptotic normality of $\Phi(\hat{F}_n)$ in B_2. This is described in Gill (1989).

The first of these, asymptotic normality of the input estimators, refers to an empirical central limit theorem for the input empirical distribution functions. This is a statement of the form

$$\sqrt{n}(\hat{F}_n - F) \to_d Z \text{ as } n \to \infty \text{ in } D(\phi), \tag{3.1}$$

where $\to_d$ denotes convergence in distribution (see below), Z is a zero mean Gaussian process (cf Donsker's Theorem, see Billingsley (1968)). We follow the theory of convergence in distribution in nonseparable metric spaces discussed in Pollard (1984), while noting that this is not the most recent form of such a theory. Briefly, we give $D(\phi)$ its open ball (or projection) σ-field $\mathcal{P}(\phi)$, which is strictly contained in its Borel σ-field. Define a random element of $D(\phi)$ to be an $\mathcal{A}$-$\mathcal{P}(\phi)$ measurable map from Ω to $D(\phi)$. For random elements $\{Y_n\}$ and Y in $D(\phi)$, we say Y_n converges in distribution to Y in $D(\phi)$, $Y_n \to_d Y$ in $D(\phi)$, if $E(f(Y_n)) \to E(f(Y))$ for all bounded continuous $\mathcal{P}(\phi)$-$\mathcal{B}$ measurable f from $D(\phi)$ to $\mathbf{R}$, where $\mathcal{B}$ is the Borel σ-field of $\mathbf{R}$. In the following, we omit discussion of measurability, see Gill (1989), Pollard (1984), Pitts (1994a) and Pitts *et al.* (1996).

To obtain (3.1), we use a weighted empirical central limit theorem that holds for random variables uniformly distributed on $(0, 1)$ under certain conditions on the weight function (Pyke and Shorack (1968), Shorack and Wellner (1986), Section 3.7.1). The limiting Gaussian process is a standard Brownian bridge. Rescaling by F, we obtain for $\phi = \phi_r$ in Example I, for $s > 2r \geq 0$,

$$E(\exp(sX_1)^+) < \infty \Rightarrow \sqrt{n}(\hat{F}_n - F) \to_d B \circ F \text{ as } n \to \infty \text{ in } D_r, \tag{3.2}$$

where $(B \circ F)(t, \omega) = B(F(t), \omega)$ for t in $\mathbf{R}$, ω in Ω, so that $\operatorname{cov}((B \circ F)(t, \cdot), (B \circ F)(s, \cdot)) = F(s \wedge t) - F(s)F(t)$, where $x \wedge y = \min\{x, y\}$.

Similarly for Example II, we have for $0 \leq 2\beta < \gamma$,

$$E(|X_1 \vee 0|^\gamma) < \infty \Rightarrow \sqrt{n}(\hat{F}_n - F) \to_d B \circ F \text{ in } D_{0\beta}. \tag{3.3}$$

The appropriate differentiability property of Φ required by the delta method is *Hadamard differentiability*, and this is now defined. The map Φ is Hadamard differentiable at f in B_1 if there exists a bounded linear map $\Phi'_f : B_1 \to B_2$ such that, if $h_n \to h$ in B_1 and $t_n \to 0$ in $\mathbf{R}$, then

$$\frac{1}{t_n}\left(\Phi(f + t_n h_n) - \Phi(f)\right) \to \Phi'_f(h) \text{ in } B_2.$$

If B_1 is a weighted D-space and F_n and F are distribution functions in the domain of Φ satisfying $\sqrt{n}(F_n - F) \to h$ in B_1, then, putting $f = F$, $h_n = \sqrt{n}(F_n - F)$ and $t_n = n^{-1/2}$, Hadamard differentiability of Φ at F implies

$$\sqrt{n}(\Phi(F_n) - \Phi(F)) \to \Phi'_F(h) \text{ in } B_2. \tag{3.4}$$

See Gill (1989) for a discussion of Hadamard and other kinds of differentiability. The delta method (Gill (1989), Theorem 3) says that if $\Phi : B_1 \to B_2$ is Hadamard differentiable at f in B_1, and X_n, $n = 1, 2, \ldots$, and Z are random elements in B_1 such that $\sqrt{n}(X_n - f) \to_d Z$ in B_1, where the distribution of Z concentrates on a separable subset of B_1, then

$$\sqrt{n}\left(\Phi(X_n) - \Phi(f)\right) \to_d \Phi'_f(Z) \text{ in } B_2.$$

To apply the delta method in a particular case involves establishing Hadamard differentiability for the functional in question. For Example I, the functional Ξ taking $1 - F$ to R satisfies a stronger differentiability property than Hadamard differentiability (for details of the analysis, see Pitts *et al.* (1996)). From this we obtain a version of (3.4). Let $r > R$ and put $R_n = \Xi(1 - F_n)$ and suppose that F is the step distribution function for a stable queue and is such that there exists $R > 0$ with $M(R) = 1$. Then

$$\sqrt{n}(F_n - F) \to g \text{ in } D_r \Rightarrow \sqrt{n}(R_n - R) \to \Xi'_{1-F}(-g),$$

where $\Xi'_{1-F}(-g) = \left(\int Re^{Rx} g(x)dx\right) / (M'(R))$. Using this and (3.2), the delta method implies the following asymptotic normality result for $\hat{R}_n$ (Pitts *et al.* (1996), Theorem 1.1).

Theorem 3.1 *Assume that F is as above and $M(s) < \infty$ for some $s > 2R$. Then*

$$\sqrt{n}(\hat{R}_n - R) \to_d Z_\Xi,$$

where Z_Ξ is normally distributed with mean zero and variance

$$\sigma^2 = (M(2R) - 1) / \left((M'(R))^2\right).$$

The variance of the limiting distribution is obtained by direct calculation.

Asymptotic normality of empirical estimators of R under various model assumptions (e.g., Poisson arrivals with either known or unknown arrival rate) has also been discussed in Gaver and Jacobs (1988), Csörgő and Teugels (1990), and Grandell (1991). See also Embrechts and Mikosch (1991), Deheuvels and Steinebach (1990), Csörgő and Steinebach (1991), and Herkenrath (1986).

For Example II, the functional $\Psi : (1_{[0,\infty)} - F_S, 1_{[0,\infty)} - F_T) \to F_W$ is much more complicated and difficult to analyse than the functional Ξ. Nevertheless, it is possible to establish an appropriate differentiability property for Ψ, given in the proposition below. For the (lengthy) details, see Pitts (1994a).

Proposition 3.2 *Assume that $\int x^\gamma F_{S,n}(dx) < \infty$, $\int x^\gamma F_{T,n}(dx) < \infty$ for $n = 1, 2, 3, \ldots$, and $\int x^\gamma F_S(dx) < \infty$ and $\int x^\gamma F_T(dx) < \infty$ for some $\gamma > 2$. Assume further that for some continuous g_S and g_T in $D_{0\gamma}$,*

$$\|\sqrt{n}(F_{S,n} - F_S) - g_S\|_{0\gamma} \to 0 \text{ as } n \to \infty$$

$$\|\sqrt{n}(F_{T,n} - F_t) - g_T\|_{0\gamma} \to 0 \text{ as } n \to \infty.$$

Then

$$\|\sqrt{n}(F_{W,n} - F_W) - \Psi'_{(1_{[0,\infty)} - F_S, 1_{[0,\infty)} - F_T)}(-g_S, -g_T)\|_\infty \to 0 \text{ as } n \to \infty,$$

where $\Psi'_{(1_{[0,\infty)} - F_S, 1_{[0,\infty)} - F_T)}$ is a bounded linear map from $D_{0\gamma} \times D_{0\gamma}$ to D_∞.

By the delta method, (3.3) and Proposition 3.2, we obtain the following asymptotic normality result for $\hat{F}_{W,n}$.

Theorem 3.3 *Assume that $\int x^{2\gamma} F_S(dx) < \infty$ and $\int x^{2\gamma} F_T(dx) < \infty$ for some $\gamma > 2$. Then*

$$\sqrt{n}(\hat{F}_{W,n} - F_W) \to_d Z_\Psi \text{ in } D_\infty,$$

where Z_Ψ is a zero mean Gaussian process.

The Gaussian process Z_Ψ has a complicated covariance structure, the details of which are given in Pitts (1994a).

4 Confidence regions

A natural next step is to find confidence regions for the unknown output quantity $\Phi(F)$. We consider various methods in this section, concentrating on the bootstrap, where we follow Gill (1989) and Grübel and Pitts (1993).

In this section, suppose that the functional Φ maps B_1 to B_2, where B_1 is a weighted D-space and $B_2 = \mathbf{R}$ as in Example I. Let

$$G_n(z) = P\big(\sqrt{n}(\Phi(\hat{F}_n) - \Phi(F)) \leq z\big),$$

so that G_n is the (unknown) distribution function of $\sqrt{n}(\Phi(\hat{F}_n) - \Phi(F))$. Provided that Φ is Hadamard differentiable at F and we have asymptotic normality of $\hat{F}_n$, then, using the delta method, $G_n \to G$ in D_∞, where G is the distribution function of $\Phi'_F(B \circ F)$, a normally distributed random variable (see Corollary 1 in Gill (1989)). Suppose that the variance of σ^2 of the limiting distribution G can be consistently estimated by $\hat{\sigma}_n^2$ say. Let

$$C_n = \Big[\Phi(\hat{F}_n) - \frac{\hat{\sigma}_n z_\alpha}{\sqrt{n}}, \infty\Big),$$

where z_α is the α-quantile of the standard normal distribution. Then C_n satisfies

$$P(C_n \ni \Phi(F)) \to \alpha \text{ as } n \to \infty,$$

and so C_n has asymptotically the correct coverage rate.

Example For Example I, we know that

$$\sigma^2 = \sigma^2(F) = (M(2R) - 1)/(M'(R)^2),$$

and so an obvious estimator for σ^2 is $\hat{\sigma}^2_{n,P} = \sigma^2(\hat{F}_n)$. This can be shown to be consistent for σ^2 (Pitts *et al.* (1996)). Another possibility is to use the jackknife estimator

$$\hat{\sigma}^2_{n,J} = (n - 1) \sum_{i=1}^{n} \left[\hat{R}_{ni} - \frac{1}{n} \sum_{k=1}^{n} \hat{R}_{nk} \right]^2,$$

where $\hat{R}_{ni}$ is $\Xi(1 - \hat{F}_{ni})$, and $\hat{F}_{ni}$ is the empirical distribution function for the original sample with the ith observation omitted. Shao (1993) shows that, if certain conditions are satisfied, including a particular differentiability condition for the functional, then the jackknife estimator for σ^2 is consistent. In Pitts *et al.* (1996) this particular differentiability property is established for Ξ, and the other conditions are shown to hold, so that confidence intervals constructed using $\hat{\sigma}_{n,J}$ have asymptotically correct coverage rates.

In the above, we use $G_n \to G$ and then estimate the variance of the limiting distribution. An alternative approach uses the bootstrap estimator G_n^* of G_n. For z in $\mathbf{R}$, this is given by

$$G_n^*(z) = \frac{1}{n^n} \sum_{\mathbf{i} \in \mathcal{I}_n} 1_{(-\infty, z]} \left(\sqrt{n} \big(\Phi(F_{ni}^*) - \Phi(\hat{F}_n) \big) \right),$$

where $\mathbf{i} = (i_1, \dots, i_n)$ is in $\mathcal{I}_n = \{1, 2, \dots, n\}^n$, and F_{ni}^* is the empirical distribution function based on $\{X_{i_1}, \dots, X_{i_n}\}$. Thus G_n^* bears the same relationship to $\hat{F}_n$ as G_n bears to F. The function G_n^* is knowable in the sense that it does not depend on unknown quantities. However, in practice direct calculation of G_n^* is not feasible, and often Monte-Carlo methods are used to approximate G_n^*.

To say that the bootstrap works almost surely means that G_n^* converges to G almost surely, so that with probability one G_n^* is eventually close to G_n. Gill (1989), Theorem 4, shows that the bootstrap works almost surely if Φ satisfies another differentiability condition, namely that if $f_n \to f$ and $h_n \to h$ in B_1 and $t_n \to 0$ in $\mathbf{R}$, then

$$\frac{1}{t_n} \left(\Phi(f_n + t_n h_n) - \Phi(f_n) \right) \to \Phi'_f(h) \text{ as } n \to \infty. \tag{4.1}$$

(Gill (1989) shows this for general B_2.)

Example In Pitts *et al.* (1996), (4.1) is shown for Ξ and so the bootstrap works almost surely for Example I. Let $q_n^*(\alpha)$ be the α-quantile of G_n^*, so that

$q_n^*(\alpha) = G_n^{*-1}(\alpha) = \inf\{y : G_n^*(y) \geq \alpha\}$. Then we obtain the $100\alpha\%$ one-sided bootstrap confidence interval for R

$$\left[\hat{R}_n - \frac{q_n^*(\alpha)}{\sqrt{n}}, \infty\right).$$

As a consequence of Gill's Theorem above, such intervals have asymptotically correct coverage rates almost surely.

We illustrate the case where B_2 is not $\mathbf{R}$, but is itself a D-space, with reference to Example II.

Example For Example II, the estimator $\hat{F}_{W,n}$ and the unknown F_W are elements of the function space D_∞, and we aim to find confidence regions in this space for F_W. For z in $\mathbf{R}$, let

$$H_n(z) = P\left(\|\sqrt{n}(\hat{F}_{W,n} - F_W)\|_\infty \leq z\right).$$

The distribution function H_n is not known. Putting $H(z) = P(\|Z_\Psi\|_\infty \leq z)$, Theorem 3.3 implies that $H_n(z) \to H(z)$ for all continuity points z of H. In contrast to the corresponding quantity in Example I, H does not have a simple dependence on the underlying F_S and F_T. This makes bootstrap confidence regions the more practicable option in this case. With $\mathbf{i}$ and $\mathcal{I}_n$ as for Example I, define

$$
\begin{aligned}
H_n^*(z) = \ & \frac{1}{n^n}\sum_{\mathbf{i}\in\mathcal{I}_n} 1_{(-\infty,z]}\big(\sqrt{n}\big\|\Psi(1_{[0,\infty)} - F_{S,\mathbf{i}}^*, 1_{[0,\infty)} - F_{T,\mathbf{i}}^*) \\
& - \Psi(1_{[0,\infty)} - \hat{F}_{S,n}, 1_{[0,\infty)} - \hat{F}_{T,n})\big\|_\infty\big),
\end{aligned}
$$

where $F_{S,\mathbf{i}}^*$ is the empirical distribution function based on $\{S_{i_1}, \ldots, S_{i_n}\}$ and similarly for $F_{T\mathbf{i}}^*$. In aiming to show that the bootstrap works in this case, we hope to apply the differentiability of the functional as we did in Example I. We have seen that Ψ satisfies the weaker property given in Proposition 3.2; this is not enough to apply Theorem 4 in Gill (1989). An easy adaptation of the proof of Proposition 3.15 in Grübel and Pitts (1993) yields that, under the assumptions of Theorem 3.3,

$$H_n^* \to_d H \text{ in } D_\infty.$$

As in Lemma 3.16 in the same paper, this implies that bootstrap confidence regions, defined by

$$\hat{F}_n \pm \frac{1}{\sqrt{n}}q_n^*(\alpha),$$

where $q_n^*(\alpha)$ is the α-quantile of H_n^*, have asymptotically correct coverage rates (see also Gill (1989), Theorem 5). In practice a Monte-Carlo approximation to H_n^* is obtained using B resamples of size n chosen uniformly and with replacement from each of $\{S_1, \ldots, S_n\}$ and $\{T_1, \ldots, T_n\}$.

For estimation of equivalent quantities arising in insurance mathematics, see Hipp (1989), and Frees (1986).

5 Concluding remarks

The methods described here, with similarly defined spaces, have also been applied to quantities of interest in other stochastic models. These include the renewal function, in Grübel and Pitts (1993) (see also Grübel (1989) for a functional approach to renewal models), and compound distribution functions, in Pitts (1994b). In principle, the same methods may be considered for application to other stochastic models and to other quantities of interest, the practicability of such application depending mainly on the feasibility of the analysis for the relevant functional. It is hoped that the illustrative examples above may provide some insight into the practicalities of applying these techniques to statistical estimation problems in stochastic models.

Bibliography

1. Asmussen, S. (1987). *Applied Probability and Queues*. Wiley, New York.

2. Basawa, I. V. and Prabhu, N. U. (1988). Large sample inference from single server queues. *Queueing Systems*, **3**, 289–304.

3. Billingsley, P. (1968). *Convergence of Probability Measures*. Wiley, New York.

4. Chow, Y. S. and Teicher, H. (1988). *Probability Theory, Independence, Interchangeability, Martingales* (2nd edn). Springer-Verlag, New York.

5. Csörgő, M. and Steinebach, J. (1991). On the estimation of the adjustment coefficient in risk theory via intermediate order statistics. *Insurance: Mathematics and Economics*, **10**, 37–50.

6. Csörgő, S. and Teugels, J. (1990). Empirical Laplace transform and approximation of compound distributions. *Journal of Applied Probability*, **27**, 88–101.

7. Deheuvels, P. and Steinebach, J. (1990). On some alternative estimators of the adjustment coefficient in risk theory. *Scandinavian Actuarial Journal*, 135–59.

8. Duffield, N. G., Lewis, J. T., O'Connell, N., Russell, R. and Toomey, F. (1995). Entropy of ATM traffic streams: a tool for estimating QoS parameters. *IEEE JSAC*, **13**, 981–90.

9. Embrechts, P. and Mikosch, T. (1991). A bootstrap procedure for estimating the adjustment coefficient. *Insurance: Mathematics and Economics*, **10**, 181–90.

10. Frees, E. W. (1986). Nonparametric estimation of the probability of ruin. *ASTIN Bulletin*, **16S**, 81–90.

11. Gaver, D. P. and Jacobs, P. A. (1988). Nonparametric estimation of the probability of a long delay in the $M/G/1$ queue. *Journal of the Royal Statistical Society B*, **50**, 393–402.

12. Gill, R. D. (1989). Non- and semi-parametric maximum likelihood estimators and the von Mises method. *Scandinavian Journal of Statistics*, **16**, 97–128.

13. Grandell, J. (1991). *Aspects of Risk Theory*. Springer-Verlag, New York.

14. Grübel, R. (1989). Stochastic models as functionals: some remarks on the renewal case. *Journal of Applied Probability*, **26**, 296–303.

15. Grübel, R. and Pitts, S. M. (1993). Nonparametric estimation in renewal theory I: the empirical renewal function. *Annals of Statistics*, **21**, 1431–51.

16. Herkenrath, U. (1986). On the estimation of the adjustment coefficient in risk theory by means of stochastic approximation procedures. *Insurance: Mathematics and Economics*, **5**, 305–13.

17. Heyde, C. C. (1988). Asymptotic efficiency results for the method of moments with application to estimation for queueing processes. In *Queueing Theory and its Applications: Liber Amicorum for J. W. Cohen*, O. J. Boxma and R. Syski, eds, 405–11. North-Holland, Amsterdam.

18. Hipp, C. (1989). Estimators and bootstrap confidence intervals for ruin probabilities. *ASTIN Bulletin*, **19**, 57–70.

19. Hipp, C. (1994). Pollard's CLT and the delta method. Technical Report, University of Karlesruhe.

20. Mammitzsch, V. (1986). A note on the adjustment coefficient in ruin theory. *Insurance: Mathematics and Economics*, **5**, 147–9.

21. Pitts, S. M. (1994a). Nonparametric estimation of the stationary waiting time distribution function for the $GI/GI/1$ queue. *Annals of Statistics*, **22**, 1428–46.

22. Pitts, S. M. (1994b). Nonparametric estimation of compound distributions with applications in insurance. *Annals of the Institute of Statistical Mathematics*, **46**, 537–55.

23. Pitts, S. M., Grübel, R. and Embrechts, P. (1996). Confidence bounds for the adjustment coefficient. *Advances in Applied Probability*, to appear.

24. Pollard, D. (1984). *Convergence of Stochastic Processes*. Springer-Verlag, New York.

25. Pyke, R. and Shorack, G. R. (1968). Weak convergence of a two-sample empirical process and a new approach to Chernoff-Savage theorems. *Annals of Mathematical Statistics*, **39**, 755–71.

26. Shao, J. (1993). Differentiability of statistical functionals and consistency of the jackknife. *Annals of Statistics*, **21**, 61–75.

27. Shorack, G. R. and Wellner, J. (1986). *Empirical Processes with Applications to Statistics*. Wiley, New York.

28. Widder, D. (1946). *The Laplace Transform*. Princeton University Press, Princeton.

11

The asymptotic behaviour of large loss networks

Stan Zachary

Heriot-Watt University

Abstract

We study the limit behaviour of controlled loss networks as capacity and offered traffic are allowed to increase in proportion, reviewing and extending recent work based on the functional law of large numbers of Hunt and Kurtz. We consider in detail single and two-resource networks.

1   Introduction

In this paper we study large loss networks in which the offered traffic is subject to acceptance controls. We review recent work of Hunt and Kurtz (1994), who established rigorous results for the asymptotic dynamics of such networks as capacity and offered traffic are allowed to increase in proportion, and we relate these results to asymptotic equilibrium behaviour. We further study the detailed behaviour of networks with at most two resources, extending results of Bean *et al.* (1994b, 1995), and giving some additional results.

The asymptotic results considered here have important applications to the control of modern communications networks, which are typically large and which may simultaneously carry traffic with very different capacity requirements and holding times. A failure to apply effective controls in such networks can lead to a serious degradation in performance.

The results also remain qualitatively correct for smaller capacity networks. Bean *et al.* (1994a, 1995) and Moretta (1995) derive refinements which permit more accurate modelling of the quantitative behaviour of networks of all capacities.

The mathematical framework is the same as that of Hunt and Kurtz (1994). Consider a sequence of loss networks, indexed by a scale parameter N. All members of the sequence are identical except in respect of capacities and call arrival rates (which, as defined more precisely by eqn (1.1) below, are essentially proportional to N), and are identically controlled. Resources (or links) are indexed in a finite set $\mathcal{J}$ and call types in a finite set $\mathcal{R}$. For the Nth member of the sequence, each resource $j \in \mathcal{J}$ has integer capacity $C_j(N)$, and calls of each type $r \in \mathcal{R}$ arrive as a Poisson process of rate $\kappa_r(N)$. Each such call, if accepted, simultaneously requires an integer A_{jr} units of the capacity of each resource j for the duration of its holding time, which is exponentially distributed with mean $1/\mu_r$. All arrival streams and holding times are independent.

Let $n^N(t) = (n_r^N(t), \; r \in \mathcal{R})$, where $n_r^N(t)$ is the number of calls of type r in progress at time t, and let $m^N(t) = (m_j^N(t), \; j \in \mathcal{J})$ where $m_j^N(t) = C_j(N) - \sum_{r \in \mathcal{R}} A_{jr} n_r^N(t)$ is the free capacity of resource j at time t. A call of type r arriving at time t is accepted if and only if $m^N(t-)$ belongs to some acceptance region $\mathcal{A}_r$, which we formally regard as a subset of the space $E = (\mathbb{Z}_+ \cup \{\infty\})^J$, where $J = |\mathcal{J}|$. (Of course, the process $m^N(\cdot)$ only takes values in $\mathbb{Z}_+^J$.) We further require that each set $\mathcal{A}_r$ is well-behaved in the sense that its indicator function $I_{\mathcal{A}_r}$ is continuous, where the topology of E is the product of the topology of the one-point compactification of $\mathbb{Z}_+$.

This framework permits the modelling of a wide variety of control mechanisms, including most of those, such as fixed routing, trunk reservation and alternative routing, employed in practical applications to communications networks. For details see Hunt and Kurtz (1994).

Suppose that, as $N \to \infty$, for all $j \in \mathcal{J}$, $r \in \mathcal{R}$,

$$\frac{1}{N} C_j(N) \to C_j, \qquad \frac{1}{N} \kappa_r(N) \to \kappa_r. \tag{1.1}$$

Then, under appropriate initial conditions, the normalized process $x^N(\cdot) = n^N(\cdot)/N$ might reasonably be expected to converge to a 'fluid limit' process $x(\cdot)$ taking values in the space $X = \{x \in \mathbb{R}_+^R : \sum_r A_{jr} x_r \le C_j \text{ for all } j \in \mathcal{J}\}$, where $R = |\mathcal{R}|$. (See, for example, Kelly, 1991.)

To make this idea precise, for each $x \in X$, let $m_x(\cdot)$ be the Markov process on E with transition rates given by

$$m \to \begin{cases} m - A_r & \text{at rate } \kappa_r I_{\{m \in \mathcal{A}_r\}} \\ m + A_r & \text{at rate } \mu_r x_r, \end{cases} \tag{1.2}$$

where A_r denotes the vector $(A_{jr}, j \in \mathcal{J})$ and $\infty \pm a = \infty$ for any $a \in \mathbb{Z}_+$. Note that the process $m_x(\cdot)$ is reducible, and so does not always have a unique invariant distribution. Hunt and Kurtz (1994, Theorem 3) show that, provided the distribution of $x^N(0)$ converges weakly to that of $x(0)$, the sequence of processes $x^N(\cdot)$ is relatively compact in $D_{\mathbb{R}^R}[0, \infty)$ and any weakly convergent subsequence has a limit $x(\cdot)$ which obeys the relation

$$x_r(t) = x_r(0) + \int_0^t (\kappa_r \pi_u(\mathcal{A}_r) - \mu_r x_r(u)) du, \tag{1.3}$$

where, for each t, π_t is some invariant distribution of the Markov process $m_{x(t)}(\cdot)$ and additionally satisfies, for all j,

$$\pi_t\{m : m_j = \infty\} = 1 \text{ if } \sum_{r \in \mathcal{R}} A_{jr} x_r(t) < C_j. \tag{1.4}$$

Thus, at each time t, the invariant distribution π_t acts as a control for the asymptotic process $x(\cdot)$, corresponding to a limiting acceptance rate for calls of each type. For a discussion of this result, which involves a separation, in the limit, of the time scales of the processes $x^N(\cdot)$ and $m^N(\cdot)$, see Hunt and Kurtz (1994) and Bean *et al.* (1995).

Of particular interest is the case where there exists a function π' on X (each value of which is a probability distribution on E) with the property that, for *all* convergent subsequences, we may take $\pi_t = \pi'_{x(t)}$ in eqn (1.3). We may then define a *velocity field* $v = (v_r, \ r \in \mathcal{R})$ on X by

$$v_r(x) = \kappa_r \pi'_x(\mathcal{A}_r) - \mu_r x_r, \tag{1.5}$$

so that eqn (1.3) becomes

$$x_r(t) = x_r(0) + \int_0^t v_r(x(u))du. \tag{1.6}$$

It will then generally be the case that, for all t, $x_r(t)$ is uniquely determined by $x_r(0)$, so that the convergence asserted above takes place in the entire sequence of networks.

Further, when such a velocity field may be defined, it is usually possible to show that, for all t, $x(t)$ is a continuous function of $x(0)$. Since X is compact, the argument of Theorem 3.3 of Bean *et al.* (1995) then applies equally to the present, more general, situation to show that there is at least one *fixed point* $\bar{x} \in X$ such that $v(\bar{x}) = 0$; that is, satisfying the fixed point equations

$$\kappa_r \pi'_x(\mathcal{A}_r) = \mu_r x_r, \quad r \in \mathcal{R}. \tag{1.7}$$

It is scarcely surprising (but for a formal proof see Bean *et al.*, 1994b) that when this fixed point $\bar{x}$ is unique, and further is such that all trajectories of $x(\cdot)$ converge to it, then the invariant distribution of the process $x^N(\cdot)$ converges weakly to the distribution concentrated on the single point $\bar{x}$, while the invariant distribution of the 'free capacity' process $m^N(\cdot)$ converges weakly to $\pi'_{\bar{x}}$. In particular, for each r, $\pi'_{\bar{x}}(\mathcal{A}_r)$ is the limiting equilibrium acceptance probability for calls of type r.

When the process $x(\cdot)$ possesses more than one fixed point, each may be associated, for any large N, with some 'quasi-equilibrium' regime of the process $x^N(\cdot)$, maintained over some extended period of time—as in the example which we discuss in Section 2.

Where there does *not* exist a function π' on X such that $\pi_t = \pi'_{x(t)}$, so that it is impossible to define a velocity field on X, then behaviour in the associated sequence of networks is typically highly pathological. Examples of such behaviour are given by Hunt (1995).

The remainder of this paper is primarily concerned with the identification of conditions under which a velocity field *may* be defined, and with the determination of the resulting dynamics and fixed points of the process $x(\cdot)$. In Section 2 we review results for single resource networks (where a velocity field may always be defined), and in Section 3 we study two-resource networks. Finally, in Section 4 we discuss briefly the general case.

However, it is convenient to make a number of further definitions at this point.

Partition the set X by defining, for each $S \subseteq \mathcal{J}$, $X_S = \{x \in X: \sum_r A_{jr} x_r(t) = C_j$ if and only if $j \in S\}$. We shall find it convenient to write X_j for $X_{\{j\}}$, and shall make similar obvious notational simplifications elsewhere.

For each subset S of $\mathcal{J}$, let $E_S = \{m \in E: m_j < \infty$ if and only if $j \in S\}$. We assume that the matrix of capacity requirements (A_{jr}) and the acceptance regions $\mathcal{A}_r$ are such that, for each $x \in X$ and $S \subseteq \mathcal{J}$, there is *at most* a single invariant distribution π_x^S of the Markov process $m_x(\cdot)$ on E which assigns probability one to the set E_S. (The distribution π_x^S may also be thought of as the invariant distribution of the obvious projection of the process $m_x(\cdot)$ onto $\mathbb{Z}_+^S$.) There is no loss of generality in this irreducibility assumption—for a discussion see again Hunt and Kurtz (1994). Note that the distribution $\pi_x^\emptyset$ exists for all $x \in X$, assigning probability one to the single point $(\infty, \dots, \infty)$ of the set $E_\emptyset$.

Then, from the above results of Hunt and Kurtz, it follows that there exist nonnegative functions $\lambda^S(\cdot)$, $S \subseteq \mathcal{J}$, summing to one, such that, for almost all t,

$$\pi_t = \sum_{S \subseteq \mathcal{J}} \lambda^S(t) \pi_{x(t)}^S \tag{1.8}$$

where, from (1.4),

$$\lambda^S(t) = 0 \text{ if } \sum_{r \in \mathcal{R}} A_{jr} x_r(t) < C_j \text{ for any } j \in S, \tag{1.9}$$

and where additionally we make the convention that $\lambda^S(t) = \lambda^S(t) \pi_{x(t)}^S = 0$ if $\pi_{x(t)}^S$ does not exist. Identification of π_t, $t \geq 0$, thus reduces to identification of the functions $\lambda^S(\cdot)$.

Finally, define also, for each x, each $S \subseteq \mathcal{J}$ such that π_x^S exists, and each $j \in \mathcal{J}$,

$$\alpha_j^S(x) = \sum_{r \in \mathcal{R}} A_{jr} \{\kappa_r \pi_x^S(\mathcal{A}_r) - \mu_r x_r\}. \tag{1.10}$$

The quantity $\alpha_j^S(x)$ will play an important role in subsequent analysis. Note in particular that

$$\alpha_j^S(x) = 0 \text{ if } j \in S. \tag{1.11}$$

This follows from the observation that, in equilibrium, the jth component of the restriction of the process $m_x(\cdot)$ to E_S has zero drift for each $j \in S$. A formal proof may be given analogously to that of Lemma 4 of Hunt and Kurtz (1994).

2   Single resource networks

We now consider further the single resource case $\mathcal{J} = \{1\}$. It is convenient to write C for C_1, A_r for A_{1r}, and $\alpha^S(x)$ for $\alpha_1^S(x)$.

Here the compactified space $E = \mathbb{Z}_+ \cup \{\infty\}$ and the requirement that, for each r, the indicator function I_{A_r} of the acceptance region $\mathcal{A}_r$ be continuous at ∞ implies that there is some *finite* $M \in E$ such that, again for each r, either $m \in \mathcal{A}_r$ for all $m > M$ (including $m = \infty$) or $m \notin \mathcal{A}_r$ for all $m > M$. Define

$\mathcal{R}^* = \{r \in \mathcal{R}: \infty \in \mathcal{A}_r\}$. Thus $\mathcal{R}^*$ is the set of call types which are accepted for all sufficiently large values of the free capacity in the network.

In most applications we might expect $\mathcal{R}^* = \mathcal{R}$. However, there are practical circumstances where this might not be the case—for example, a call type which was to be allocated less resource when the network was nearly full might be modelled as two call types with disjoint acceptance regions.

Now note that, for all x,

$$\pi_x^\emptyset(\mathcal{A}_r) = I_{\{r \in \mathcal{R}^*\}} \tag{2.1}$$

(where again I is the indicator function) and so, from eqn (1.10), $\alpha^\emptyset(x) = \sum_r \mathcal{A}_r\{\kappa_r I_{\{r \in \mathcal{R}^*\}} - \mu_r x_r\}$. This quantity is also the drift rate *towards* the origin of the process $m_x(\cdot)$ while in the set $[M+1, \infty)$ and so elementary Lyapounov techniques for such processes (see, for example, Fayolle *et al.*, 1995) show that the restriction of this process to $\mathbb{Z}_+$ is ergodic—and so the distribution $\pi_x^1 (= \pi_x^{\{1\}})$ exists—if and only if $\alpha^\emptyset(x) > 0$.

Let $X_1^+ = \{x \in X_1: \alpha^\emptyset(x) > 0\}$ and let $X_1^- = X_1 \setminus X_1^+$. It is then straightforward to show that a velocity field for the limit process $x(\cdot)$ may be defined everywhere on X, the function π_x' being given by

$$\pi_x' = \begin{cases} \pi_x^\emptyset & \text{if } x \in X_\emptyset \cup X_1^-, \\ \pi_x^1 & \text{if } x \in X_1^+. \end{cases} \tag{2.2}$$

In the case $x \in X_\emptyset$ this result follows from eqn (1.4) (or equivalently from eqn (1.9)), while in the case $x \in X_1^-$ it is immediate from the above criterion for the existence of π_x^1. To prove the remaining case note, from eqns (1.3), (1.8), (1.10), and (1.11), together with the condition $\lambda^\emptyset(t) + \lambda^1(t) = 1$, we have easily that

$$\sum_{r \in \mathcal{R}} \mathcal{A}_r x_r(t) = \sum_{r \in \mathcal{R}} \mathcal{A}_r x_r(0) + \int_0^t \lambda^\emptyset(u)\alpha^\emptyset(x(u))du. \tag{2.3}$$

Since necessarily $\sum_r \mathcal{A}_r x_r(t) \leq C$ for all t, it follows that $\lambda^\emptyset(t) = 0$ for (almost) all t with $x(t) \in X_1^+$.

The simple idea underlying this argument—that the process $x(\cdot)$ must remain within X—is due to Hunt (1990). Hunt and Kurtz (1994) prove the above result in the case $\mathcal{R}^* = \mathcal{R}$. A slightly more formal version of the present argument is given by Bean *et al.* (1994b).

It is now readily verified that, for each r, $\pi_x'(\mathcal{A}_r)$ is Lipschitz continuous on $X_\emptyset \cup X_1^-$ (trivially) and also on the set X_1 (see Bean *et al.*, 1995). Hence trajectories of the process $x(\cdot)$ are well-defined functions of their positions at time 0 and discontinuities in the velocity of any trajectory occur only at times of passage from $X_\emptyset$ to X_1^+. (Passage from X_1^+ to $X_\emptyset$ is impossible by the continuity of the function $\alpha^\emptyset$ on X and the relation (2.3)). It follows from standard arguments for dynamical systems that, for each t, $x(t)$ is a continuous function of $x(0)$ and so, as indicated earlier, the process $x(\cdot)$ has at least one fixed point.

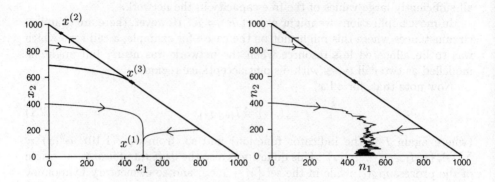

FIG. 1. Analytical and simulation results for numerical example

In the case where

$$\sum_{r \in \mathcal{R}} A_r \kappa_r I_{\{r \in \mathcal{R}^*\}}/\mu_r \leq C, \tag{2.4}$$

define $\hat{x} \in X$ by $\hat{x}_r = \kappa_r I_{\{r \in \mathcal{R}^*\}}/\mu_r$. Then $\alpha^{\emptyset}(\hat{x}) = 0$ and so $\hat{x} \in X_{\emptyset} \cup X_1^-$. It follows from eqns (1.7), (2.1), and (2.2) that $\hat{x}$ is the unique fixed point of the process $x(\cdot)$ in the set $X_{\emptyset} \cup X_1^-$. If $\mathcal{R}^* = \mathcal{R}$ then it follows easily from the eqns (1.7) that there is no further fixed point in X^+.

When the condition (2.4) holds but $\mathcal{R}^* \neq \mathcal{R}$, then there may be more than one fixed point. Bean *et al.* (1994b) give a numerical example with two call types. All calls require a single unit of resource and $C = 1000$, $\kappa_1 = 500$, $\kappa_2 = 700$, $\mu_1 = 1.0$, $\mu_2 = 0.1$. The acceptance regions are given by $\mathcal{A}_1 = \{m: m > 0\}$ and $\mathcal{A}_2 = \{m: 0 < m < 5\}$ (so that here $\mathcal{R}^* = \{1\}$). They show that the process $x(\cdot)$ possesses three distinct fixed points $x^{(1)}$, $x^{(2)}$, $x^{(3)}$. Every trajectory of the process $x(\cdot)$ tends to one of these points, although the point $x^{(3)}$ is unstable in the sense of possessing a domain of attraction of Lebesgue measure zero in X. The limit behaviour of the corresponding sequence of networks is therefore essentially bistable. The left panel of Fig. 1 shows sample trajectories of the process $x(\cdot)$—the thick line separates the domains of attraction of $x^{(1)}$ and $x^{(2)}$ and is of course itself a trajectory of the system, tending to $x^{(3)}$. The right panel shows simulated trajectories of the process $x^1(\cdot)$ ($= n^1(\cdot)$) in the associated sequence of networks. Here C is sufficiently large that the process $x^1(\cdot)$ should be reasonably well-approximated by $x(\cdot)$ and indeed the bistable behaviour of $x^1(\cdot)$ is clearly evident. However, this process is of course ergodic, so that, over sufficiently long time periods, it alternates between typically lengthy residences in the neighbourhoods of $x^{(1)}$ and $x^{(2)}$.

In the case where the relation (2.4) does *not* hold the fixed points of the process $x(\cdot)$ necessarily lie in X_1^+. Where, additionally, $A_r = 1$ for all r, an argument of Bean *et al.* (1995) for the case $\mathcal{R}^* = \mathcal{R}$ extends unchanged to the present case to show that there is a unique fixed point $\bar{x} \in X_1^+$. Provided only that all trajectories of $x(\cdot)$ then converge to $\bar{x}$ (this is difficult to show formally except in the case $R = 2$), identification of this point via the equations (1.7) permits the determination of limiting equilibrium behaviour—in particular limiting call acceptance probabilities—for the associated sequence of networks.

3 Two-resource networks

We now study the two-resource case $\mathcal{J} = \{1, 2\}$. Here some distinctly pathological behaviour is possible, as is shown by the example of Hunt (1995), which we discuss briefly below. We require conditions under which such pathological behaviour may not occur.

For any x and j, the restriction of the process $m_x(\cdot)$ to E_j is essentially one-dimensional and it follows, as in the previous section, that the distribution π_x^j exists if and only if $\alpha_j^\emptyset(x) > 0$. It again follows as there, and using the condition (1.4), that when $x(t) \notin X_{12}$ then $\pi_t = \pi'_{x(t)}$ where π'_x is given by

$$\pi'_x = \begin{cases} \pi_x^\emptyset & \text{if } x \in X_\emptyset \cup X_1^- \cup X_2^-, \\ \pi_x^j & \text{if } x \in X_j^+, \end{cases} \tag{3.1}$$

and where, for each j, $X_j^+ = \{x \in X_j \colon \alpha_j^\emptyset(x) > 0\}$, $X_j^- = X_j \setminus X_j^+$. It remains to consider the identification of π_t in the case where $x(t) \in X_{12}$. The key here is again given by the functions α_j^S.

For either $j \in \mathcal{J}$, let j' denote its complement in $\mathcal{J}$. For each j, define the function β_j on X by

$$\beta_j(x) = \begin{cases} \alpha_j^{j'}(x) & \text{if } \alpha_{j'}^\emptyset(x) > 0, \\ \alpha_j^\emptyset(x) & \text{if } \alpha_{j'}^\emptyset(x) \le 0. \end{cases} \tag{3.2}$$

Recall that $\alpha_j^{j'}(x)$ is defined if and only if $\alpha_{j'}^\emptyset(x) > 0$. The quantity $\beta_j(x)$ also has an informal interpretation in terms of the restriction of the process $m_x(\cdot)$ to $E_{12} = \mathbb{Z}_+^2$. In the case $\alpha_{j'}^\emptyset(x) > 0$, suppose that the component j of this restricted process is far from 0 but the component j' is in equilibrium; then $\beta_j(x)$ is the averaged (negative) drift rate of the component j. In the case $\alpha_{j'}^\emptyset(x) \le 0$, a similar but simpler interpretation holds. These ideas may be formalized as, for example, by Fayolle *et al.* (1995), but for our purposes a formal definition is more easily made as above in terms of the invariant distributions associated with the restrictions of the process $m_x(\cdot)$ to $E_{j'}$ or $E_\emptyset$ as appropriate.

Define subsets of X_{12} as follows. Let

$$\begin{aligned} U &= \{x \in X_{12} \colon \beta_1(x) \wedge \beta_2(x) > 0\}, \\ V_j &= \{x \in X_{12} \colon \beta_j(x) > 0, \ \beta_{j'}(x) \le 0\}, \quad j = 1, 2, \\ W &= \{x \in X_{12} \colon \beta_1(x) \vee \beta_2(x) \le 0\}, \\ W^- &= \{x \in W \colon \alpha_1^\emptyset(x) \vee \alpha_2^\emptyset(x) \le 0\}, \end{aligned}$$

$$W^+ = \{x \in W: \alpha_1^{\emptyset}(x) \wedge \alpha_2^{\emptyset}(x) > 0\}.$$

Note that it follows from the definition (3.2) that $W = W^- \cup W^+$, so that the above sets form a partition of X_{12}.

Bean *et al.* (1994b) show that, under the condition

$$(\infty, \infty) \in \mathcal{A}_r \text{ for all } r \in \mathcal{R}, \tag{3.3}$$

for (almost) all t,

$$\pi_t = \begin{cases} \pi_{x(t)}^{12} & \text{if } x(t) \in U, \\ \pi_{x(t)}^{j} & \text{if } x(t) \in V_j, \quad j = 1, 2, \\ \pi_{x(t)}^{\emptyset} & \text{if } x(t) \in W^-. \end{cases} \tag{3.4}$$

We shall discuss below the necessity of the condition (3.3) for the result (3.4), and also the remaining case, $x(t) \in W^+$. However, note that when the result (3.4) holds and W^+ is empty (as will usually be the case in applications), it is again possible to define a velocity field for the process $x(\cdot)$ everywhere on X.

The result (3.4) is proved using essentially the same arguments as those used to establish the result (2.2) in the single resource case. We give here an outline. Note first that, under the condition (3.3), it follows from the definitions (1.10) and (3.2), that, for all x,

$$\beta_j(x) \leq \alpha_j^{\emptyset}(x), \quad j = 1, 2. \tag{3.5}$$

Further, again under this condition (3.3), standard results for Markov chains on $\mathbb{Z}_+^2$ with partial spatial homogeneity (see Fayolle *et al.*, 1995, or Zachary, 1995) show that, for all x,

$$\pi_x^{12} \text{ exists if and only if } \beta_1(x) \wedge \beta_2(x) > 0. \tag{3.6}$$

Note also that, analogously to eqn (2.3), and by again using in particular the result (1.11), we have that for each j,

$$\sum_{r \in \mathcal{R}} A_{jr} x_r(t) = \sum_{r \in \mathcal{R}} A_{jr} x_r(0) + \int_0^t \{\lambda^{\emptyset}(u)\alpha_j^{\emptyset}(x(u)) + \lambda^{j'}(u)\alpha_j^{j'}(x(u))\} du. \tag{3.7}$$

From eqn (3.5), for t with $x(t) \in U$ and each j, $\alpha_j^{\emptyset}(x(t)) > 0$ and $\alpha_j^{j'}(x(t)) > 0$. It follows from eqn (3.7), arguing as in the single resource case, that, for (almost) all t with $x(t) \in U$, $\lambda^{\emptyset}(t) = \lambda^1(t) = \lambda^2(t) = 0$ and so $\pi_t = \pi_{x(t)}^{12}$ as required.

The remaining cases of the result (3.4) are proved similarly, on making use also of the result (3.6).

It seems likely that the result (3.6) continues to hold in the absence of the condition (3.3) (here the only doubt in the existing literature lies with the boundary case $\beta_1(x) \wedge \beta_2(x) = 0$) in which case a relatively straightforward variation of the above argument may be used to show that the result (3.4) also continues to hold. Thus, under what are at worst mild regularity conditions, and certainly in the case where the condition (3.3) does hold, a velocity field for the process $x(\cdot)$

may be defined everywhere on the set $X \setminus W^+$. When W^+ is empty, again only mild regularity conditions are required to show that the trajectories of the process $x(\cdot)$ are well-defined and that, for each t, $x(t)$ is a continuous function of $x(0)$. It follows that, in this case, there exists at least one fixed point for the process $x(\cdot)$.

Hunt (1995) gives an example in which the set W^+ is nonempty. Here, and in general, for t such that $x(t) \in W^+$,

$$\pi_t = \lambda^1(t)\pi^1_{x(t)} + \lambda^2(t)\pi^2_{x(t)}, \tag{3.8}$$

where as usual $\lambda^1(t)$ and $\lambda^2(t)$ are positive and sum to one. However, beyond this, the behaviour of the process $x(\cdot)$ within the set W^+ is indeterminate, corresponding to the fact that here the sequence of processes $x^N(\cdot)$ may have different limits in different subsequences. Trajectories of two such limits may agree up to the time of entrance into W^+, but behave quite differently thereafter.

It is therefore important for the control of networks to have conditions which ensure that the set W^+ is empty. For each $r \in \mathcal{R}$, let $\mathcal{J}_r = \{j \in \mathcal{J}: A_{jr} > 0\}$. Extend the definition of $\mathcal{R}^*$ given in the previous section to two- (and more) resource networks by letting

$$\mathcal{R}^* = \{r \in \mathcal{R}: E_S \subset \mathcal{A}_r \text{ for all } S \text{ with } S \cap \mathcal{J}_r = \emptyset\}. \tag{3.9}$$

Thus, using also the continuity of $I_{\mathcal{A}_r}$, calls of type $r \in \mathcal{R}^*$ are accepted for all sufficiently large values of the free capacities of those resources in $\mathcal{J}_r$—regardless of the state of the remaining resources. In a variation of Conjecture 5 of Hunt and Kurtz (1994) we conjecture that a sufficient condition for W^+ to be empty is given by $\mathcal{R}^* = \mathcal{R}$. (This of course implies in particular the condition (3.3).)

The following theorem shows this to be the case where $A_{1r} = A_{2r}$ for those call types r such that $A_{1r} \wedge A_{2r} > 0$. It generalizes a result of Moretta (1995).

Theorem 3.1 *Suppose that $\mathcal{R}^* = \mathcal{R}$ and that*

$$A_{1r} = A_{2r} \text{ for all } r \text{ with } \mathcal{J}_r = \mathcal{J}. \tag{3.10}$$

Then W^+ is empty.

Proof Suppose there exists $x \in W^+$. Then, for each j, $\alpha^\emptyset_j(x) > 0$ and so the distribution π^j_x exists. Thus, again for each j, $\alpha^{j'}_j(x) = \beta_j(x) \leq 0$, and since also (by the result (1.11)) $\alpha^j_j(x) = 0$, it follows that

$$\sum_{r \in \mathcal{R}} A_{jr}\kappa_r\{\pi^{j'}_x(\mathcal{A}_r) - \pi^j_x(\mathcal{A}_r)\} \leq 0. \tag{3.11}$$

For r such that $A_{jr} > 0$, either $\mathcal{J}_r = \{j\}$ or $\mathcal{J}_r = \mathcal{J}$. In the former case the condition $\mathcal{R}^* = \mathcal{R}$ implies that $\pi^{j'}_x(\mathcal{A}_r) = 1$, while $\pi^j_x(\mathcal{A}_r) < 1$ (since necessarily $E_j \not\subset \mathcal{A}_r$). Hence

$$\sum_{r: \mathcal{J}_r = \mathcal{J}} A_{jr}\kappa_r\{\pi^{j'}_x(\mathcal{A}_r) - \pi^j_x(\mathcal{A}_r)\} \leq 0 \tag{3.12}$$

with strict inequality if $\mathcal{J}_r = \{j\}$ for at least one r.

The irreducibility assumption of Section 1 implies that there is at least one call type r such that $\mathcal{J}_r = \{1\}$ or $\mathcal{J}_r = \{2\}$. It follows, on interchanging j and j' in eqn (3.12), and using the condition (3.10), that the eqn (3.12) is self-contradictory.   □

As usual the fixed points of the process $x(\cdot)$ are determined by solution of the equations (1.7). In the case where, for either j, $\sum_r A_{jr}\kappa_r/\mu_r \leq C_j$, we may in effect replace C_j by ∞ and consider the single resource j'. Otherwise, and when the set U is nonempty, the analysis may be more complicated, involving in particular the (nontrivial) determination of two-dimensional invariant distribution π_x^{12} for $x \in U$.

We know of no example in which $\mathcal{R}^* = \mathcal{R}$ and there is more than one fixed point. Moretta (1995) considers the case where $\mathcal{R}^* = \mathcal{R}$ and the matrix $(A_{jr},\ j \in \mathcal{J},\ r \in \mathcal{R})$ is given by

$$A_{jr} = \begin{pmatrix} 1 & 0 & 1 \\ 0 & 1 & 1 \end{pmatrix}.$$

He uses a coupling argument to show that here, if there is more than one fixed point, then all fixed points necessarily lie in U. If therefore an (essentially straightforward single resource) analysis identifies a fixed point outside this set, there will be no further fixed point within it. Moretta also presents compelling evidence that, for this model, there is only ever one fixed point.

Moretta also considers the problem of the determination of the invariant distribution π_x^{12} for $x \in U$, and that of determining more refined approximations to call acceptance probabilities in networks whose capacities are insufficiently large to justify direct application of the above asymptotic theory.

4   General networks

In the previous two sections we have outlined an essentially complete theory for the identification of the 'driving' distribution π_t of eqn (1.3) in the case of single and two-resource networks. This has used little more than Hunt's elementary observation that the process $x(\cdot)$ must remain within the set X. (Only for t such that $x(t)$ belongs to the set W^+, defined in the previous section, is a more careful argument required, and this too is due to Hunt (1995).)

For networks with more than two resources, the identification of π_t is very much more complex. For $x \in X$, define a set $\mathcal{S} \subseteq \mathcal{J}$ to be *blocking* with respect to x if $\pi_x^{\mathcal{S}}$ exists and $\sum_r A_{jr}x(t)_r = C_j$ for all $j \in \mathcal{S}$. One very reasonable conjecture is that, for any t, $\pi_t = \pi_{x(t)}^{\mathcal{S}}$ whenever there exists a 'maximal' blocking set $\mathcal{S}$ with respect to $x(t)$ containing every other such blocking set.

Again as remarked earlier, we are particularly interested in the identification of conditions under which a velocity field may be defined for the process $x(\cdot)$. We hesitate to make any conjectures here, but merely observe that for none of the 'pathological' examples of Hunt (1995) is the condition $\mathcal{R}^* = \mathcal{R}$ satisfied.

Acknowledgements   The author is grateful to many people for their contributions to this paper—especially to Nigel Bean, Richard Gibbens, Phil Hunt, Frank Kelly and Brian Moretta, and to Edward Ionides for assistance in constructing the simulations.

Bibliography

1. Bean, N.G., Gibbens, R.J., and Zachary, S. (1994a). The performance of single resource loss systems in multiservice networks. In Jacques Labetoulle and James W. Roberts (eds), *The Fundamental Role of Teletraffic in the Evolution of Telecommunications Networks*, Proceedings of the 14th International Teletraffic Congress, pp. 13–21, Elsevier Science B.V.

2. Bean, N.G., Gibbens, R.J., and Zachary, S. (1994b). Dynamic and equilibrium behaviour of controlled loss networks. Research Report #94-31, Statistical Laboratory, The University of Cambridge, 16 Mill Lane, Cambridge, CB2 1SB, U.K.

3. Bean, N.G., Gibbens, R.J., and Zachary, S. (1995). Asymptotic analysis of large single resource loss systems under heavy traffic, with applications to integrated networks. *Adv. Appl. Probab.*, **27**, 273–292.

4. Fayolle, G., Malyshev, V.A., and Menshikov, M.V. (1995). *Topics in the Constructive Theory of Countable Markov Chains*. Cambridge University Press.

5. Hunt, P.J. (1990). *Limit theorems for stochastic loss networks*. Ph.D. dissertation, University of Cambridge.

6. Hunt, P.J. (1995). Pathological behaviour in loss networks. *J. Appl. Probab.*, **32**, 519–533.

7. Hunt, P.J. and Kurtz, T.G. (1994). Large loss networks *Stochastic Process. Appl.*, **53**, 363–378.

8. Kelly, F.P. (1991). Loss networks. *Ann. Appl. Probab.* **1**, 319–378.

9. Moretta, B. (1995). *Behaviour and control of single and two resource loss networks*. Ph.D. dissertation, Heriot-Watt University.

10. Zachary, S. (1995). On two-dimensional Markov chains in the positive quadrant with partial spatial homogeneity. *Markov Process. Relat. Fields*, **1**, 267–280.

12

Admission controls for loss networks with diverse routing

Iain MacPhee

University of Durham

Ilze Ziedins

University of Auckland

Abstract

This paper studies control policies for large symmetric star-shaped networks, in which arrivals request capacity on one or two links. We formulate the optimization problem of interest for the network, and consider the limit as the number of links in the network becomes large while the arrival rate and capacities at any single link are held fixed. The optimal solution of the time-independent version of the deterministic problem obtained in this limit is to use a threshold policy, also known as trunk reservation. This work complements that of Hunt and Laws for fully connected networks.

1   Introduction

In this paper we study control policies for a highly symmetric star-shaped loss network. Loss networks are often used as models of certain types of telecommunication networks (such as circuit-switched networks), and we will freely use the terminology associated with that application. The review paper by Kelly (1991) gives an excellent introduction to and overview of results for loss networks.

The star-shaped loss network consists of K links, each link connecting an external node to a central node or hub (see Fig. 1). We assume that each link consists of C circuits, that is it has capacity C. We allow two types of route on this network. Single link routes connect an external node to the central hub. Two-link routes connect two external nodes via the central hub. There are thus K single link routes and $\binom{K}{2}$ two-link routes consisting of the pairs $\{(i,j) : 1 \leq i,j \leq K, i \neq j\}$. We assume that calls at the routes arrive as independent Poisson processes. The arrival rate at each one-link route is ν. The arrival rate at each two-link route is $\lambda_K = \frac{\lambda}{K-1}$. There are exactly $K-1$ two-link routes passing through each link, and so this scaling ensures that the total traffic offered to a single link is held fixed at $\nu + \lambda$, regardless of the size of the network.

When a call arrives, it requests one unit of capacity (or circuit) from all the

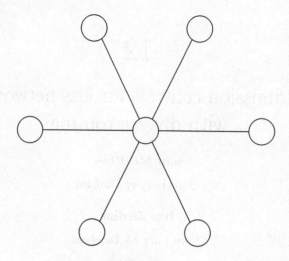

FIG. 1. A star network, $K = 6$.

links along its route. If that capacity is not immediately available, the call is blocked and lost. Otherwise, if the requested capacity is available, the call seizes it simultaneously on all the links along its route, and releases it all after an exponentially distributed period of time which does not depend on the type of the call. Without loss of generality we assume that the call duration has mean 1. Arrival processes and call durations are all independent of one another.

We assume that single link calls give a return of α if they are accepted and two-link calls a return of β. We are interested in the structure of acceptance/rejection (or admission) policies that maximize the expected return per unit time from the network in equilibrium.

We will be studying this network as $K \to \infty$, with the capacities and arrival rates at each link held fixed. This limiting regime is one where asymptotically, for K large, the occupancies of the links are approximately independent. Such a limiting regime is known as a diverse routing limit, since for K large each link has a large number of multi-link routes with very low arrival rates passing through it.

The symmetric star-shaped network has been previously considered by Whitt (1985), Ziedins and Kelly (1989) and Hunt (1992, 1995). Whitt (1985) obtained limiting results, as $K \to \infty$, for the uncontrolled case, where calls are always accepted if there is capacity to carry them. Ziedins and Kelly (1989) studied the stationary distribution directly. Hunt (1995, 1992) obtained weak convergence results for a more general symmetric star-shaped network that we will be applying later. More recently, Antoniu (1995) and Antoniu and MacPhee (1996) have studied star-shaped symmetric and asymmetric networks numerically.

In this paper, we will be restricting our attention to *local* admission policies, where the decision whether to accept or reject a call depends only on the state of the links along which a call is routed. Of course, for a single link call, such a policy depends only on the occupancy of the link along which it is routed. For two-link calls, we distinguish three classes of such local policies.

I The decision whether to accept or reject a two-link call depends on the state of both links, but both links decide independently whether or not to accept the call. This is a one-dimensional policy.

II Two-link calls are directed from one external node to another. The admission decision for a two-link call is based only on the state of the *first* link through which it passes. If a call is accepted on its first link, it is then rejected only if the second link is full. Since the two-link routes in this model are now directed, the total number of two-link routes has doubled, and we therefore assume that the arrival rate on any directed two-link route is now $\frac{\lambda}{2(K-1)}$. Again, this is a one-dimensional policy.

III Finally, the admission decision for a two-link call depends on the state of *both* links on which it is to be carried. However, interaction is allowed, so that the links on a route do not decide independently of one another whether to carry an offered call. This is a fully two-dimensional policy.

For a K-link network, the exactly optimal policy can be found using the theory of Markov decision processes. The optimal policy will, of course, depend on the state of the whole network, and so local policies such as those we have described above will not, in general, be optimal. However, Antoniu (1995), and Antoniu and MacPhee (1996) give a number of examples showing that the increase in returns obtained by using a global policy (rather than the suboptimal local policies we consider) may be negligible, even for relatively small K and C – in examples up to size $K = 5$, $C = 3$ over a range of offered traffics in cases where single link calls are more valuable per link than two-link calls (chapter 3 of Antoniu 1995) the increase is always less than 1%. Thus we restrict our attention here to attempting to characterize the optimal policy within the class of local policies that we are considering.

One of the aims of this work is to study the optimality or otherwise of trunk reservation policies. Trunk reservation policies are threshold policies. Calls are accepted on a link when the amount of free capacity on that link is greater than a trunk reservation parameter r. This parameter may depend both on the link (in an asymmetric network) and on the type of call (higher-valued calls in general having smaller reservation parameters). Threshold policies have been widely studied, e.g. Kelly (1990). It is known that they are optimal under various assumptions (such as Poisson arrivals, exponential holding times) if the network consists of just one link (Lippman 1975). In general such a policy is *not* optimal. For instance, Key (1990) gives an example of a two-link network, with single link and two-link traffic, where the rejection region for two-link calls is concave. However, threshold policies may be very close to optimal. Certainly, the optimal policies of class I and class II (the one-dimensional policies) are of threshold form, although, perhaps unexpectedly, they may be randomized – calls are accepted with some probability θ if the amount of free capacity is exactly r.

The results presented in this paper complement those of Hunt and Laws (1991) and many of the proofs of our results follow theirs. Hunt and Laws consider a symmetric fully connected network, with K links, and capacity C on each link. Offered calls may be carried either on the direct single link route or via an alternative two-link path (there are $K - 2$ such two-link alternatives to each direct single link route). The aim is to find the policy that minimizes the blocking probability (the probability that a call cannot be carried). Hunt and Laws consider the network as $K \to \infty$, and find that the asymptotically optimal policy is least busy alternative routing with trunk reservation. Under this policy a call is routed directly if possible. If not, it is routed via the least busy two-link alternative route, as long as both links on the alternative route have no more than v circuits in use, for some fixed v. Note that the limiting regime under which they consider their fully connected network also gives a form of diverse routing, since the arrival rate on any given two-link alternative route will become negligible as the network becomes large, and there are an ever-increasing number of such two-link alternative routes passing through each link. Our approach is very similar to theirs – we state the necessary weak convergence results, giving a deterministic motion in the limit, and then study the behaviour of the deterministic limit.

In section 2 we outline (without proof) the theoretical argument for the class I policies. Section 3 gives examples illustrating some interesting features of the optimal policies. This paper covers results that will be described in greater detail in two forthcoming papers.

2 Optimality conditions

In this section we discuss the treatment of policies of class I in detail. Consider the K-link network. Let $Kx_j^K(t)$ be the number of links with j circuits occupied at time t (so there are $C - j$ circuits available to carry calls on those links). We consider a sequence of networks, indexed by K, the Kth network operating under any admissible policy of class I. Then the following theorem is a special case of Theorem 2 in Hunt (1992).

Theorem 2.1 *Consider a sequence of networks, indexed by K, the K-th network operating under any admissible policy of class I. Then $\{x^K(\cdot)\}$ is relatively compact and $x(\cdot)$, the limit as $K \to \infty$ of any convergent subsequence, has the following properties:*

(i) $\sum_{j=0}^{C} x_j(t) = 1$ *for all $t \geq 0$.*

(ii) $x_j(t) \geq 0$ *for $0 \leq j \leq C$ and all $t \geq 0$.*

(iii) *There exist $\psi(\cdot)$ and $\eta(\cdot)$ such that almost surely, for almost all t, for $i = 0, 1, \ldots, C$,*

$$\psi_i(t), \psi_{ib}(t), \eta_i(t), \eta_{ib}(t) \geq 0, \tag{2.1}$$
$$\psi_i(t) + \psi_{ib}(t) = \nu x_i(t), \tag{2.2}$$

$$\eta_i(t) + \eta_{ib}(t) = \lambda x_i(t); \qquad (2.3)$$

and for $i = 0, 1, \ldots, C - 1,$

$$\frac{d}{dt}\left(\sum_{j=0}^{i} x_j(t)\right)(t) = (i+1)x_{i+1}(t) - \psi_i(t) - \frac{1}{\lambda}\eta_i(t)1 \cdot \eta(t) \qquad (2.4)$$

where we have defined

$$1 \cdot \eta(t) \equiv \sum_{i=0}^{C-1} \eta_i(t).$$

Conditions (i) and (ii) of the theorem are obvious. In the third condition, the $\psi(\cdot)$ and $\eta(\cdot)$ have a natural interpretation. The quantity $\nu x_i(t)$ is the arrival rate for single link calls at a link multiplied by the probability that the link has occupancy i at time t – this is split into $\psi_{ib}(t)$, corresponding to those calls that are blocked, and $\psi_i(t)$, corresponding to those that are accepted. The quantity $\lambda x_i(t)$ is split in a similar fashion. It will often be the case that one or other of $\psi_i(t)$ and $\psi_{ib}(t)$ is equal to 0, corresponding to single links being blocked or accepted when the occupancy is i. A similar observation holds for the $\eta(\cdot)$. The differential equation (2.4) is a time-dependent version of the detailed balance equations. Calls depart from links that have occupancy $i+1$ at rate $i+1$ (since the mean call duration is 1), and so the proportion of links in state i increases at rate $(i+1)x_{i+1}(t)$ due to call departures. It decreases at rate $\psi_i(t)$ due to single link arrivals that are accepted and at rate $\frac{1}{\lambda}\eta_i(t)1 \cdot \eta(t)$ due to two-link arrivals that are accepted on links that are in state i.

The conditions of the theorem now give us the constraints for the formulation of our time-dependent optimization problem. Let $\pi_j(i)$ be the probability that a j-link call, arriving to a link in state i, is accepted on that link, $i = 0, \ldots, C$, $j = 1, 2$. Our objective is to maximize the expected average return per unit time. That is, we wish to

$$\text{maximize } \limsup_{T \to \infty} \frac{1}{T}\int_0^T \left\{\alpha \sum_{i=0}^{C-1} \psi_i(t) + \frac{\beta}{2}\frac{(1 \cdot \eta(t))^2}{\lambda}\right\} dt$$

subject to the conditions (obtained in Theorem 2.1 above)

$$\psi_i(t) + \psi_{ib}(t) = \nu x_i(t), \qquad \eta_i(t) + \eta_{ib}(t) = \lambda x_i(t),$$

$$\frac{d}{dt}\left(\sum_{j=0}^{i} x_j(t)\right) = (i+1)x_{i+1}(t) - \psi_i(t) - \frac{\eta_i(t)1 \cdot \eta(t)}{\lambda}$$

$$\text{and } \sum_{i=0}^{C} x_i(t) = 1.$$

Note that $\psi_i(t) = \nu\pi_1(i)x_i(t)$ and $\eta_i(t) = \lambda\pi_2(i)x_i(t)$.

Rather than considering the time-dependent problem outlined above, we consider a simpler time-independent version:

$$\text{maximize } \alpha \sum_{i=0}^{C-1} \psi_i + \frac{\beta}{2\lambda}(1 \cdot \eta)^2$$

subject to the constraints

$$\psi_i + \psi_{ib} = \nu x_i, \qquad \eta_i + \eta_{ib} = \lambda x_i, \qquad 0 \leq i \leq C \tag{2.5}$$

$$(i+1)x_{i+1} = \psi_i + \frac{1}{\lambda}\eta_i 1 \cdot \eta, \qquad 0 \leq i \leq C-1 \tag{2.6}$$

with of course $\sum_{i=0}^{C} x_i = 1$ and all variables non-negative. The exact relationship between the two models depends on the convergence behaviour of the continuous time model which we discuss in more detail at a later point.

Define $y \cdot z \equiv \sum_{j=0}^{C-1} y_j z_j$. Then the Lagrangian for the time-independent problem is

$$\begin{aligned}
L &= \alpha \sum_{i=0}^{C-1} \psi_i + \frac{\beta}{2\lambda}(1 \cdot \eta)^2 + \sum_{i=0}^{C} a_i(\psi_i + \psi_{ib} - \nu x_i) + \sum_{i=0}^{C} b_i(\eta_i + \eta_{ib} - \lambda x_i) \\
&\quad + \sum_{i=0}^{C-1} d_i\left((i+1)x_{i+1} - \psi_i - \frac{\eta_i}{\lambda}1 \cdot \eta\right) + e\left(\sum_{i=0}^{C} x_i - 1\right) \\
&= \sum_{i=0}^{C} \psi_i(\alpha + a_i - d_i) + \sum_{i=0}^{C} \psi_{ib} a_i + \sum_{i=0}^{C}(\eta_i + \eta_{ib})b_i - \frac{1}{\lambda}(d \cdot \eta)(1 \cdot \eta) \\
&\quad + \frac{\beta}{2\lambda}(1 \cdot \eta)^2 + \sum_{i=0}^{C} x_i(id_{i-1} - \nu a_i - \lambda b_i + e) - e.
\end{aligned}$$

As the variables are all non-negative, at any maximum of L we have

$$\text{either } \frac{\partial L}{\partial y} = 0 \quad \text{or} \quad y = 0, \frac{\partial L}{\partial y} \leq 0 \tag{2.7}$$

for each variable y. The partial derivatives are given by

$$\frac{\partial L}{\partial \psi_{ib}} = a_i, \qquad \frac{\partial L}{\partial \psi_i} = \alpha + a_i - d_i, \tag{2.8}$$

$$\frac{\partial L}{\partial \eta_{ib}} = b_i, \qquad \lambda\frac{\partial L}{\partial \eta_i} = \lambda b_i - d_i 1 \cdot \eta - d \cdot \eta + \beta 1 \cdot \eta, \tag{2.9}$$

$$\frac{\partial L}{\partial x_i} = id_{i-1} - \nu a_i - \lambda b_i + e. \tag{2.10}$$

Then we have the following:

Theorem 2.2 *The optimal policy for the time-independent problem has the following properties:*

(i) $\pi_j(i-1) \geq \pi_j(i)$, $j = 1, 2$, $0 \leq i \leq C$.

(ii) $\pi_1(C-1) = 1$ *and/or* $\pi_2(C-1) = 1$.

(iii) *For* $j = 1, 2$ *there exists* $r_j \in (0, 1, \ldots, C-1)$ *such that*

$$\pi_j(i) = 1, \qquad i < C - r_j$$
$$\pi_j(C - r_j) = \theta, \qquad where \ \ \theta \in (0, 1]$$
$$\pi_j(i) = 0, \qquad i > C - r_j.$$

Note that condition (ii) ensures that at least one of r_1, r_2 equals 0. We see that the optimal policy is of trunk reservation form, possibly with randomization at the boundary. The proof of this proceeds by appealing directly to the conditions on the partial derivatives in (2.7). A direct consequence of these conditions is that the quantity d_i which appears as a multiplier in the Lagrangian can be seen as the cost of accepting a call on a link when it is in state i. For instance, suppose that the optimal policy has $\pi_1(i) = 1$, with $\psi_{ib} = 0$ and $\psi_i = \nu x_i > 0$. Then we must also have, applying (2.7) and (2.8),

$$\frac{\partial L}{\partial \psi_i} = \alpha + a_i - d_i = 0 \qquad and \qquad \frac{\partial L}{\partial \psi_{ib}} = a_i \leq 0,$$

and so $\alpha \geq d_i$. If the optimal policy has $\pi_1(i) = 0$ we have $\psi_{ib} = \nu x_i > 0$, $\psi_i = 0$ and now $\frac{\partial L}{\partial \psi_i} \leq 0$ and $\frac{\partial L}{\partial \psi_{ib}} = a_i = 0$ imply that $\alpha \leq d_i$. Thus single link calls are accepted on a link in state i if $\alpha \geq d_i$. Similarly, applying the conditions (2.7) and (2.9) we can show that two-link calls are accepted on a link in state i if

$$\beta \geq d_i + \frac{\sum_{j=0}^{C-1} d_j x_j \pi_2(j)}{\sum_{j=0}^{C-1} x_j \pi_2(j)}. \tag{2.11}$$

The first term in (2.11), d_i, is the cost of accepting on the link in question, and the remaining term is the cost of accepting on the second link (of which we do not know the state), conditional on the call being accepted. Thus the **d** have a natural interpretation as the implied or shadow costs of using capacity.

Once this observation has been made, the proof proceeds by assuming that at the maximum the d_i are increasing in i, and then showing that all of the necessary conditions for a maximum are satisfied under this assumption. Finally, we observe that if the d_i are increasing at the optimum point, then by the discussion in the previous paragraph, this implies that the optimal policy is of trunk reservation form.

It can be shown that a unique equilibrium distribution, $\hat{\mathbf{x}}$, exists when the policy is of trunk reservation form. The $\hat{x}_i$ satisfy the detailed balance conditions, obtained from (2.6): $(i+1)\hat{x}_{i+1} = R_i \hat{x}_i$, for $0 \leq i < C$ where $R_i = \nu \pi_1(i) + \lambda \pi_2(i) \sum_{k=0}^{C-1} \pi_2(k) \hat{x}_k$ and so

$$\hat{x}_i = \hat{x}_0 \frac{\Pi_{j=0}^{i-1} R_j}{i!}, \qquad 0 < i \leq C.$$

It is also possible to show (using arguments similar to those of Hunt and Laws (1991), Theorem 5) that under certain conditions when a threshold control is applied to the time-dependent model the limiting deterministic motion $x(t)$ of Theorem 2.1 satisfies $x(t) \to \hat{x}$. We are currently able to do this only under restricted starting conditions, e.g. $x_0(0) = 1, x_i(0) = 0, 1 \le i \le C$, or $x(0)$ 'close to' $\hat{x}$. A proof for general initial conditions would allow the argument of the appendix to the paper of Hunt and Laws (1991) to be adapted to apply to this problem in which case the optimal control for the time-independent problem is also average cost optimal for the time-dependent problem. At the moment the possibility of a limit cycle being optimal for the time-dependent problem for some starting configuration $\mathbf{x}(0)$ cannot be ruled out. Thus we can certainly conclude that the optimal trunk reservation policy above is at least a locally optimal policy (within class I) for the time-dependent problem and we conjecture that it is also globally optimal.

The development for policies of class II is very similar to that for class I policies. The time-independent optimization problem becomes

$$\text{maximize} \quad \alpha \sum_{i=0}^{C-1} \psi_i \; + \; \beta \sum_{i=0}^{C-1} \eta_i \left(1 - x_C \right)$$

subject to the linear constraints

$$\psi_i + \psi_{ib} = \nu x_i, \qquad \eta_i + \eta_{ib} = \frac{1}{2} \lambda x_i,$$

for $i = 0, 1, \ldots, C$ and the quadratic constraints

$$(i+1)x_{i+1} = \psi_i + \eta_i \left(1 - x_C \right) + x_i \sum_{j=0}^{C-1} \eta_j$$

for $i = 0, 1, \ldots, C - 1$. We also require $\sum_{i=0}^{C} x_i = 1$ and all of the variables to be non-negative. The policy obtained from this optimization problem has the same threshold form as for the policies of class I.

3   Examples

The behaviour of the limiting model changes considerably over the range of values of its various parameters. In this section we consider two examples illustrating some interesting features of this behaviour.

In the case where $2\alpha = \beta$ so that both call types produce the same revenue per link used, we might expect to be indifferent to the type of any calls that request capacity and thus admit both types of call whenever there is free capacity. This is indeed optimal for small capacities, for instance when $C = 5$, but when $C = 20$ we find that for some ranges of offered traffic we should block some calls.

Figure 2 plots the ratio of d_{19}, the 'price' of using the last of the 20 circuits, to α when the admission policy is to admit all calls. The surface gives the value of the ratio and the contour below identifies the region of rates where the ratio

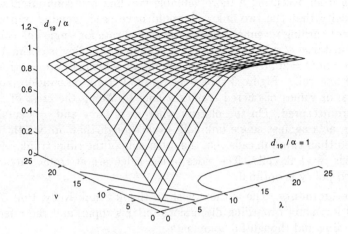

FIG. 2. Shadow price, d_{19}, of last circuit when $C = 20$.

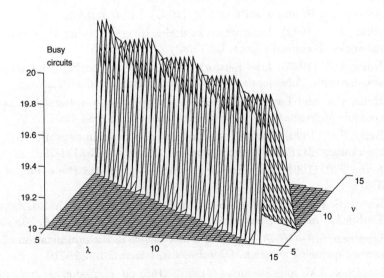

FIG. 3. Optimal trunk reservation parameters. $C = 20$, $\alpha = 1$, $\beta = 1.6$.

is greater than 1. We see that we should block single link calls when ν is small and λ is large. Two-link calls are accepted as the values d_i satisfy inequality

2.11. Our heuristic explanation for rejecting one-link calls is that, with these arrival rates, accepting a one-link call on a link which is almost at capacity may prevent us from accepting a more valuable two-link call soon after, while the extra capacity that the two-link call would have used remains empty (and is therefore not earning revenue) because the arrival rate for single link calls is low.

As an instance of a situation where a randomized policy is optimal we offer the case where $C = 20$, $\alpha = 1$ and $\beta = 1.6$, i.e. two-link calls are worth less than two single link calls. Figure 3 shows the optimal trunk reservation parameters for a range of values of offered traffics ν and λ. Along the crest of the ridge all calls are accepted. On the plateau where λ is large and ν is small, trunk reservation acts against single link calls even though they are worth more per circuit used than two-link calls. On the other side of the ridge trunk reservation acts against two-link calls. The sides of the ridge are at traffic rates where a randomized policy is optimal.

Acknowledgements The authors wish to thank Frank Kelly, Phil Hunt and Neil Laws for many rewarding discussions on this topic, and the referee for a careful reading and thoughtful comments.

Bibliography

1. Antoniu, N.H. (1995). Optimal admission policies for small star networks. PhD Thesis. University of Durham.

2. Antoniu, N.H. and MacPhee, I.M. (1996). In preparation.

3. Hunt, P.J. (1992). Loss networks under diverse routing II: *L*-symmetric networks. Technical report, University of Cambridge.

4. Hunt, P.J. (1995). Loss networks under diverse routing: the symmetric star network. *Advances in Applied Probability*, **25**, 255–272.

5. Hunt, P.J. and Laws, C.N. (1991). Asymptotically optimal loss network control. *Mathematics of Operations Research*, **18**, 880–900.

6. Kelly, F.P. (1990). Routing and capacity allocation in networks with trunk reservation. *Mathematics of Operations Research*, **15**, 771-793.

7. Kelly, F.P. (1991). Loss networks. *Annals of Applied Probability*, **1**, 319–378.

8. Key, P.B. (1990). Optimal control and trunk reservation in loss networks. *Probability in the Engineering and Informational Sciences*, **4**, 203–242.

9. Lippman, S.A. (1975). Applying a new device in the optimization of exponential queueing systems. *Operations Research*, **23**, 687–710.

10. MacPhee, I.M. and Ziedins, I. (1996). Optimal admission controls for loss networks with diverse routing I. Working paper.

11. Whitt, W. (1985). Blocking when service is required from several facilities simultaneously. *A.T. & T. Technical Journal*, **64**, 1807–1856.

12. Ziedins, I.B. and Kelly, F.P. (1989). Limit theorems for loss networks with diverse routing. *Advances in Applied Probability*, **21**, 804–830.

13

On load balancing in Erlang networks

Murat Alanyali and Bruce Hajek

University of Illinois at Urbana Champaign

Abstract

This chapter summarizes our recent work on the dynamic resource allocation problem. The question of interest is the performance of simple allocation strategies which can be implemented on-line. The chapter focuses on the least load routing policy. The analysis is based on fluid limit equations and the theory of large deviations for Markov processes with discontinuous statistics.

1 Introduction

The dynamic resource allocation problem arises in a variety of applications with load sharing features (see Ganger *et al.* (1993), and Willebeck-LeMair and Reeves (1993) for some examples). The generic resource allocation setting involves a number of locations containing resources. The dynamic aspect of the problem is the arrivals of consumers, each of which requires a certain amount of service from the resources, and the control variable of the problem is the "allocation policy", which specifies at which location each consumer is to be served. Oftentimes in applications, locations contain finitely many resources; hence the main objective of the allocation policy is to guarantee low blocking probability. On the other hand, in some applications such as spread spectrum mobile radio networks, there are no sharp capacity constraints, and the goal becomes to dynamically balance the load. In either case, one wants the allocation policy to have low complexity, require little information about the system state, and be robust to changes in the traffic parameters.

An instance of resource allocation arises in the wireless network pictured in Fig. 1. The network consists of a number of base stations and users. The users require communication channels that are available at the base stations, whereas each station may serve the users within its geographical range. The resource allocation problem in this setting concerns the question of station selection.

Our mathematical abstraction of a load sharing network is a triple (U, V, N), where U is a finite set of *consumer types*, V is a finite set of *locations*, and $(N(u) \subset V : u \in U)$ is a set of *neighborhoods* (see Fig. 2 for examples). A

The work in this paper was sponsored in part by the National Science Foundation under Contract NSF NCR 93 14253 and by a TUBITAK NATO Fellowship.

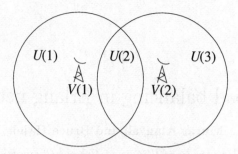

FIG. 1.   A typical wireless network with overlapping neighborhoods.

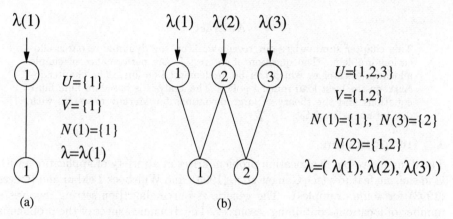

FIG. 2. Two simple load sharing networks: (a) the single-location network, and (b) the W network.

demand for this network is a vector $(\lambda(u) : u \in U)$ of positive numbers, where $\lambda(u)$ denotes the arrival rate of *type u consumers*. Each consumer is served, starting immediately upon its arrival, for the duration of its *holding time*. The neighborhood $N(u)$ denotes the locations that are available to type u consumers, in the sense that each such consumer can be served only at a location within $N(u)$. An *allocation policy* is an algorithm which assigns consumers to locations within their respective neighborhoods. The *load* at location $v \in V$ at a given time t, $X_t(v)$, is the number of consumers at v at time t. The allocation policy, together with the consumer arrival and departure times and an initial condition, determine the load process $X = (X_t : t \geq 0)$.

Load balancing is a possible guiding principle for resource allocation, whereby the load is allocated across locations as evenly as possible. It is well known that load balancing can be an effective allocation strategy, when the cost is convex (or the reward concave) as a function of the allocated loads. For example, $x(1)^2 + x(2)^2$ is minimized over probability vectors $(x(1), x(2))$ by $x(1) = x(2) = 1/2$. This is connected with the convexity of the function $f(x) = x^2$.

A reasonable allocation strategy for dynamic load balancing is the *least load routing* (LLR) policy, under which each arriving consumer is assigned to a location with the least load in the associated neighborhood. We focus on the analysis of the LLR policy under the following stochastic description of the network dynamics, indexed by a positive scalar γ. For each $u \in U$, consumers of type u arrive according to a Poisson process of rate $\gamma\lambda(u)$, the processes for different types of arrivals being independent. The holding time of each consumer is exponentially distributed with unit mean, independent of the past history. Given a positive number κ, we say that the network *overflows* when the load of some location exceeds its designated *capacity* $\lfloor \gamma\kappa \rfloor$.

Under the LLR policy, the load process is Markov with an explicit generator; therefore in principle one can learn much about the load process by computing its equilibrium distribution. Nevertheless, computation of the equilibrium distribution appears intractable for arbitrary network topologies. An alternative approach is to approximate the typical behavior of the network load by *fluid limit approximations*, which are deterministic weak limits of the network load, suitably normalized, for large values of γ.

Certain events of interest, such as network overflow for large enough values of κ, correspond to large deviations of the network from its typical behavior, and hence are not described accurately by the fluid limit approximation. Although network overflow is a rare event, it has a strong impact on network performance; it is thus desirable to estimate its probability and mode accurately. An appropriate tool for this purpose is large deviations theory, which may be employed to study network overflow in terms of *overflow exponents*: the overflow exponent of the network, $F^{LLR}(\kappa)$, and the overflow exponent of a location $v \in V$, $F^{LLR}(v, \kappa)$, under policy LLR are defined as

$$F^{LLR}(\kappa) = -\lim_{T\to\infty} \lim_{\gamma\to\infty} \gamma^{-1} \log P(\text{Network overflow time} \leq T | X_0 = 0)$$

$$F^{LLR}(v, \kappa) = -\lim_{T\to\infty} \lim_{\gamma\to\infty} \gamma^{-1} \log P(\text{Overflow time of location } v \leq T | X_0 = 0).$$

Note that $F^{LLR}(\kappa) = \min_{v\in V} F^{LLR}(v, \kappa)$ whenever the above quantities exist.

This paper concerns both the fluid limit approximations and the overflow exponents of load sharing networks operating under various allocation policies. Section 2 identifies the fluid limit approximations of the network load under the LLR policy as solutions of certain integral equations with boundary constraints. Section 3 provides the overflow exponents for the two simple networks of Fig. 2 under LLR, and conjectures the form of the overflow exponents for networks with arbitrary topologies. Finally, extensions of Section 2 to networks with either migration of load or finite capacity constraints, and extensions of Section 3 to two alternative allocation policies, are discussed in Section 4.

2 The fluid limit approximation

The fluid limit approximation is a description of the network behavior which is asymptotically exact in heavy traffic. To observe the typical behavior of the network load under the LLR policy, consider the *W network* of Fig. 2.b with the

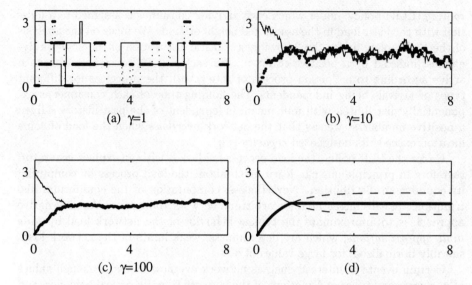

FIG. 3. The normalized load of the W network under the LLR policy.

demand $\gamma\lambda$ where $\lambda = (1,1,1)$, so that consumers of each type arrive at rate γ. Suppose that initially location 1 has zero load, whereas location 2 has load 3γ. Figures 3.a–3.c depict typical sample paths of the *normalized load* (load divided by γ) $X_t^\gamma = (X_t^\gamma(1), X_t^\gamma(2))$ for $\gamma = 1, 10, 100$, during the time interval $[0,8]$. As γ tends to infinity, the sample paths converge to the limiting trajectory depicted in Fig. 3.d.

Roughly speaking, the sequence of processes $(X^\gamma : \gamma > 0)$ is uniformly stochastically equicontinuous. It can thus be shown using methods of weak convergence (see Ethier and Kurtz (1986)) that if the sequence of initial conditions $(X_0^\gamma : \gamma > 0)$ is tight, then $(X^\gamma : \gamma > 0)$ is tight. Therefore, for any sequence $\gamma_n \to \infty$, there is a subsequence γ_{n_k} such that the distribution of $X^{\gamma_{n_k}}$ converges in Prohorov metric. This is equivalent to weak convergence of $X^{\gamma_{n_k}}$, and the weak limit x, together with some A, satisfies the following equations (Lemma 3.3 of Alanyali and Hajek (1995a)):

$$x_t(v) = x_0(v) + \sum_{u \in N^{-1}(v)} A_{u,v}(t) - \int_0^t x_s(v)ds, \qquad (2.1)$$

$$A_{u,v}(0) = 0, \quad A_{u,v}(t) \text{ nondecreasing}, \qquad (2.2)$$

$$\sum_{v \in N(u)} A_{u,v}(t) = \lambda(u)t, \qquad (2.3)$$

$$\int_0^t I\left\{x_s(v) > \min_{v' \in N(u)} x_s(v')\right\} dA_{u,v}(s) = 0. \qquad (2.4)$$

In the fluid analogy, $A_{u,v}(t)$ corresponds to the amount of type u fluid received by location v up until time t. By (2.1), $x_t(v)$ is the load at location v at time t under assignment regime A and linear discharge rate. Condition (2.3) is a "conservation of fluid" equation, stating that the total type u fluid arrived by time t is equal to the total type u fluid that is assigned to locations by time t. Note that Conditions (2.1)–(2.3) would be satisfied under any allocation policy. Condition (2.4) is the impact of the LLR policy, which implies that location v does not receive any type u fluid unless the load at v is equal to the minimum load in the neighborhood of u.

Solutions of a static optimization problem play an important role in the characterization of the fluid limit approximations. Let an *assignment* a, given by $(a_{u,v} : u \in U, v \in V)$, be called *admissible* if $a \geq 0$, and $a_{u,v} = 0$ whenever $v \notin N(u)$. An admissible assignment a *satisfies* demand λ if $\sum_v a_{u,v} = \lambda(u)$ for all $u \in U$. The *load* at location $v \in V$ corresponding to assignment a is given by $q(v) = \sum_u a_{u,v}$, and $q = (q(v) : v \in V)$ is called the *load vector*. Let $\mathcal{A}_\lambda$ denote the set of admissible assignments that satisfy demand λ, and let $\Phi : R^V \to R$ be a strictly convex, differentiable function which is symmetric in its arguments. The *static load balancing (SLB) problem* is defined as

$$SLB(\lambda, \Phi): \; Minimize(\Phi(q) \; : \; a \in \mathcal{A}_\lambda).$$

Lemma 2.1 *(Alanyali and Hajek (1995a)) There exists a solution to $SLB(\lambda, \Phi)$. An assignment $a \in \mathcal{A}_\lambda$ is a solution if and only if for all $u \in U$ and all $v \in N(u)$*

$$a_{u,v} = 0 \quad whenever \quad q(v) > \min_{v' \in N(u)} q(v').$$

Furthermore, all such assignments yield the same load vector.

The unique load vector corresponding to the solutions of $SLB(\lambda, \Phi)$ is called the *balanced load vector*. One connection between the SLB problem and load balancing is the fact that the balanced load vector minimizes the maximum load $\max_{v \in V} q(v)$ over the admissible assignments satisfying the demand λ (Corollary 3 of Hajek (1990)).

Solutions of the fluid equations (2.1)–(2.4) have the following properties:

Monotonicity: If (x, A) and $(\tilde{x}, \tilde{A})$ are two solutions to the fluid equations with $x_0(v) \geq \tilde{x}_0(v)$ for all $v \in V$, then $x_t(v) \geq \tilde{x}_t(v)$ for all $v \in V$ and $t \geq 0$.

Uniqueness: If (x, A) and $(\tilde{x}, \tilde{A})$ are two solutions to the fluid equations with $x_0 = \tilde{x}_0$, then $x_t = \tilde{x}_t$ for all $t \geq 0$. This follows easily by applying the monotonicity property twice with $\tilde{x}_0 \leq x_0$ and $\tilde{x}_0 \geq x_0$.

Insensitivity to initial state: Let x denote the unique process such that for some A, (x, A) is a solution to the fluid equations with initial state x_0, where x_0 is arbitrary. Then, $\lim_{t \to \infty} x_t = q$, where q is the balanced load vector for $SLB(\lambda, \Phi)$.

By the uniqueness property, if X_0^γ converges weakly to x_0, then X^γ converges weakly to the process x that solves the fluid equations (2.1)–(2.4) with the initial state x_0. This implies the convergence of X^γ to x on finite time intervals, whereas x converges to the balanced load vector q as time tends to infinity. The following lemma establishes the weak limit of X^γ in steady state. The proof relies on convergence of X^γ to x over large, fixed-length time intervals, uniformly over the initial times of such intervals.

Lemma 2.2 *The steady state distribution of the process X^γ converges weakly, as γ tends to infinity, to the deterministic distribution concentrated at the balanced load vector q.*

Given an allocation policy π, define for each γ,

$$J_\gamma^\pi = \liminf_{T \to \infty} E^\pi \left[\frac{1}{T} \int_0^T \sum_v X_t(v)^2 dt \,|\, X_0 = x_0 \right].$$

Note that the function $\Phi(x) = \sum_v x(v)^2$ is convex and symmetric in its coordinates; hence a smaller value of J_γ^π indicates a more balanced load, on average. Lemma 2.2 describes the limiting equilibrium distribution of X^γ, in sufficient detail to imply the following optimality property of the LLR policy:

Theorem 2.3 *For any allocation policy π,*

$$\liminf_{\gamma \to \infty} \gamma^{-2} J_\gamma^\pi \geq \lim_{\gamma \to \infty} \gamma^{-2} J_\gamma^{LLR} = \sum_v q(v)^2.$$

Theorem 2.3 shows that, to first order, the LLR policy achieves asymptotically optimal balancing. Other strategies, such as Bernoulli splitting covered in Section 4, have the same property. The large deviations analysis described below can distinguish between the strategies.

3   Overflow exponents

Identifying the overflow exponent of a load sharing network entails establishing a large deviations principle regarding the network load. Let $\tilde{X}^\gamma$ denote the piecewise linearization of the normalized load X^γ at the jump instants. The sequence $(\tilde{X}^\gamma : \gamma > 0)$ is said to *satisfy the large deviations principle* (LDP) in $C_{[0,T]}(R_+^V)$ with the rate function $\Gamma : C_{[0,T]}(R_+^V) \times R_+^V \to R_+ \cup \{+\infty\}$ if for each $x_0 \in R_+^V$ the function $\Gamma(\cdot, x_0)$ is lower semicontinuous, and for any sequence $(x^\gamma : \gamma > 0)$ such that $\lim_{\gamma \to \infty} x^\gamma = x_0$ and Borel measurable $S \subset C_{[0,T]}(R_+^V)$,

$$\limsup_{\gamma \to \infty} \gamma^{-1} \log P_{x^\gamma}\left((\tilde{X}_t^\gamma : 0 \leq t \leq T) \in S \right) \leq - \inf_{\phi \in \overline{S}} \Gamma(\phi, x_0)$$

$$\liminf_{\gamma \to \infty} \gamma^{-1} \log P_{x^\gamma}\left((\tilde{X}_t^\gamma : 0 \leq t \leq T) \in S \right) \geq - \inf_{\phi \in S^o} \Gamma(\phi, x_0),$$

where $\overline{S}$ and S^o denote respectively the closure and the interior of S. This definition of LDP is somewhat weaker than the uniform (in initial conditions) LDP stated in Freidlin and Wentzell (1984), page 92.

Given that $(\tilde{X}^\gamma : \gamma > 0)$ satisfies the large deviations principle in $C_{[0,T]}(R_+^V)$ with the rate function Γ, the overflow exponents are given by the solutions of the following variational optimization problems:

$$F^{LLR}(\kappa) = \inf\{\Gamma(\phi, 0) : \phi \in C_{[0,T]}(R_+^V), \ \phi_0 = 0, \ \max_{v \in V} \phi_T(v) = \kappa, \ T > 0\}$$

$$F^{LLR}(v, \kappa) = \inf\{\Gamma(\phi, 0) : \phi \in C_{[0,T]}(R_+^V), \ \phi_0 = 0, \ \phi_T(v) = \kappa, \ T > 0\}.$$

This section identifies the overflow exponents for the two basic networks of Fig. 2 under the LLR policy, and conjectures their general form for networks with arbitrary topologies.

3.1 The single-location network

Large deviations of the single-location network of Fig. 2.a have been studied extensively. In particular Section 12 of Shwartz and Weiss (1995) establishes a large deviations principle for the network load, which can be employed to obtain the following theorem:

Theorem 3.1 *The overflow exponent of the single-location network exists and is given by* $H_{\lambda(1)}(0, \kappa)$, *where*

$$H_{\lambda(1)}(x, y) = \int_x^y \left(\log\left(\frac{z}{\lambda(1)}\right)\right)_+ dz, \qquad y \geq x \geq 0.$$

Intuitively, for $x < y$, $H_{\lambda(1)}(x, y)$ is a measure of how improbable it is for the normalized load process, starting at x, to make a transition to y within a fixed, long time interval. Note that $H_{\lambda(1)}(x, y) = 0$ for $0 \leq x < y \leq \lambda(1)$. The time-dependent extremal trajectories associated with the most likely mode of transition are not indicated directly in $H_{\lambda(1)}$, but they are as follows: if $\lambda(1) < x < y$, the extremal trajectory from x to y satisfies the simple differential equation $\phi' = \phi - \lambda(1)$. This trajectory is the time reversal of the fluid limit trajectory from y to x (see Shwartz and Weiss (1995) for details).

3.2 The W network

Stochastic ordering arguments provide two upper bounds on the overflow exponent of the W network of Fig. 2.b under *any* allocation policy: *1) The single-location bound:* The load at location 1 is stochastically larger than the load of a single-location network with demand $\gamma\lambda(1)$. Therefore the overflow time of a single-location network with capacity $\lfloor \gamma\kappa \rfloor$ and demand $\gamma\lambda(1)$ dominates the overflow time of location 1, which in turn dominates the overflow time of the W network. *2) Pooling bound:* The network necessarily overflows if the total load exceeds $\lfloor 2\gamma\kappa \rfloor$. Thus the overflow time of the W network is dominated by the overflow time of a single-location network with capacity $\lfloor 2\gamma\kappa \rfloor$ and demand $\gamma(\lambda(1) + \lambda(2) + \lambda(3))$.

The transition mechanism of the load process under LLR changes discontinuously along the boundary $(x \in R_+^2 : x(1) = x(2))$. Large deviations of Markov processes with discontinuous transition mechanisms have been studied by a number of authors. The pioneering paper of Dupuis and Ellis (1992) concerns a Markov process with constant transition mechanism in each of two half spaces, and gives an explicit representation of the rate function for the process observed at a fixed point in time. Subsequently Blinovskii and Dobrushin (1994), and Ignatyuk, Malyshev, and Scherbakov (1994) derived process-level large deviations principles for the case of constant transition mechanism in each half space, using different approaches. In their book Shwartz and Weiss (1995) considered processes on a half space with a flat boundary which cannot be crossed. The recent paper by Dupuis and Ellis (1995) establishes large deviations principles for Markov processes with transition mechanisms that are continuous over facets generated by a finite number of hyperplanes. While in general the paper does not identify the rate function explicitly, it does give an explicit integral representation for the case of a single hyperplane of discontinuity. The paper by Alanyali and Hajek (1995c) concentrates on processes with a single hyperplane of discontinuity, and contains the same explicit form of the rate function under somewhat relaxed technical conditions, using a different approach.

The large deviations principle satisfied by the load process in the W network is established in Alanyali and Hajek (1995b). The proof is based on Alanyali and Hajek (1995c) regarding the discontinuity of the transition mechanism, and on the techniques used in Section 12.6 of Shwartz and Weiss (1995) regarding the flat boundaries of the state space R_+^2, with departure rates tending to zero at the boundaries.

Let $q = (q(1), q(2))$ denote the balanced load vector for the W network with demand λ. More explicitly,

$$q(1) = [(\lambda(1) + \lambda(2) + \lambda(3))/2]_{\lambda(1)}^{\lambda(1)+\lambda(2)}, \quad q(2) = [(\lambda(1) + \lambda(2) + \lambda(3))/2]_{\lambda(3)}^{\lambda(3)+\lambda(2)},$$

where for real x, a, b such that $a \leq b$, $[x]_a^b$ denotes the number in the interval $[a, b]$ that is closest to x. We shall assume without loss of generality that $\lambda(1) \geq \lambda(3)$, which in turn implies that $q(1) \geq q(2)$. The following theorems of Alanyali and Hajek (1995b) identify the overflow exponents of the W network. Stephen Turner of Cambridge University mentioned to us at the workshop that he also had found the result in Theorem 3.2 in independent, unpublished work. He outlined a proof that is similar to ours which relies on the existence of an LDP for the W network, but which does not require an explicit representation of the LDP rate function.

Theorem 3.2 *The overflow exponent of the W network under the LLR policy exists and is given by*

$$F^{LLR}(\kappa) = \begin{cases} H_{\lambda(1)+\lambda(2)+\lambda(3)}(0, 2\kappa) & \text{if } \kappa \leq \kappa_* \\ H_{\lambda(1)+\lambda(2)+\lambda(3)}(0, 2\kappa_*) + H_{\lambda(1)}(\kappa_*, \kappa) & \text{if } \kappa > \kappa_*, \end{cases} \tag{3.1}$$

where $\kappa_* = q(1)q(2)/\lambda(1)$.

Here we shall give an intuitive explanation for the formulas appearing in Theorem 3.2. In the remainder of this section, the "load" at a location will be

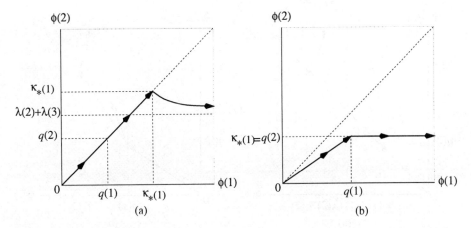

FIG. 4. The most likely scenario for the overflow of location 1 in the W network
for the cases (a) $q(1) = q(2)$, (b) $q(1) > q(2)$.

understood to be the normalized load for some suitably large value of γ. Refer to
Fig. 4, which depicts the extremal trajectories associated with Theorem 3.2. (The
constant $\kappa_*(1)$ in the figure is the same as κ_*.) Consider first the case $q(1) = q(2)$.
If $\kappa \leq q(1) = q(2)$, $F^{LLR}(\kappa) = 0$, which is expected since network overflow is not
a rare event for such κ. If $q(1) = q(2) < \kappa \leq \kappa_*$, then overflow typically occurs
because the whole network becomes overloaded, and locations 1 and 2 maintain
roughly equal loads and act as a single pooled location. For larger values of
κ, the most likely scenario is that first the loads at the two locations together
build up to level κ_*, and then the load at location 1 continues to grow to level
κ. The given value of κ_* minimizes the right hand side of (3.1) for all κ. Finally,
consider the case that $q(1) > q(2)$. Then $q(1) = \lambda(1) > \lambda(2) + \lambda(3) = q(2)$, and
it follows that $F^{LLR}(\kappa) = H_{\lambda(1)}(0, \kappa)$ for all values of κ. In this case, the typical
scenario for network overflow is that the load at location 1 reaches κ, while the
load at location 2 remains near its mean $q(2)$.

Theorem 3.3 *For $v = 1, 2$, the overflow exponent of location v in the W network under LLR exists, and is given by*

$$F^{LLR}(v, \kappa) = \begin{cases} H_{q(1)}(0, \kappa) + H_{q(2)}(0, \kappa) & \text{if } \kappa \leq \kappa_*(v) \\ H_{q(1)}(0, \kappa_*(v)) + H_{q(2)}(0, \kappa_*(v)) \\ \quad + H_{\lambda(2v-1)}(\kappa_*(v), \kappa) & \text{if } \kappa > \kappa_*(v), \end{cases}$$

where $\kappa_*(v) = q(1)q(2)/\lambda(2v - 1)$.

Let us explain the intuition behind Theorem 3.3. First, the very definitions
imply that when the overflow exponents exist, $F^{LLR}(1, \kappa) = F^{LLR}(\kappa)$. To see
why the given expressions in fact satisfy this relation, note that $\kappa_*(1) = \kappa_*$, and
(*i*) if $q(1) = q(2)$ then $H_{q(1)}(0, \kappa) + H_{q(2)}(0, \kappa) = H_{\lambda(1)+\lambda(2)+\lambda(3)}(0, 2\kappa)$ for all κ,
(*ii*) if $q(1) > q(2)$ then $\kappa_* = \kappa_*(1) = q(2)$, and hence $H_{q(1)}(0, \kappa) + H_{q(2)}(0, \kappa) =
H_{\lambda(1)+\lambda(2)+\lambda(3)}(0, 2\kappa) = 0$ whenever $\kappa \leq \kappa_*$. Thus, the expressions given for

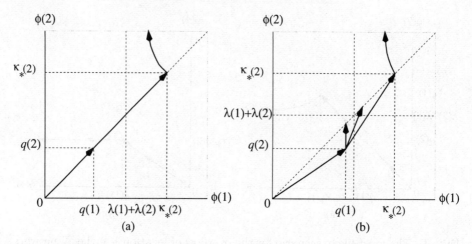

FIG. 5. The most likely scenario for the overflow of location 2 in the W network for the cases (a) $q(1) = q(2)$, (b) $q(1) > q(2)$.

$F^{LLR}(1, \kappa)$ and $F^{LLR}(\kappa)$ are indeed equivalent. The intuitive explanation given for Theorem 3.2 thus also applies to explain the expression for $F^{LLR}(1, \kappa)$.

Finally, let us give an intuitive explanation for the expression for the overflow exponent $F^{LLR}(2, \kappa)$. Consult Fig. 5. If $q(1) = q(2)$, the explanation is similar to that for $F^{LLR}(1, \kappa)$, so assume that $q(1) > q(2)$. If $\kappa \leq q(2)$ then $F^{LLR}(2, \kappa) = 0$, which makes sense since for such κ network overflow is not a rare event. If $q(2) < \kappa \leq q(1)$, then the load at location 2 can grow to level κ, while, even without any large deviation occurring at location 1, all type 2 arrivals are assigned to location 2. Thus, it makes sense that $F^{LLR}(2, \kappa) = H_{q(2)}(0, \kappa)$ for such values of κ. Finally, if $\kappa > q(1) > q(2)$, as the load at location 2 begins to build beyond $q(2)$, the load at location 1 begins to build beyond $q(1)$, even though the two loads are not equal. In that way, all type 2 consumers are assigned to location 2, even after the load at location 2 exceeds $q(1)$. Eventually the loads at the two locations simultaneously become approximately equal to $\kappa \wedge \kappa_*(2)$. If $\kappa > \kappa_*(2)$, then the load at location 2 unilaterally continues to increase to level κ. It is interesting to note that the initial segments of the most likely trajectories depend on κ as κ ranges over $\kappa > q(1) > q(2)$, as illustrated by the multiple trajectories in Fig. 5.b.

To close this section we compare the network overflow exponent of LLR to the single-location and pooling upper bounds for a numerical example. Specifically, consider the W network with demand $\lambda = (1 - \alpha, 2\alpha, 1 - \alpha)$, where $0 \leq \alpha \leq 1$. The network overflow exponent under LLR, $F^{LLR}(\kappa)$, and the two bounds are plotted in Fig. 6 for $\alpha = 0.5$. Also depicted are the overflow exponents of two other policies mentioned in Section 4.3. For the values of $\kappa \leq \kappa_*$ the simple LLR policy performs as well as any other policy, in the sense that $F^{LLR}(\kappa)$ achieves the pooling bound. For larger values of κ, however, the suboptimality of LLR reveals itself.

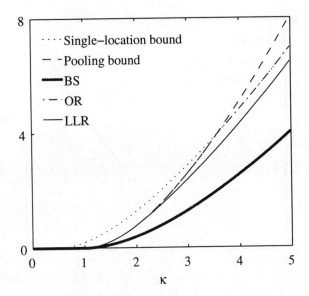

FIG. 6. The overflow exponents of some allocation policies, along with the single-location and pooling bounds, for $\alpha = 0.5$.

Larger values of α correspond to increased load sharing capability of the network for the same total demand. Figure 7 plots $F^{LLR}(\kappa)$ for different values of α. Note that when $\alpha = 0$ and $\alpha = 1$, $F^{LLR}(\kappa)$ achieves the single-location and pooling bounds respectively.

3.3 Arbitrary topologies

Establishing explicit large deviations principles for arbitrary load sharing networks operating under the LLR policy appears difficult. Nevertheless, the forms of the overflow exponents given in Theorems 3.2 and 3.3, and the corresponding most likely scenarios for network overflow, suggest the following conjecture on the overflow exponents for networks with arbitrary topologies:

Conjecture For each $v \in V$ and $\kappa \geq 0$, $F^{LLR}(v, \kappa)$ can be identified as follows. Let S range over the set of set-valued functions of the form $S = (S(x) : 0 \leq x \leq \kappa)$, where $v \in S(x) \subset V$ for $0 \leq x \leq \kappa$, and $S(x) \subset S(x')$ for $x \geq x'$. Associated with each such S and $0 \leq x \leq \kappa$, let $R(x) = \{u \in U : N(u) \subset S(x) \cup \{v' : q(v') > x\}\}$, and let $(q(v', x) : v' \in S(x))$ denote the balanced load vector in the subnetwork $(R(x), S(x), N(x))$ with demand $(\lambda(u) : u \in R(x))$, where $N(x, u) = N(u) \cap S(x)$ for $u \in R(x)$. Then

$$F^{LLR}(v, \kappa) = \inf_S \int_0^\kappa \sum_{v' \in S(x)} \left(\log\left(\frac{x}{q(v', x)} \right) \right)_+ dx.$$

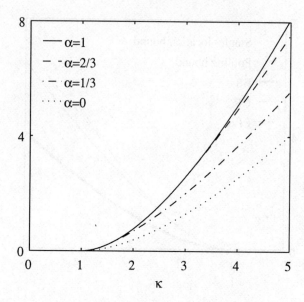

FIG. 7. $F^{LLR}(\kappa)$ for several values of α.

An intuitive justification of the conjecture is as follows. The set $S(x)$ denotes the set of locations with load that would increase to at least level x. The set $R(x)$ is the set of consumer types which would have exceptionally large numbers of arrivals in order to cause the large deviation of the load. The quantity $q(v', x)$ is the nominal arrival rate at location v' at the time the load at v' crosses level x. The conjecture is consistent with Theorem 3.3.

4   Extensions

The first two extensions below deal with the fluid limit analysis, and the third deals with the overflow exponents based on the theory of large deviations.

4.1   The migration model

Consider the wireless network of Section 1, where the type of a user is determined by its geographical location. If the users of the network are mobile, it is conceivable that a user can visit different neighborhoods while in the system. In our load sharing network abstraction, this corresponds to migration of consumers among different *types*, and is not covered by the basic dynamic model of Section 1. The migration model is a generalization of the basic model, and incorporates consumer migrations as well as type-dependent departure rates from the network.

The analytical description of the migration model involves a *routing matrix* $R = [r_{u,u'}]_{U \times U}$, where $r_{u,u'} \geq 0$ for $u \neq u'$, and $\sum_{u' \in U} r_{u,u'} \leq 0$ for all $u \in U$. For $u, u' \in U$, such that $u' \neq u$, each type u consumer transforms into a type u' consumer with rate $r_{u,u'}$, or departs from the system with rate $-\sum_{u' \in U} r_{u,u'}$.

We assume that R is nonsingular, so that every consumer eventually departs from the system. Define the *effective demand*, ρ, as $\rho = -\lambda R^{-1}$ so that $\gamma\rho(u)$ is the mean number of type u consumers in the network in equilibrium. We assume that each new consumer is assigned to a location via the LLR policy. In addition, when a consumer changes its type, it is reassigned using LLR. Its location may or may not change.

Let $C_t(u, v)$ denote the number of type u consumers at location v at time t. Define $C_t^\gamma(u, v) = \gamma^{-1} C_t(u, v)$, and note that $X_t^\gamma(v) = \sum_u C_t^\gamma(u, v)$. If c is a weak limit of a subsequence of C^γ, then (c, A) for some process A satisfies the following fluid equations:

$$c_t(u, v) = c_0(u, v) + A_{u,v}(t) + \int_0^t r_{u,u} c_s(u, v) ds, \tag{4.1}$$

$$A_{u,v}(0) = 0, \quad A_{u,v}(t) \text{ nondecreasing}, \tag{4.2}$$

$$\sum_{v \in N(u)} A_{u,v}(t) = \lambda(u)t + \sum_{u' \neq u} \int_0^t l_s(u') r_{u',u} ds, \tag{4.3}$$

$$\int_0^t I\left\{ x_s(v) > \min_{v' \in N(u)} x_s(v') \right\} dA_{u,v}(s) = 0, \tag{4.4}$$

where $x_t(v) = \sum_{u \in N^{-1}(v)} c_t(u, v)$, and $l_t(u) = \sum_{v \in N(u)} c_t(u, v)$.

In this more general setting, the initial state c_0 and the fluid equations (4.1)–(4.4) do not necessarily determine the load trajectory x uniquely, and hence do not necessarily identify a unique weak limit for the normalized load process X^γ. However, as t tends to infinity, the load trajectories determined by (4.1)–(4.4) converge to the balanced load vector for problem $SLB(\rho, \Phi)$. This suffices to establish the convergence of the equilibrium distribution of X^γ; hence Theorem 2.3 continues to hold under the migration model (Section 5 of Alanyali and Hajek (1995a)). The overflow exponents for the migration model have not been investigated.

4.2 Finite capacities

Another variation of the basic model of Section 1 involves incorporation of finite capacity constraints on the locations. Namely, given a nonnegative capacity vector $\kappa = (\kappa(v) : v \in V)$, we consider the case in which for each location v, the number of consumers assigned to the location cannot exceed its capacity $\lfloor \gamma\kappa(v) \rfloor$. A consumer is *lost* if upon its arrival all the locations in its neighborhood are already loaded to capacity. The goal of the allocation policy is to minimize the fraction of consumers lost in the system.

Define $\mathcal{B}_{\lambda,\kappa}$ as the set of admissible assignments a such that $\sum_v a_{u,v} \leq \lambda(u)$ for all $u \in U$, and $q(v) \leq \kappa(v)$ for all $v \in V$, where q denotes the load vector determined by a. The *static load packing* (*SLP*) *problem* is defined as

$$SLP(\lambda, \kappa) : Maximize\left(\sum_v q(v) \; : \; a \in \mathcal{B}_{\lambda,\kappa}\right),$$

and provides the following lower bound for the consumer loss probability under *any* allocation policy:

Lemma 4.1 *For any assignment policy π and $\gamma > 0$, the probability of consumer loss satisfies*

$$P_\gamma^\pi(Loss) \geq 1 - \frac{\sum_v q(v)}{\sum_u \lambda(u)},$$

where q is a load vector corresponding to a solution of $SLP(\lambda, \kappa)$.

The *least ratio routing* (LRR) policy is a variation of the LLR policy whereby an arriving consumer is assigned to an admissible location with the least *relative load*, $f(x, v)$, defined by $f(x, v) = x(v)/\kappa(v)$. The LRR policy can be analyzed within the framework of lossless networks by introducing a *loss location* v_L of infinite capacity, to which consumers are assigned if they would be lost otherwise. Under the LRR policy, weak limits of the normalized load process satisfy the following fluid equations together with some A:

$$x_t(v) = x_0(v) + \sum_{u \in N^{-1}(v)} A_{u,v}(t) - \int_0^t x_s(v) ds, \tag{4.5}$$

$$A_{u,v}(0) = 0, \;\; A_{u,v}(t) \text{ nondecreasing}, \tag{4.6}$$

$$0 \leq x_t(v) \leq \kappa(v), \quad \sum_{v \in N(u) \cup v_L} A_{u,v}(t) = \lambda(u)t, \tag{4.7}$$

$$\int_0^t I\left\{f(x_s, v) > \min_{v' \in N(u)} f(x_s, v')\right\} dA_{u,v}(s) = 0 \quad v \neq v_L, \tag{4.8}$$

$$\int_0^t I\left\{\min_{v \in N(u)} f(x_s, v) < 1\right\} dA_{u,v_L}(s) = 0. \tag{4.9}$$

The solutions (x, A) of equations (4.5)–(4.9) enjoy the monotonicity and thus the uniqueness properties of x. Furthermore $\lim_{t \to \infty} x_t = q$ where q is the load vector corresponding to a certain solution of $SLP(\lambda, \kappa)$, so that the mechanism of Section 2 can be applied to establish the following theorem (Section 4 of Alanyali and Hajek (1995a)):

Theorem 4.2 *For any allocation policy π,*

$$\liminf_{\gamma \to \infty} P_\gamma^\pi(Loss) \geq \lim_{\gamma \to \infty} P_\gamma^{LRR}(Loss) = 1 - \frac{\sum_v q(v)}{\sum_u \lambda(u)}.$$

4.3   Optimal repacking and Bernoulli splitting

Optimal repacking (OR) and *Bernoulli splitting* (BS) are alternative allocation policies. The OR policy is a "brute force" approach whereby at each time t, the consumers in the network are repacked so as to solve the problem $SLB(L_t, \Phi)$, where for each $u \in U$, $L_t(u)$ denotes the number of type u consumers in the network at time t. The BS policy is a randomized nonrepacking policy under which each arriving type u consumer is assigned to location v with probability $a_{u,v}/\lambda(u)$, where a is an optimal assignment for the static problem $SLB(\lambda, \Phi)$. It is easy to see that both OR and BS policies enjoy the same optimality properties as the LLR policy that are implied by the fluid limit approximations. However, the network overflow exponents under the three policies differ rather drastically.

The process L is a vector of independent single-location network loads with demand vector $\gamma\lambda$. The large deviations principle for the single-location network, together with the Contraction Principle (Theorem 4.2.1 of Dembo and Zeitouni (1992)), yield the large deviations principle for the network load under the OR policy. Solution of the relevant variational optimization problem yields that the network overflow exponent under the OR policy can be expressed as $\min_{F \subset V} H_{\lambda(F)}(0, |F|\kappa)$ where $\lambda(F) = \sum_{u:N(u) \subset F} \lambda(u)$.

Under the BS policy the load process is a vector of independent single-location network loads with the demand vector γq, where q is the balanced load vector for problem $SLB(\lambda, \Phi)$. The techniques mentioned in the above paragraph yield that the network overflow exponent under the BS policy is given by $\min_{v \in V} H_{q(v)}(0, \kappa)$.

Figure 6 plots the network overflow exponents of the OR and BS policies for the W network with demand $\lambda = (0.5, 1, 0.5)$ on the same scale as the LLR policy. The OR policy minimizes the maximum load in the network at all times; hence its overflow exponent dominates the overflow exponent of any allocation policy. Furthermore the OR policy achieves the smaller of the single-location and pooling bounds. The BS policy only exerts open-loop control, and its performance is significantly worse than that of LLR for the whole range of capacities as illustrated in Fig. 6.

Bibliography

1. Alanyali, M. and Hajek, B. (1995a). Analysis of simple algorithms for dynamic load balancing. Submitted to *Mathematics of Operations Research*.

2. Alanyali, M. and Hajek, B. (1995b). On large deviations in load sharing networks. Submitted to *The Annals of Applied Probability*.

3. Alanyali, M. and Hajek, B. (1995c). On large deviations for Markov processes with discontinuous statistics. Submitted to *The Annals of Applied Probability*.

4. Blinovskii, V.M. and Dobrushin, R.L. (1994). Process level large deviations for a class of piecewise homogeneous random walks. In *The Dynkin Festschrift: Markov Processes and their Applications*, 1–59. Birkhauser, Boston.

5. Dembo, A. and Zeitouni, O. (1992). *Large deviations techniques and applications*. Jones and Bartlett, Boston.

6. Dupuis, P. and Ellis, R.S. (1992). Large deviations for Markov processes with discontinuous statistics, II: random walks. *Probability Theory and Related Fields*, **91**, 153–194.

7. Dupuis, P. and Ellis, R.S. (1995). The large deviation principle for a general class of queueing systems I. *Transactions of the American Mathematical Society*, **347**, 2689–2751.

8. Dupuis, P., Ellis, R.S., and Weiss, A. (1991). Large deviations for Markov processes with discontinuous statistics, I: general upper bounds. *Annals of Probability*, **19**, 1280–1297.

9. Ethier, S. and Kurtz, T. (1986). *Markov Processes: Characterization and convergence*. Wiley, New York.

10. Freidlin, M.I. and Wentzell, A.D. (1984). *Random Perturbations of Dynamical Systems*. Springer-Verlag, New York.

11. Ganger, G.R., Worthington, B.L., Hou, R.Y., and Patt, Y.N. (1993). Disk subsystem load balancing: disk striping vs. conventional data placement. *Proceedings of 26th Hawaii International Conference on System Sciences*. **1**, 40–49.

12. Hajek, B. (1990). Performance of global load balancing by local adjustment. *IEEE Transactions on Information Theory*, **36**, 1398–1414.

13. Ignatyuk, I.A., Malyshev, V., and Scherbakov, V.V. (1994). Boundary effects in large deviation problems. *Russian Mathematical Surveys*, **49**, 41–99.

14. Shwartz, A. and Weiss, A. (1995). *Large deviations for performance analysis, queues, communication and computing*. Chapman & Hall, London.

15. Willebeck-LeMair, M.H. and Reeves, A.P. (1993). Strategies for dynamic load balancing on highly parallel computers. *IEEE Transactions on Parallel and Distributed Systems*, **9**, 979–993.

14

Analysing system behaviour on different time scales

Suzanne P. Evans

Birkbeck College, University of London

Abstract

Events in an integrated services telecommunications network occur on multiple time scales and careful analysis of time dependent behaviour is important to derive appropriate system descriptions at different levels. This paper analyses time dependent behaviour expressed in terms of the eigenstructure of a Markov generator. Identifying terms that are dominant on different time scales is not always straightforward and we discuss a number of technical conditions required to guarantee a separation of time scales. These conditions are shown to have implications for the appropriate representation of fast and slow time scale dynamics and for the sensitivity of these representations to errors in the model parameters.

1   Introduction

The work described here was motivated by consideration of an integrated services telecommunications network where events occur on multiple time scales and the system is, or can be, controlled at a number of different levels. A close understanding of the transient and time dependent behaviour of the system helps to derive appropriate and simple system descriptions at different levels corresponding to different time scales. It may also suggest different approaches to control at these levels.

We shall consider systems which can be modelled as a finite state continuous time Markov chain with infinitesimal generator Q. Transitions between states are triggered by events that occur on different time scales and therefore the elements of Q will range over several orders of magnitude. We are interested in the structure of Q in terms of the different time scales and in the consequent time dependent behaviour of the process. As we shall see, the results cover discrete time systems as well.

In many cases the time scale structure of the system is well understood and states can be appropriately grouped so that the corresponding structure of Q is clear. A more systematic procedure is sometimes required when the structure is not obvious, or when it is desirable to analyse it more closely. For example, for a sequence of time thresholds, the structure of the chain can be analysed by computing the strong components of the graph whose directed arcs denote state transitions that occur on a time scale faster than the given threshold. The sequence of thresholds, together with the corresponding components and the induced partial ordering on the components, define an hierarchical ordering, a grouping of states and hence a structure for Q.

The simplest and most frequently studied Q structure of this type is that governing a nearly completely decomposable (NCD) Markov chain. Here states are grouped into larger aggregates of states in such a way that transitions between states within the same aggregate occur on a much faster time scale than transitions between states belonging to different aggregates. The corresponding Q matrix has a near block diagonal structure. The time dependent behaviour generally expected of such a process can be described by a slow moving, almost Markovian process between aggregates and by approximate stationary distributions over states within aggregates. Thus, as the process approaches equilibrium, the transient behaviour can be divided into a fast transient, which vanishes quickly and represents the fast approach to equilibrium within each aggregate, and a slow transient representing the slow approach to equilibrium of the process moving between aggregates.

The double limit behaviour described above for NCD chains is well known. It has been used, for example, to develop effective iterative techniques for computing the stationary distribution of an NCD chain, where standard algorithms are ineffective. It has also been used to develop various approximations to the stationary distribution, together with corresponding error bounds. Less formally it is often used in the analysis of large, complex stochastic systems to justify a decomposition into approximate submodels for which exact or approximate stationary distributions are obtainable. This may then allow the overall system to be solved approximately via a higher level model linking the submodel solutions. Many of these more *ad hoc* procedures have been very successful in obtaining good, usable, approximate solutions to large and complex systems. However, even in the simplest two-level NCD structure, there are some subtleties in any mathematical justification of the decomposition approach that are not always appreciated, and some sensitivities as well. This can lead to, at the least, a less than full understanding of the status of the approximations with possibly misleading interpretation and use.

The purpose of this paper is to explore some of the technical conditions that guarantee the separation of transient behaviour in an NCD chain into a fast and a slow component, and which also guarantee the accuracy of certain approximations. In particular we give conditions and carry out a full analysis for an approximation based on product quasi-stationary distributions, which have certain desirable properties. We also consider the sensitivity of our approximations to errors in transition rates and the implications for their use.

The structure of the paper is as follows. Section 2 defines a continuous time NCD model and an equivalent discrete time model which we shall find more convenient to analyse. It introduces the general approach which we shall take to the analysis of an NCD chain. Section 3 contains the mathematical definitions and results required in the remainder of the paper. Section 4 introduces the product quasi-stationary distributions on which we shall base our approximations, and considers conditions for their stability. Section 5 gives a detailed analysis of the spectral structure of an NCD chain and the associated time dependent behaviour. We note the conditions under which our approximation is accurate and the resulting implications for the sensitivity of such approximations to errors in the transition rates. Section 6 contains a summary of our results and a brief discussion of some telecommunications applications.

2 Time dependent behaviour and spectral structure

In our applications, we are usually concerned with a *continuous* time Markov chain with finite state space. However, we can equivalently analyse a *discrete* time process by making use of the uniformization procedure relating a continuous time chain governed by infinitesimal generator Q to a discrete time chain governed by the stochastic matrix $A = \nu^{-1}Q + I$ for a suitably chosen $\nu > 0$. If Q is irreducible and ν is sufficiently large, the matrix A is stochastic, irreducible and aperiodic, the discrete time chain governed by A will be ergodic and A^k, which governs the time dependent behaviour of the discrete process, will have the same ergodic properties as e^{Qt}. Therefore, without loss of generality, we shall consider only discrete time NCD chains.

2.1 The NCD chain

A nearly completely decomposable, discrete time Markov chain, with finite state space, is one whose $n \times n$ probability transition matrix can be written in the form

$$A = \begin{pmatrix} D_{11} & E_{12} & \cdots & E_{1L} \\ E_{21} & D_{22} & \cdots & E_{2L} \\ \vdots & \vdots & & \vdots \\ E_{L1} & E_{L2} & \cdots & D_{LL} \end{pmatrix} \tag{2.1}$$

where the off-diagonal blocks $E_{k\ell}$ are small compared with the diagonal blocks $D_{\ell\ell}$. We quantify this by letting $\varepsilon = \| A - diag(D_{\ell\ell}) \|$ where $\| \ \|$ denotes the spectral norm and ε is called the coupling parameter.

Thus the n possible states of the chain are grouped into L aggregates, each aggregate ℓ containing n_ℓ states. Events occurring on the fast time scale are represented by transitions between states within the same aggregate and events occurring on the slow time scale are represented by transitions between states in different aggregates. We assume that the process governed by A is irreducible and ergodic and we wish to characterize its time dependent behaviour.

2.2 Time dependent behaviour in NCD chains

From the Perron–Frobenius theorem A has the simple eigenvalue one, with corresponding left and right eigenvectors $\mathbf{y}_1 > 0$ and $\mathbf{x}_1 > 0$, and all other eigenvalues have modulus less than one. If $\mathbf{y}_1$ is normalized so that $\mathbf{y}_1^T \mathbf{x}_1 = 1$, then $\mathbf{y}_1$ is the unique $n \times 1$ stationary probability vector associated with the chain. If $\mathbf{x}_1$ is normalized so that $\mathbf{y}_1^T \mathbf{x}_1 = 1$ then, since A is stochastic, $\mathbf{x}_1 = \mathbf{1}$. We now have

$$A^k \to \mathbf{1}\mathbf{y}_1^T \quad \text{as} \quad k \to \infty \tag{2.2}$$

and if $\mathbf{z}_0$ is a vector of initial state probabilities

$$\mathbf{z}_k^T = \mathbf{z}_0^T A^k \to \mathbf{y}_1^T \quad \text{as} \quad k \to \infty, \tag{2.3}$$

where $\mathbf{z}_k$ is the probability vector at time k. We are interested in the *transient* behaviour of the NCD chain, that is the way in which the limits are reached in eqns (2.2) and (2.3). This depends on the complete spectral structure of A. In particular we want to know what happens as ε goes to zero. We shall see that allowing ε to become small is not sufficient to characterize transient behaviour.

2.3   Spectral structure

In examining the spectral structure of A, and relating it to time dependent behaviour, we shall follow McAllister *et al.* (1984) and base our analysis on a spectral decomposition in terms of (well conditioned) invariant *subspaces* of A, whose contributions disappear on different time scales. This means that the matrices involved are not required to have complete sets of eigenvectors. Although we are unlikely in practice to encounter a transition matrix which is degenerate in this way, we may well encounter one which is close to degeneracy and results cast in the form of a more detailed spectral decomposition are likely to be unstable. We seek a spectral decomposition of the form

$$A = \mathbf{1}\mathbf{y}_1^T + X_2 A_2 Y_2^T + X_3 A_3 Y_3^T, \tag{2.4}$$

that is

$$(\mathbf{y}_1 \ Y_2 \ Y_3)^T A (\mathbf{1} \ X_2 \ X_3) = diag(1, A_2, A_3) \tag{2.5}$$

where $(\mathbf{y}_1 \ Y_2 \ Y_3)^T = (\mathbf{1} \ X_2 \ X_3)^{-1}$, so that eqn (2.5) represents a similarity transform of A, the columns of $Y_1(X_1)$ span an $(L-1)$-dimensional left (right) invariant subspace of A and the columns of $Y_2(X_2)$ span an $(n-L)$-dimensional left (right) invariant subspace of A. The idea is to construct the decomposition so that *all* the eigenvalues of A_2 have a modulus within $O(\varepsilon)$ of one and *all* the eigenvalues of A_3 are bounded away from one uniformly in ε. If this can be done then, if $\mathbf{z}_0$ represents the initial state probability vector, the probability vector at time k is given by

$$\mathbf{z}_k^T = \mathbf{y}_1^T + \mathbf{z}_0^T X_2 A_2^k Y_2^T + \mathbf{z}_0^T X_3 A_3^k Y_3^T = \mathbf{y}_1^T + \mathbf{g}_k^T Y_2^T + \mathbf{h}_k^T Y_3^T \tag{2.6}$$

and as $\mathbf{z}_k$ approaches $\mathbf{y}_1$, the transient in the column space of Y_3 will vanish quickly while the transient in the column space of Y_2 will take a long time to disappear. If this is so we may ignore the space spanned by Y_3 but will be very interested in the structure of the space spanned by Y_2. A classic spectral decomposition of this type for an NCD stochastic matrix is contained in McAllister *et al.* (1984). The construction is built around $\mathbf{y}_1$, and L conditional probability vectors derived from $\mathbf{y}_1$. The authors discuss the structure of the Y_i and the X_i

in detail, as well as the conditions required for the decomposition to work. We shall prefer to take a different approach, since y_1 is generally unknown and we are not necessarily interested in the long term stationary behaviour of the chain if the time scale for reaching stationarity is very long. It is enough if the approximations we use are appropriate on the time scale of interest and are stable. The conditions for our approach to work turn out to be slightly different from those of McAllister *et al.* (1984). In particular, we dispense with their requirement that the long term stationary probability of being in aggregate ℓ should be bounded away from zero uniformly in ε for $\ell = 1, \ldots, L$. However, to obtain an approximation to y_1 we need a new condition, and this turns out to have a strong bearing on the stability of such approximations.

3 Mathematical preliminaries

3.1 Constructing and bounding invariant subspaces

In the process of constructing our spectral decomposition of A, we often start with a pair of vectors y and x, scaled so that $y^T x = 1$, and representing a left and right pair of eigenvectors of some matrix. We then look for matrices Y and X such that $(y\,Y)^T = (x\,X)^{-1}$, and seek to establish bounds concerning them. To do this we require the following lemma.

Lemma 3.1 *Let A be a non-negative, irreducible and aperiodic matrix. If β is the Perron root of A and $y > 0$ and $x > 0$ are the corresponding left and right eigenvectors scaled so that $1^T y = 1$ and $y^T x = 1$, then there exist matrices J and K such that:*

(1) $(y\,K)^T = (x\,J)^{-1}$

(2) $(y\,K)^T A\,(x\,J) = \begin{pmatrix} \beta & 0^T \\ 0 & C \end{pmatrix}$

(3) $\| J \| = 1$, $\| K \| = \| x \| \| y \| = \sec\theta$, θ *being the angle between y and x*

(4) $n^{-1/2} \le \| y \| \le 1$, $\| x \| \le n^{1/2} \sec\theta$

(5) $\sec\theta \le \left\{ 1 + (\| A \| / \delta(\beta, C))^2 \right\}^{1/2}$ *where* $\delta(\beta, C) = \| (\beta I - C)^{-1} \|^{-1}$.

Proof These results are a variant on standard results for a simple eigenvalue and its eigenvectors. See, for example, Stewart (1983).   □

Note From the preceding lemma we may note the dependence of the bounds for invariant vectors and subspaces on the quantity $\delta(\beta, C)$. This is a distance measure which appears also in the next theorem and will be formally defined and discussed in Section 3.2.

Corollary If $y > 0$ and $x = 1$ above, then K, x and y are bounded above by numbers that depend only on n, the length of x and y. Specifically
$n^{-1/2} \le \| y \| \le 1$, $\| x \| = n^{1/2}$, $\| K \| = \| x \| \| y \| \le n^{1/2}$.

We now require results concerning *approximations* to pairs of left and right invariant subspaces of a matrix A. In particular we need sufficient conditions for such approximations to be 'close' to exact invariant subspaces. Such conditions relate to the *stability* of invariant subspaces.

Suppose we have $(Y_1 \ Y_2)$ and $(X_1 \ X_2)$ such that $(Y_1 \ Y_2)^T = (X_1 \ X_2)^{-1}$. Suppose further that the pairs Y_1, X_1 and Y_2, X_2 are 'nearly' corresponding left and right invariant spaces of some matrix A so that

$$(Y_1 \ \ Y_2)^T A (X_1 \ \ X_2) = \begin{pmatrix} B & H \\ G & C \end{pmatrix}$$

where G and H are 'small'. We look for a matrix U, close to I, such that

$$U^{-1} \begin{pmatrix} B & H \\ G & C \end{pmatrix} U = \begin{pmatrix} \overline{B} & 0 \\ 0 & \overline{C} \end{pmatrix} \tag{3.1}$$

$$\Rightarrow \quad (Y_1^* \ \ Y_2^*)^T A (X_1^* \ \ X_2^*) = \begin{pmatrix} \overline{B} & 0 \\ 0 & \overline{C} \end{pmatrix} \tag{3.2}$$

where $Y_1^* = Y_1 U^{-T}$, $X_1^* = X_1 U$ and $Y_2^* = Y_2 U^{-T}$, $X_2^* = X_2 U$ are exact pairs of left and right invariant subspaces, which are close to Y_1, X_1 and Y_2, X_2 insofar as U is close to I. If we choose

$$U = \begin{pmatrix} I & -Q^T \\ P & I \end{pmatrix} \begin{pmatrix} (I + Q^T P)^{-1} & 0 \\ 0 & (I + P Q^T)^{-1} \end{pmatrix} \tag{3.3}$$

$$U^{-1} = \begin{pmatrix} I & Q^T \\ -P & I \end{pmatrix} \tag{3.4}$$

then it is easy to show that eqn (3.1) will be satisfied if

$$PB - CP = G - PHP \quad \text{and} \quad BQ^T - Q^T C = H - Q^T G Q^T. \tag{3.5}$$

Thus, to obtain U satisfying eqn (3.1), and hence exact left and right invariant subspace pairs Y_1^*, X_1^* and Y_2^*, X_2^*, we need to solve eqns (3.5) for P and Q. In these equations we expect both G and H to be small and hope to find correspondingly small P and Q. Theorem 3.2 gives a sufficient condition for the existence of solutions to eqns (3.5) and gives bounds for their spectral norms.

Theorem 3.2 *Consider the matrix* $\begin{pmatrix} B & H \\ G & C \end{pmatrix}$ *and let*

$$\delta(B, C) = \min \left\{ \inf_{\|P\|=1} \| PB - CP \|, \ \inf_{\|Q\|=1} \| BQ^T - Q^T C \| \right\}.$$

If $\dfrac{\| H \| \| G \|}{(\delta(B,C))^2} < \dfrac{1}{4}$, *then we can find P and Q such that*

$$\| P \| < \frac{2 \| G \|}{\delta(B,C)}, \quad \| Q \| < \frac{2 \| H \|}{\delta(B,C)} \quad \text{and} \quad U^{-1} \begin{pmatrix} B & H \\ G & C \end{pmatrix} U = \begin{pmatrix} \overline{B} & 0 \\ 0 & \overline{C} \end{pmatrix}$$

$$\text{where} \quad \begin{aligned} \overline{B} &= (I + Q^T P)(B + HP)(I + Q^T P)^{-1} \\ \overline{C} &= (I + PQ^T)(C - GQ^T)(I + PQ^T)^{-1}. \end{aligned}$$

Proof McAllister *et al.* (1984) give a proof that applies in the more general setting of bounded linear operators on Banach space. □

Note 1 From this theorem, and the preceding discussion, we see that if G, H, P and Q are all $O(\varepsilon)$, then $\overline{B} = B + O(\varepsilon^2)$ and $\overline{C} = C + O(\varepsilon^2)$. In this case the eigenvalues of our original matrix A are the eigenvalues of $\overline{B}$ and $\overline{C}$, which are within $O(\varepsilon^2)$ of the eigenvalues of B and C. However, the exact invariant subspace pairs Y_1^*, X_1^* and Y_2^*, X_2^* are only within $O(\varepsilon)$ of the approximate pairs Y_1, X_1 and Y_2, X_2.

Note 2 The quantity $\delta(B, C)$ is a generalization of the distance measure in Lemma 3.1. From Theorem 3.2 we see that the stability of invariant subspaces depends on this measure.

3.2 The distance measure $\delta(.,.)$

An important concept throughout this work is the distance measure $\delta(.,.)$, between square matrices. It is related to the distance between the eigenvalues of the two matrices and, as we have already seen, is of fundamental importance in establishing stability and bounds for invariant subspaces.

Let B and C be square matrices. Then $\delta(B, C)$ is defined as

$$\delta(B, C) = \min \left\{ \inf_{\|P\|=1} \| PB - CP \|, \inf_{\|Q\|=1} \| BQ^T - Q^T C \| \right\}.$$

The following properties hold; see Stewart (1973) and McAllister *et al.* (1984).

(1) $\delta(B, C) \leq \min \{|\beta - \gamma| : \beta$ an eigenvalue of B and γ an eigenvalue of $C\}$
(2) $\delta(XBX^{-1}, YCY^{-1}) \geq \dfrac{\delta(B, C)}{\kappa(X)\kappa(Y)}$, where $\kappa(X)$, $\kappa(Y)$ are the condition
   numbers of X, Y, namely $\kappa(X) = \| X \| \| X^{-1} \|$, $\kappa(Y) = \| Y \| \| Y^{-1} \|$
(3) $\delta(XBX^T, YCY^T) = \delta(B, C)$, if X and Y are orthogonal matrices
(4) $|\delta(B + E, C + F) - \delta(B, C)| \leq \| E \| \| F \|$
(5) $\delta(\beta, C) = \| (\beta I - C)^{-1} \|^{-1}$
(6) $\delta = \delta(diag(B_1, \ldots, B_p), diag(C_1, \ldots, C_q))$ satisfies $k_1 \delta_{\min} \leq \delta \leq k_2 \delta_{\min}$,
   where $\delta_{\min} = \min\{\delta(B_i, C_j)\ i = 1, \ldots, p;\ j = 1, \ldots, q\}$ and $k_1, k_2 > 0$ are
   constants depending only on the dimensions of the B_i and C_j.

Note 1 (1) relates $\delta(B, C)$ to the distance between eigenvalues of B and C. However, the similarity transforms of (2), which leave the eigenvalues of B and C unaltered, may substantially change $\delta(.,.)$.

Note 2 (4) is a continuity property not shared by the simple minimum difference measure given on the right hand side of (1). It is crucial in our analysis.

In the next section we begin the process of finding a suitable approximation to a spectral decomposition of the form (2.5) for the stochastic matrix A governing our NCD chain. We then need to establish sufficient conditions to ensure that our approximate decomposition is within $O(\varepsilon)$ of the exact decomposition, to establish the separation of transient behaviour into a fast and a slow transient, and to characterize the slow transient space spanned by columns of Y_2.

4 Behaviour within aggregates

The stochastic matrix A governing the discrete time NCD chain has the form $A = diag(D_{\ell\ell}) + O(\varepsilon)$, where the coupling parameter, ε, is assumed small. We assume, henceforth, that each $D_{\ell\ell}$ is irreducible and aperiodic. Then, for each ℓ, let $\overline{\mathbf{y}}_\ell$ and $\overline{\mathbf{x}}_\ell$ be the left and right eigenvectors of $D_{\ell\ell}$ corresponding to the Perron root ρ_ℓ. The elements of $\overline{\mathbf{y}}_\ell$ and $\overline{\mathbf{x}}_\ell$ are strictly positive and we normalize so that $\mathbf{1}_\ell^T \overline{\mathbf{y}}_\ell = 1$ and $\overline{\mathbf{y}}_\ell^T \overline{\mathbf{x}}_\ell = 1$, where $\mathbf{1}_\ell$ will be used to represent a column vector of ones of length n_ℓ, the number of states in aggregate ℓ. Clearly $D_{\ell\ell}$ is stochastic to within $O(\varepsilon)$ and so, by a standard property of non-negative matrices,

$$\rho_\ell = 1 - O(\varepsilon). \tag{4.1}$$

Darroch and Seneta (1965) define the *product* quasi-stationary distribution

$$\overline{\mathbf{z}}_\ell = (\overline{x}_{1\ell}\,\overline{y}_{1\ell}, \ldots, \overline{x}_{n_\ell\ell}\,\overline{y}_{n_\ell\ell})^T = X_\ell \overline{\mathbf{y}}_\ell \tag{4.2}$$

where X_ℓ is the diagonal matrix, whose diagonal elements are the elements of $\overline{\mathbf{x}}_\ell$. Then $\overline{\mathbf{z}}_\ell$ $(= X_\ell \overline{\mathbf{y}}_\ell)$ and $\mathbf{1}_\ell$ $(= X_\ell^{-1} \overline{\mathbf{x}}_\ell)$ are corresponding left and right eigenvectors of the similarity transform of $D_{\ell\ell}$, $X_\ell^{-1} D_{\ell\ell} X_\ell$, corresponding to its Perron root ρ_ℓ and scaled so that $\mathbf{1}_\ell^T \overline{\mathbf{z}}_\ell = 1$.

4.1 Choice of quasi-stationary distribution

Both $\overline{\mathbf{y}}_\ell$ and $\overline{\mathbf{z}}_\ell$ are candidates for the quasi-stationary distribution over states of aggregate ℓ on which to base an approximate spectral decomposition of A. Their properties are discussed in some detail in Darroch and Seneta (1965). However, the following limiting properties establish the essential difference between them. Both limits assume that the process was in aggregate ℓ at time 0 and give a limiting probability of being in state i of aggregate ℓ:

$$\overline{y}_{i\ell} = \lim_{n \to \infty} P(\text{process in state } i \text{ of aggregate } \ell \text{ at time } n \,| \tag{4.3}$$
$$\text{process has not left } \ell \text{ by time } n),$$

$$\overline{z}_{i\ell} = \lim_{m \to \infty} \lim_{n \to \infty} P(\text{process in state } i \text{ of aggregate } \ell \text{ at time } m < n \,| \tag{4.4}$$
$$\text{process has not left } \ell \text{ by time } n).$$

Both $\overline{\mathbf{y}}_\ell$ and $\overline{\mathbf{z}}_\ell$ are independent of the initial state in aggregate ℓ. However, only $\overline{\mathbf{z}}_\ell$ is independent of how the process eventually exits from ℓ. We prefer to use $\overline{\mathbf{z}}_\ell$ to represent quasi-stationary behaviour within ℓ, thus keeping clear of any dependence on the way in which the process moves between aggregates.

4.2 Spectral structure of $D_{\ell\ell}$

By definition, $\overline{\mathbf{y}}_\ell$ and $\overline{\mathbf{x}}_\ell$ are left and right eigenvectors of $D_{\ell\ell}$ corresponding to its Perron root ρ_ℓ and normalized so that $\mathbf{1}_\ell^T \overline{\mathbf{y}}_\ell = 1$ and $\overline{\mathbf{y}}_\ell^T \overline{\mathbf{x}}_\ell = 1$. Thus, from Lemma 3.1, there exist matrices J_ℓ and K_ℓ such that

$$(\overline{\mathbf{y}}_\ell \quad K_\ell)^T = (\overline{\mathbf{x}}_\ell \quad J_\ell)^{-1} \tag{4.5}$$

$$\| J_\ell \| = 1, \quad n^{-1/2} \leq \| \overline{\mathbf{y}}_\ell \| \leq 1 \tag{4.6}$$

$$(\overline{\mathbf{y}}_\ell \quad K_\ell)^T D_{\ell\ell}(\overline{\mathbf{x}}_\ell \quad J_\ell) = \begin{pmatrix} \rho_\ell & \mathbf{0}^T \\ \mathbf{0} & C_\ell \end{pmatrix}. \tag{4.7}$$

It also follows from Lemma 3.1 that $\overline{\mathbf{x}}_\ell$ and K_ℓ will be bounded above, uniformly in ε, provided that $\delta(\rho_\ell, C_\ell)$ is bounded away from zero, uniformly in ε. This introduces an essential condition for the separation of time scales to work.

Condition 1 $\varepsilon/\delta(\rho_\ell, C_\ell) = O(\varepsilon)$ $\ell = 1, \ldots, L$

A number of nice results follow if Condition 1 is assumed.

Lemma 4.1 *If Condition 1 is satisfied then, for $\ell = 1, \ldots, L$,*

(1) $\| \overline{\mathbf{y}}_\ell \|$, $\| \overline{\mathbf{x}}_\ell \|$, $\| J_\ell \|$ *and* $\| K_\ell \|$ *are all bounded above, uniformly in ε.*

(2) $\varepsilon/\delta(1, C_\ell) = O(\varepsilon)$ *so that $\delta(1, C_\ell)$ is bounded away from zero uniformly in ε. That is, the subdominant eigenvalue of $D_{\ell\ell}$ is bounded away from 1, uniformly in ε.*

(3) $\overline{\mathbf{x}}_\ell = \mathbf{1}_\ell + \mathbf{a}_\ell$, *where* $\| \mathbf{a}_\ell \| = O(\varepsilon)$.

Proof Result (1) follows immediately from Condition 1 and Lemma 3.1. Result (2) follows from eqn (4.1), Condition 1 and the continuity property of $\delta(.,.)$. The proof of (3) is rather longer and is omitted since it follows a standard pattern which we shall see in later proofs. It uses the fact that the *substochastic* matrix $D_{\ell\ell}$ is within $O(\varepsilon)$ of a primitive *stochastic* matrix, whose Perron root, 1, has associated right eigenvector $\mathbf{1}$. Condition 1 and (1) above ensure that the conditions of Theorem 3.2 are satisfied and that the $O(\varepsilon)$ shift from a stochastic matrix to the substochastic matrix $D_{\ell\ell}$ is accompanied by an $O(\varepsilon)$ shift in the right eigenvector corresponding to the Perron root. The result follows. $\square$

We now consider our similarity transform of $D_{\ell\ell}$, the matrix $X_\ell^{-1} D_{\ell\ell} X_\ell$, and its left and right eigenvectors $\overline{\mathbf{z}}_\ell$, $\mathbf{1}_\ell$ corresponding to its Perron root, ρ_ℓ. We define

$$\overline{K}_\ell = X_\ell K_\ell \quad \text{and} \quad \overline{J}_\ell = X_\ell^{-1} J_\ell. \tag{4.8}$$

Then, clearly

$$(\overline{\mathbf{z}}_\ell \quad \overline{K}_\ell)^T = (\mathbf{1}_\ell \quad \overline{J}_\ell)^{-1} \tag{4.9}$$

$$(\overline{\mathbf{z}}_\ell \quad \overline{K}_\ell)^T D_{\ell\ell}(\mathbf{1}_\ell \quad \overline{J}) = \begin{pmatrix} \rho_\ell & \mathbf{0}^T \\ \mathbf{0} & \overline{C}_\ell \end{pmatrix}. \tag{4.10}$$

From Lemma 4.1(3),

$$X_\ell = I + O(\varepsilon) \tag{4.11}$$

and therefore the following lemma is immediate.

Lemma 4.2 *If Condition 1 is satisfied then, for $\ell = 1, \ldots, L$,*

(1) $\| \overline{\mathbf{z}}_\ell \|$, $\| \mathbf{1}_\ell \|$, $\| \overline{J}_\ell \|$ *and* $\| \overline{K}_\ell \|$ *are all bounded above, uniformly in ε.*

(2) $X_\ell^{-1} D_{\ell\ell} X_\ell = D_{\ell\ell} + O(\varepsilon)$ *and* $X_\ell D_{\ell\ell} X_\ell^{-1} = D_{\ell\ell} + O(\varepsilon)$ *so that*

(3) $\overline{C}_\ell = C_\ell + O(\varepsilon)$ *and therefore*

(4) $\delta(\rho_\ell, \overline{C}_\ell) = \delta(\rho_\ell, C_\ell) + O(\varepsilon)$ *and* $\delta(1, \overline{C}_\ell) = \delta(\rho_\ell, C_\ell) + O(\varepsilon)$.

5 The spectral decomposition of A

5.1 The first stage

Define

$$T_z = \begin{pmatrix} \overline{z}_1 & 0 & \cdots & 0 \\ 0 & \overline{z}_2 & \cdots & 0 \\ \vdots & \vdots & & \vdots \\ 0 & 0 & \cdots & \overline{z}_L \end{pmatrix} \quad \text{and} \quad T_1 = \begin{pmatrix} 1_1 & 0 & \cdots & 0 \\ 0 & 1_2 & \cdots & 0 \\ \vdots & \vdots & & \vdots \\ 0 & 0 & \cdots & 1_L \end{pmatrix}. \tag{5.1}$$

Then clearly

$$(T_z \quad \overline{K})^T = (T_1 \quad \overline{J})^{-1} \tag{5.2}$$

and we form the following similarity transform of A:

$$(T_z \quad \overline{K})^T A (T_1 \quad \overline{J}) = \begin{pmatrix} \overline{B} & H \\ G & \overline{C} \end{pmatrix}. \tag{5.3}$$

Assume that Condition 1 holds. We shall show that the conditions of Theorem 3.2 hold for ε sufficiently small and that therefore T_z, T_1 and $\overline{K}$, $\overline{J}$ are within $O(\varepsilon)$ of left and right invariant subspaces of A. Consider the matrices G, H, $\overline{C}$ and $\overline{B}$ in turn.

$$G = \overline{K}^T A T_1 = \begin{pmatrix} g_1 & g_{12} & \cdots & g_{1L} \\ g_{21} & g_2 & \cdots & g_{2L} \\ \vdots & \vdots & & \vdots \\ g_{L1} & g_{L2} & \cdots & g_L \end{pmatrix}$$

where for $k \neq \ell$, $g_{\ell k} = \overline{K}_\ell^T E_{\ell k} 1_k = O(\varepsilon)$, from Lemma 4.2(1), and where

$$\begin{aligned} g_\ell &= \overline{K}_\ell^T D_{\ell\ell} 1_\ell \\ &= K_\ell^T X_\ell^T D_{\ell\ell} X_\ell^{-1} \overline{x}_\ell \\ &= K_\ell^T D_{\ell\ell} \overline{x}_\ell + O(\varepsilon) \quad \text{from Lemma 4.2, results (1) and (2)} \\ &= O(\varepsilon). \end{aligned}$$

Similar results may be obtained for H and therefore

$$\| G \| = O(\varepsilon) \quad \text{and} \quad \| H \| = O(\varepsilon). \tag{5.4}$$

Note also that

$$G1 = \overline{K}^T A T_1 1 = \overline{K}^T A 1 = \overline{K}^T 1 = 0. \tag{5.5}$$

Now consider $\overline{C}$.

$$\overline{C} = \overline{K}^T A \overline{J} = \begin{pmatrix} \overline{C}_1 & \overline{C}_{12} & \cdots & \overline{C}_{1L} \\ \overline{C}_{21} & \overline{C}_2 & \cdots & \overline{C}_{2L} \\ \vdots & \vdots & & \vdots \\ \overline{C}_{L1} & \overline{C}_{L2} & \cdots & \overline{C}_L \end{pmatrix}$$

where, for $k \neq \ell$, $\overline{C}_{\ell k} = \overline{K}_\ell^T E_{\ell k} \overline{J}_k = O(\varepsilon)$ from Lemma 4.2(1). Thus

$$\overline{C} = diag(\overline{C}_\ell) + O(\varepsilon) = diag(C_\ell) + O(\varepsilon) \quad \text{from Lemma 4.2(3).} \tag{5.6}$$

Finally we consider the matrix $\overline{B}$.

$$\overline{B} = T_z^T A T_1 = \begin{pmatrix} b_1 & b_{12} & \cdots & b_{1L} \\ b_{21} & b_2 & \cdots & b_{2L} \\ \vdots & \vdots & & \vdots \\ b_{L1} & b_{L2} & \cdots & b_L \end{pmatrix}$$

where, using the boundedness results of Lemma 4.2 once more, we have for $k \neq \ell$, $b_{\ell k} = \overline{z}_\ell^T E_{\ell k} \mathbf{1}_k = O(\varepsilon)$. Also, using Lemma 4.2(2), we have $b_\ell = \overline{z}_\ell^T D_{\ell\ell} \mathbf{1}_\ell = \rho_\ell + O(\varepsilon) = 1 - O(\varepsilon)$. Thus

$$\overline{B} = diag(\rho_\ell) + O(\varepsilon) = I + O(\varepsilon). \tag{5.7}$$

It is easy to show that $\overline{B}$ is stochastic. Moreover, it is irreducible because A is and because $\overline{z}_\ell > 0$ for all ℓ. It may be interpreted as the matrix of transition probabilities of a slowly moving, approximately Markovian process *between* aggregates, provided that the separation of time scales works and that within each aggregate ℓ, the quasi-stationary distribution $\overline{z}_\ell$ is quickly reached.

From eqns (5.6) and (5.7), and the continuity property of $\delta(.,.)$, we have $\delta(\overline{B}, \overline{C}) = \delta(diag(\rho_\ell), diag(C_\ell)) + O(\varepsilon)$. Condition 1 and Property (6) of $\delta(.,.)$ then give us

$$\varepsilon/\delta(\overline{B}, \overline{C}) = O(\varepsilon). \tag{5.8}$$

We are now in a position to prove the following theorem.

Theorem 5.1 *We have*

$$(T_z \quad \overline{K})^T A (T_1 \quad \overline{J}) = \begin{pmatrix} \overline{B} & H \\ G & \overline{C} \end{pmatrix}.$$

If Condition 1 is satisfied then, for sufficiently small ε, there exist matrices P and Q, both $O(\varepsilon)$, such that

$$M^{-1} \begin{pmatrix} \overline{B} & H \\ G & \overline{C} \end{pmatrix} M = \begin{pmatrix} \tilde{B} & 0 \\ 0 & A_3 \end{pmatrix} \quad \text{where} \tag{5.9}$$

$$M = \begin{pmatrix} I & -Q^T \\ P & I \end{pmatrix} \begin{pmatrix} (I + Q^T P)^{-1} & 0 \\ 0 & (I + PQ^T)^{-1} \end{pmatrix} \tag{5.10}$$

$$M^{-1} = \begin{pmatrix} I & Q^T \\ -P & I \end{pmatrix} \tag{5.11}$$

$$\begin{aligned} \tilde{B} &= (I + Q^T P)(\overline{B} + HP)(I + Q^T P)^{-1} \\ &= \overline{B} + O(\varepsilon^2) \end{aligned} \tag{5.12}$$

$$\begin{aligned}
A_3 &= (I + PQ^T)(\overline{C} - GQ^T)(I + PQ^T)^{-1} \\
&= \overline{C} + O(\varepsilon^2) \\
&= diag(C_\ell) + O(\varepsilon)
\end{aligned} \tag{5.13}$$

$$P1 = 0. \tag{5.14}$$

Proof In the preceding discussion eqns (5.4) and (5.8) were obtained by assuming Condition 1. Thus, if ε is sufficiently small, the conditions of Theorem 3.2 will be satisfied and eqns (5.9)–(5.13) all follow immediately, with $\| P \|, \| Q \| = O(\varepsilon)$. Equation (5.14) is an easy consequence of eqn (5.5) and the requirement that $P\overline{B} - \overline{C}P = G - PHP$. □

Note 1 From eqn (5.13), we see that the dominant eigenvalue of A_3 is within $O(\varepsilon)$ of the largest eigenvalue of the C_ℓ. From Lemma 4.1(2), it is therefore bounded away from one uniformly in ε. From eqn (5.12), we see that the eigenvalues of $\tilde{B}$ are within $O(\varepsilon^2)$ of the eigenvalues of $\overline{B}$, which, from eqn (5.7), are all within $O(\varepsilon)$ of one. That is, the eigenvalues of $\tilde{B}$ are all within $O(\varepsilon)$ of one. Thus Condition 1, plus the requirement that ε be sufficiently small, is enough to guarantee the separation of time scales.

Note 2 From Theorem 5.1, the approximate left invariant spaces T_z and $\overline{K}$ have been shifted by $O(\varepsilon)$ to become exact left invariant spaces. Specifically $T_z \longrightarrow T_z + \overline{K}Q$ and $\overline{K} \longrightarrow \overline{K} - PT_z = Y_3$.

From Notes 1 and 2, it is clear that the process governed by A has a fast transient in the space spanned by the columns of $\overline{K} - PT_z$, which will become small very quickly. The remaining part will be contained within the space spanned by columns of $T_z + \overline{K}Q$ $(= T_z + O(\varepsilon))$ and will converge slowly to the stationary vector, $\mathbf{y}_1$, contained within this space. Therefore, ignoring the contribution from the fast transient, $\mathbf{y}_1$ will be made up, to within $O(\varepsilon)$, of the quasi-stationary vectors $\overline{\mathbf{z}}_\ell$ in some proportion.

Remark We should note that the analysis up to this point is robust to errors of $O(\varepsilon)$ in A. Such perturbations leave Condition 1 unchanged, so that all the spectral decompositions are stable and, in particular, the quasi-stationary distributions are accurate to within $O(\varepsilon)$.

Summarizing the first stage of similarity transforms in the spectral decomposition of A, we have

$$M^{-1} (T_z \ \ \overline{K})^T A (T_1 \ \ \overline{J}) M = \begin{pmatrix} \tilde{B} & 0 \\ 0 & A_3 \end{pmatrix} \tag{5.15}$$

$$\tilde{B} = N (\overline{B} + HP) N^{-1} \quad \text{with} \quad N = I + Q^T P \tag{5.16}$$

and expressions for M and M^{-1} are given as eqns (5.10) and (5.11). We now consider how to separate out $\mathbf{y}_1$ and Y_2 from the space $T_z + \overline{K}Q$.

5.2 The second stage

It is required, now, to decompose the matrix $\tilde{B}$ obtained in Theorem 5.1. However, the matrix we have available is $\overline{B}$, the irreducible, stochastic matrix obtained from the *approximate* invariant spaces as $\overline{B} = T_z A T_1$. As we have seen, $\overline{B} = I + O(\varepsilon)$ and is within $O(\varepsilon^2)$ of $\tilde{B}$. Thus both $\overline{B}$ and $\tilde{B}$ have the simple eigenvalue one, and all remaining eigenvalues within $O(\varepsilon)$ of one. The perturbation of $O(\varepsilon^2)$ as we move from the approximation $\overline{B}$ to $\tilde{B}$ will only be stable in terms of eigenvectors and invariant spaces if, roughly speaking, no eigenvalue of $\overline{B}$, other than the Perron root one, is within $O(\varepsilon^2)$ of one. Then no eigenvalue of $\overline{B}$, other than the Perron root, can become the Perron root of $\tilde{B}$.

Let $\mathbf{v} > \mathbf{0}$ and $\mathbf{1}$ be the left and right eigenvectors of $\overline{B}$ corresponding to the Perron root one, and scaled so that $\mathbf{v}^T \mathbf{1} = 1$. From Lemma 3.1 and its corollary there exist matrices U_2 and V_2 such that

$$(\mathbf{v} \quad V_2)^T = (\mathbf{1} \quad U_2)^{-1} \quad \text{and} \tag{5.17}$$

$$\| \mathbf{1} \| = L^{1/2}, \quad \| U_2 \| = 1, \quad \| \mathbf{v} \| \le 1, \quad \| V_2 \| \le L^{1/2}. \tag{5.18}$$

Form the similarity transform of $\overline{B}$

$$(\mathbf{v} \quad V_2)^T \overline{B} (\mathbf{1} \quad U_2) = \begin{pmatrix} 1 & \mathbf{0}^T \\ \mathbf{0} & \overline{A}_2 \end{pmatrix}. \tag{5.19}$$

We now introduce Condition 2, required to separate $\mathbf{y}_1$ from the slow transient.

Condition 2 $\varepsilon^2 / \delta(1, \overline{A}_2) = O(\varepsilon)$

Condition 2 says that $\delta(1, \overline{A}_2)$ cannot approach zero faster than $O(\varepsilon)$, which implies that the eigenvalues of $\overline{A}_2$, that is the remaining eigenvalues of $\overline{B}$, cannot approach one faster than $O(\varepsilon)$. Informally, the slow transient cannot be *too* slow.

From eqn (5.16), the matrix $\tilde{B}$ is a (well conditioned) similarity transform of $\overline{B} + HP$. We shall show that $\mathbf{v}$, $\mathbf{1}$ and V_2, U_2 are 'close' to left and right invariant vectors and subspaces of $\overline{B} + HP$. Since $P\mathbf{1} = \mathbf{0}$ (Theorem 5.1) it is easily shown that

$$(\mathbf{v} \quad V_2)^T (\overline{B} + HP)(\mathbf{1} \quad U_2) = \begin{pmatrix} 1 & \mathbf{s}^T \\ \mathbf{0} & A_2 \end{pmatrix} \tag{5.20}$$

where

$$\mathbf{s}^T = \mathbf{v}^T HP U_2 = O(\varepsilon^2) \quad \text{from eqn (5.18)} \tag{5.21}$$

$$A_2 = \overline{A}_2 + V_2^T HP U_2 = \overline{A}_2 + O(\varepsilon^2) \quad \text{from eqn (5.18).} \tag{5.22}$$

Now $\delta(1, A_2) = \delta(1, \overline{A}_2 + O(\varepsilon^2)) = \delta(1, \overline{A}_2) + O(\varepsilon^2)$. Thus, from Condition 2

$$\varepsilon^2 / \delta(1, A_2) = O(\varepsilon). \tag{5.23}$$

If we set

$$\mathbf{q}^T = \mathbf{s}^T (I - A_2)^{-1} \tag{5.24}$$

we have the following similarity transform

$$\begin{pmatrix} 1 & \mathbf{q}^T \\ 0 & I \end{pmatrix} \begin{pmatrix} 1 & \mathbf{s}^T \\ 0 & A_2 \end{pmatrix} \begin{pmatrix} 1 & -\mathbf{q}^T \\ 0 & I \end{pmatrix} = \begin{pmatrix} 1 & \mathbf{0}^T \\ 0 & A_2 \end{pmatrix} \qquad (5.25)$$

where, from eqns (5.21), (5.23) and (5.24), we have

$$\| \mathbf{q} \| \le \| (I - A_2)^{-1} \| \| \mathbf{s} \| = \frac{\| \mathbf{s} \|}{\delta(1, A_2)} = O(\varepsilon). \qquad (5.26)$$

Assuming Condition 2, we can now summarize the second stage of similarity transforms in the spectral decomposition of A. From eqns (5.16), (5.20) and (5.25), we have a succession of similarity transforms of $\tilde{B}$, giving

$$\begin{pmatrix} 1 & \mathbf{q}^T \\ 0 & I \end{pmatrix} (\mathbf{v} \ \ V_2)^T N^{-1} \tilde{B} N (1 \ \ U_2) \begin{pmatrix} 1 & -\mathbf{q}^T \\ 0 & I \end{pmatrix} = \begin{pmatrix} 1 & \mathbf{0}^T \\ 0 & A_2 \end{pmatrix} \qquad (5.27)$$

which may be expressed as the single similarity transform

$$(\mathbf{z}_1 \ \ Z_2)^T \tilde{B} (\mathbf{w}_1 \ \ W_2) \qquad (5.28)$$

where clearly $(\mathbf{z}_1 \ Z_2)^T = (\mathbf{w}_1 \ W_2)^{-1}$. A little algebra then gives us

$$\begin{aligned} \mathbf{z}_1 &= (I + P^T Q)^{-1} (\mathbf{v} + V_2 \mathbf{q}) &= \mathbf{v} + O(\varepsilon) & (5.29) \\ Z_2 &= (I + P^T Q)^{-1} V_2 &= V_2 + O(\varepsilon^2) & (5.30) \\ \mathbf{w}_1 &= \mathbf{1} & & (5.31) \\ W_2 &= U_2 - \mathbf{1}\mathbf{q}^T + Q^T P U_2 &= U_2 + O(\varepsilon). & (5.32) \end{aligned}$$

Thus, subject to Condition 2, our *approximate* invariant vectors and subspaces, $\mathbf{v}$, $\mathbf{1}$ and V_2, U_2, are, at worst, within $O(\varepsilon)$ of the exact decomposition of $\tilde{B}$.

Remark Despite the foregoing, it should be noted that this stage of the decomposition is extremely vulnerable to errors of $O(\varepsilon)$ in A, which could completely scramble the eigenvectors of $\overline{B}$ and $\tilde{B}$, as we have noted. Thus, errors of $O(\varepsilon)$ in A can lead to errors of $O(1)$ in $\mathbf{v}$, our approximation to the stationary vector of the slowly moving process between aggregates.

5.3   The complete decomposition of A

We now put together the results of Sections 5.1 and 5.2. From eqn (5.28),

$$\begin{pmatrix} \mathbf{z}_1 & Z_2 & 0 \\ 0 & 0 & I \end{pmatrix}^T \begin{pmatrix} \tilde{B} & 0 \\ 0 & A_3 \end{pmatrix} \begin{pmatrix} \mathbf{w}_1 & W_2 & 0 \\ 0 & 0 & I \end{pmatrix} = \begin{pmatrix} 1 & \mathbf{0}^T & \mathbf{0}^T \\ 0 & A_2 & 0 \\ 0 & 0 & A_3 \end{pmatrix} \qquad (5.33)$$

where, clearly,

$$\begin{pmatrix} \mathbf{z}_1 & Z_2 & 0 \\ 0 & 0 & I \end{pmatrix}^T = \begin{pmatrix} \mathbf{w}_1 & W_2 & 0 \\ 0 & 0 & I \end{pmatrix}^{-1}. \qquad (5.34)$$

Putting this together with eqn (5.15), we have the final decomposition of A.

$$\begin{pmatrix} \mathbf{z}_1 & Z_2 & 0 \\ 0 & 0 & I \end{pmatrix}^T M^{-1} \, (T_z \quad \overline{K})^T A \, (T_1 \quad \overline{J}) \, M \begin{pmatrix} \mathbf{w}_1 & W_2 & 0 \\ 0 & 0 & I \end{pmatrix}$$

$$= \begin{pmatrix} 1 & \mathbf{0}^T & \mathbf{0}^T \\ 0 & A_2 & 0 \\ 0 & 0 & A_3 \end{pmatrix}. \tag{5.35}$$

That is,

$$(\mathbf{y}_1 \quad Y_2 \quad Y_3)^T A (1 \quad X_2 \quad X_3) = diag(1, A_2, A_3) \tag{5.36}$$

where

$$(\mathbf{y}_1 \quad Y_2 \quad Y_3)^T = (1 \quad X_2 \quad X_3)^{-1}. \tag{5.37}$$

We can now collect all the results of Section 5 into the following theorem.

Theorem 5.2 *Let the matrix A satisfy Conditions 1 and 2; then for ε sufficiently small, there exist matrices P and Q and a vector $\mathbf{q}$ satisfying $\| P \|$, $\| Q \|$, $\| \mathbf{q} \| = O(\varepsilon)$, such that A has a spectral decomposition of the form (5.36)–(5.37) with*

$$
\begin{array}{llll}
(1) & \mathbf{y}_1 & = & T_z \mathbf{v} + T_z V_2 \mathbf{q} + \overline{K} Q \mathbf{v} + O(\varepsilon^2) \\
(2) & Y_2 & = & T_z V_2 + \overline{K} Q V_2 + O(\varepsilon^2) \\
(3) & Y_3 & = & \overline{K} - P T_z \\
(4) & X_2 & = & T_1 U_2 + \overline{J} P U_2 - \mathbf{1}\mathbf{q}^T \\
(5) & X_3 & = & \overline{J} + T_1 Q^T + O(\varepsilon^2) \\
(6) & A_2 & = & \overline{A}_2 + O(\varepsilon^2) \\
(7) & A_3 & = & \overline{C} + O(\varepsilon^2) & = & diag(C_\ell) + O(\varepsilon).
\end{array}
$$

Proof The proof follows from the results of Section 5 brought together in eqn (5.35), together with eqns (5.10)–(5.13) and (5.29)–(5.32). □

6 Discussion and summary

We have considered the most common model of a system that operates on more than one time scale, the case of a nearly completely decomposable Markov chain. Clearly the analysis is easily extendible to a hierarchy of time scales. Such models are often appropriate in the context of an integrated services telecommunications network. See, for example, Evans (1990) and Tse *et al.* (1995).

A standard practice when dealing with an NCD chain is to use an approximation to the stationary vector based on, first, a set of probability vectors, each representing approximately stationary behaviour within a fast time scale aggregate, and second, a single probability vector, approximating stationary behaviour of the slow moving process between aggregates. In our analysis, this corresponds to, first, finding the set of quasi-stationary vectors $\overline{\mathbf{z}}_\ell$, $\ell = 1, \ldots, L$, and second, forming the stochastic matrix, $\overline{B} = T_z^T A T_1$, and finding its left eigenvector, $\mathbf{v}$, corresponding to Perron root one, to represent stationary behaviour over aggregates. If the overall stationary vector $\mathbf{y}_1$ is required, it can then be approximated by $T_z \mathbf{v}$. The detailed analysis of this two-stage procedure exhibits some interesting features.

Provided the coupling parameter ε is sufficiently small, Condition 1 (given in Section 4.2 and used extensively in Section 5.1) guarantees the separation of time scales, regarded as typical of these chains. It will also guarantee the stability of the $\overline{\mathbf{z}}_\ell$, which, to within $O(\varepsilon)$, then span the slow transient space. This part of the process was seen to be robust to errors of $O(\varepsilon)$ in A.

Condition 2 (given in Section (5.2)) justifies the use of $\mathbf{v}$ and the approximation $\mathbf{y}_1 = T_z \mathbf{v}$. However, this part of the procedure, the analysis of the slow time scale dynamics, turns out to be extremely sensitive to errors of $O(\varepsilon)$. This suggests that caution should be exercised when analysing and using the spectral structure of $\overline{B}$. It also tends to reinforce our recommendation of the product quasi-stationary vectors $\overline{\mathbf{z}}_\ell$, which have a minimal dependence on the way the process moves between aggregates. In many applications these, rather than the overall stationary vector, may be the primary tools used to control the process and allocate resources. Situations where this may be true are discussed in both Evans (1990) and Tse *et al.* (1995), although the latter base their analysis on conditional rather than quasi-stationary distributions. The work in Tse *et al.* (1995) is particularly interesting since it looks at the limit as the coupling parameter $\varepsilon \to 0$, combined with a large deviations limit. This enables the authors to identify the conditions under which it is the fast or the slow time scale dynamics that lead to buffer overflow.

There are many matrices A that do not satisfy Conditions 1 and/or 2. An understanding of the implications for expected, typical behaviour on the different time scales should prove useful in the analysis and understanding of systems with multiple time scales.

Bibliography

1. Darroch, J. N. and Seneta, E. (1965). On quasi-stationary distributions in absorbing discrete-time finite Markov chains. *J. Appl. Probab.*, **2**, 88–100.

2. Evans, S. P. (1990). A quasi-stationary analysis of a virtual path in a B-ISDN network shared by services with very different characteristics. *Comput. Networks ISDN Syst.*, **20**, 391–399.

3. McAllister, D. F., Stewart, G. W. and Stewart, W. J. (1984). On a Rayleigh–Ritz refinement technique for nearly uncoupled stochastic matrices. *Linear Algebra Appl.*, **60**, 1–25.

4. Stewart, G. W. (1973). Error and perturbation bounds for subspaces associated with certain eigenvalue problems. *SIAM Rev.*, **15**, 727–764.

5. Stewart, G. W. (1983). Computable error bounds for aggregated Markov chains. *J. ACM*, **30** (2), 271–285.

6. Tse, D. N. C., Gallager, R. G. and Tsitsiklis, J. N. (1995). Statistical multiplexing of multiple time-scale Markov streams. *IEEE J. Sel. Areas Commun.*, **13** (6), 1028–1038.

15

On optimal returns and suboptimality bounds for systems satisfying generalised conservation laws

R. Garbe and K. D. Glazebrook

University of Newcastle

Abstract

Recent work by Bertsimas and Niño-Mora has demonstrated the power
of approaches to the analysis of controlled stochastic systems based on
(generalised) conservation laws. This work is described. The utility of this
theoretical framework is in part founded on its capacity to obtain the total
expected return from any policy in closed form. Work is described which
exploits this capacity, firstly to obtain a range of suboptimality bounds
for discounted and undiscounted branching bandits (and thereby a simple
dynamic programming proof of the optimality of Gittins index policies
for such processes). Secondly, we obtain characterisations of the optimal
return functions of controlled systems satisfying conservation laws and with
linear objectives. This work suggests the use of greedy approaches to the
development of solutions for a range of job selection/scheduling problems.

1 Introduction

Our concern is with stochastic systems in which one or more resources are available for allocation among demands upon the system (jobs, customers, projects etc.) which may be of several different types. Typically we seek a control rule for allocation of the resources which optimises some measure of overall system performance. For the most part such problems have been approached via dynamic programming, perhaps buttressed by special arguments. Of the latter, the simplest and most effective have been interchange arguments. See Chapter 8 of Walrand (1988) for a good summary.

In the main, progress in this area has been hard won. A notable success was the elucidation by Gittins and Jones (1974) of index-based solutions to the classical multi-armed bandit problem with discounted rewards earned over an infinite horizon. An extensive literature has been developed from this result and has highlighted the important role of index-based approaches. See, for example, Gittins (1979, 1989), Glazebrook (1976, 1982a, 1993), Katehakis and Veinott

This research was supported by the Engineering and Physical Sciences Research Council by
means of grant no. GR/KO3043.

(1987), Tsitsiklis (1986, 1993), Varaiya, Walrand and Buyukkoc (1985), Whittle
(1980, 1981) and Weber (1992). Much of this work has been discussed within a
dynamic programming framework (broadly conceived). The mathematical arguments which have been used have often been technically involved.

An alternative approach to scheduling control problems pioneered by Gelenbe and Mitrani (1980) is radically different from those based on dynamic
programming. Here we seek solutions by: (1) characterising the space of all
possible performances of the system and (2) optimising the overall system-wide
performance measure over this space. In this way, a range of control problems
involving multiclass queueing systems satisfying so-called conservation laws can
be solved by reformulating them as linear programs with feasible space a convex
polytope of special structure called (the base of) a polymatroid. See Federgruen
and Groenevelt (1988a, b) and Shanthikumar and Yao (1992). In a major recent
advance, Bertsimas and Niño-Mora (1996) demonstrated that this approach also
provides a powerful new way into the analysis of a general class of controlled
stochastic systems, called branching bandit processes. These are shown to satisfy generalised versions of the above conservation laws, a fact which allows the
performance space to be characterised as (the base of) an extended polymatroid. Consideration of the duals of linear programs over such spaces yields
Gittins index solutions to branching bandits with linear objectives. These ideas
are described in more detail in Section 2.

The authors and co-workers have embarked upon a major program of research
aimed at exploiting these exciting new developments and extending the theory.
In Sections 3 and 4 are accounts of the early stages of this work. Section 3
gives a simple dynamic programming proof of the optimality of Gittins index
policies for both discounted and undiscounted branching bandits and also derives
a range of suboptimality bounds for general policies. These analyses are clean
and straightforward and demonstrate the power of the Bertsimas and Niño-Mora
framework which derives from its capacity to give formulae for the total return
of a given policy in closed form. This capacity also plays a key role in Section 4,
which is broader in scope. Here the aim is to characterise the optimal returns for
systems satisfying conservation laws and having linear objectives, where these
returns are regarded as a function of the set of demands which are given access
to the system. The motivation here is to develop approaches to the solution
of problems which call for selection of the jobs to be processed in advance of
scheduling. The analyses lead us to a range of developments of the Bertsimas
and Niño-Mora framework, including the introduction of the crucial notion of
reducibility. Proposals for further work are included in Section 5.

2   Conservation laws

A general stochastic system has n job types, labelled $E = \{1, 2, \ldots, n\}$. A
scheduling control for the system is denoted u and must be nonidling and nonan-
ticipative. Control u yields *performance vector* $\underline{x}^u = \{x^u(i)\}$, $i \in E$, and $x^u(i)$,
the performance of type i jobs under u, must be an expectation.

Examples In a multiclass queueing system allow job type i to denote customer class i. In a problem in which the goal is to minimise expected holding costs we may make the choice

$$x^u(i) = \alpha(i)E_u\{N_t(i)\}, \; i \in E, \tag{2.1}$$

where $\alpha(i)$ is a constant, $N_t(i)$ denotes the number of class i customers present in the system at t and E_u is the expectation taken over realisations of the system under control u.

In a rather different vein De, Ghosh and Wells (1993) consider a selection and sequencing problem involving n jobs labelled $1, 2, \ldots, n$. Job i has processing requirement P_i, a nonnegative random variable independent of P_j, $j \neq i$. The goal is to choose a permutation of $E = \{1, 2, \ldots, n\}$ to maximise the total expected revenue earned from completing the jobs. If the revenue from completing i is $r_i e^{-\alpha T_i}$ where T_i is the time of completion then an appropriate choice of performance would be

$$x^u(i) = E\left(\int_{S_i^u}^{S_i^u + P_i} \alpha e^{-\alpha t} dt\right), \; i \in E, \tag{2.2}$$

where S_i^u denotes the time at which job i's processing begins under u.

If π is a permutation of E then π will also denote the admissible control, an absolute priority rule, which assigns job type $\pi(1)$ lowest priority, followed by $\pi(2)$ and so on. Whether such priorities are exercised in a preemptive or nonpreemptive fashion will need to be specified in each case.

Performances of the kind in (2.1) will tend to increase when job type i is given a lower priority. We shall characterise systems in relation to the behaviour of the performance of any nominated subset of job types, $S \subseteq E$, say, under controls which give that subset (collectively) lowest priority. In particular, we shall seek systems such that some linear combination of the performances in S is maximised by such a control. Since performances of the kind in (2.2) tend to decrease when i has low priority, here the goal will be to have linear combinations of the performances in nominated subset S minimised under controls which prefer job types in S^c to those in S.

Systems of the kind we seek are said to obey *generalised conservation laws* (GCL) as follows:

Definition 2.1 *(GCL) A system satisfies GCL if there exists a performance vector $\underline{x}$, a set function b (respectively f) : $2^E \to R^+$ and a matrix $A = (A_i^S)$, (respectively $F = (F_i^S)$), $i \in E$, $S \subseteq E$, satisfying*

$$A_i^S > 0, (F_i^S > 0), i \in S \text{ and } A_i^S = 0, (F_i^S = 0), i \in S^c \text{ for all } S \subseteq E$$

such that

$$b(S) = \sum_{i \in S} A_i^S x_i^\pi \text{ for all } \pi : \{\pi(1), \ldots, \pi(|S|)\} = S, \text{ for all } S \subseteq E, \tag{2.3}$$

or, respectively,

$$f(S) = \sum_{i \in S} F_i^S x_i^\pi \text{ for all } \pi : \{\pi(1), \ldots, \pi(|S|)\} = S, \text{ for all } S \subseteq E, \quad (2.4)$$

and such that for all admissible controls u we have

$$b(S) \le \sum_{i \in S} A_i^S x_i^u \text{ for all } S \subset E \text{ and } b(E) = \sum_{i \in E} A_i^E x_i^u \quad (2.5)$$

or, respectively,

$$f(S) \ge \sum_{i \in S} F_i^S x_i^u \text{ for all } S \subset E \text{ and } f(E) = \sum_{i \in E} F_i^E x_i^u. \quad (2.6)$$

Note that (in)equalities (2.4) and (2.6) would be what we might expect of a performance such as that in (2.1), since together they indicate that $\sum_{i \in S} F_i^S x_i^u$ is maximised when u is an absolute priority rule which gives the job types in S lowest priority. Similarly, (2.3) and (2.5) could be appropriate for performance (2.2). A system satisfying GCL of type 1, or GCL(1) for short, will be assumed to satisfy (2.3) and (2.5). A system satisfying GCL of type 2, or GCL(2), satisfies (2.4) and (2.6).

An important special case occurs when A_i^S or F_i^S is $I_S(i)$, the indicator function for $S, S \subseteq E$. Here we shall say that the system satisfies *strong conservation laws* (SCL) and shall adopt the usage SCL(1) or SCL(2), as above.

Examples

Strong conservation laws Several multiclass queueing systems satisfying SCL(2) are discussed by Shanthikumar and Yao (1992). In the examples cited here we will require that

- all service requirements are independent and are also independent of the arrival process;
- each customer class has a stream of arrivals independent of the arrivals of other classes;
- priorities are imposed preemptively.

Simple work conservation principles yield the conclusion that for a multiclass $G/G/1$ system the performance

$$\underline{x}^u := \left[E_u\{V_t(i)\}\right], \ i \in E, \quad (2.7)$$

satisfies SCL(2) for any $t > 0$. In (2.7), $V_t(i)$ stands for the work in the system at time t arising from customers in class i. Specialising to the $G/M/1$ case with exponential service times we conclude that (using Little's formula for (2.9))

$$\underline{x}^u := \left[E_u\{N_t(i)\}/\mu(i)\right], \ i \in E, \quad (2.8)$$

$$\underline{x}^u := \left[\rho(i)E_u\{W(i)\}\right], \ i \in E, \quad (2.9)$$

also satisfy SCL(2). In (2.8) and (2.9), $N_t(i)$ is the number of class i customers present in the system at t, $\mu(i)$ is the corresponding service rate and $\rho(i)$ is the traffic intensity. $W(i)$ denotes the response time for class i customers. The absence of a subscript indicates that the performance measure in (2.9) is steady state.

Shanthikumar and Yao (1992) give a simple argument which makes transparent that for a multiclass $G/M/c$ system with all customers sharing the same exponential service time distribution the performances

$$\underline{x}^u := \big[E_u\{N_t(i)\}\big], \ i \in E, \tag{2.10}$$

$$\underline{x}^u := \big[E_u\{V_t(i)\}\big], \ i \in E, \tag{2.11}$$

$$\underline{x}^u := \big[\lambda(i)E_u\{W(i)\}\big], \ i \in E, \tag{2.12}$$

all satisfy SCL(2). In (2.12), $\lambda(i)$ is a class i arrival rate. These conclusions extend to a network of queues variant of this system.

Generalised conservation laws Bertsimas and Niño-Mora (1996) establish that a general class of controlled stochastic systems called *branching bandit process* satisfy GCL. A branching bandit process is most helpfully thought of as a model for the scheduling of collections of projects of which there are K different types, labelled $\{1, 2, \ldots, K\} \equiv \Delta$. Each type k project can be in one of a finite number of states. The project/state pairs (k, x_k), denoting a project of type k in state x_k, are replaced by a single identifier (i, say) and are the job types in what follows. Hence $E = \{1, 2, \ldots, n\}$, the set of all possible job types, is given by

$$E = \bigcup_{k \in \Delta} E_k, \tag{2.13}$$

where E_k denotes the state space of a type k project.

At each decision epoch one job among those present is chosen for processing. The processing of a type i job takes random time v_i at the end of which i is replaced by N_{ij} type j jobs, $j \in E$. The collection $\{v_i, N_{ij}\}$, $j \in E$, has a general joint distribution, independent of all other project types and identically distributed for the same i. This construction can be used to model both Markovian transitions within projects and also exogenous arrivals. In the *discounted* case, rewards are earned at rate r_i during the processing of a type i job. The total reward under control u may be written

$$\sum_{i \in E} r_i x_i^u, \tag{2.14}$$

where

$$x_i^u = E_u\left\{ \int_0^\infty e^{-\alpha t} I_i(t) dt \right\}. \tag{2.15}$$

In (2.15), $\alpha > 0$ is a discount rate and

$$I_i(t) = \begin{cases} 1, & \text{if a type } i \text{ job is processed at time } t \\ 0, & \text{otherwise} \end{cases}$$

Bertsimas and Niño-Mora (1996) show that $\underline{x}^u := \{x_i^u\}$, $i \in E$, satisfies GCL(1). They characterise the coefficients A_i^S in (2.3) and (2.5). For now, we simply note that these are such that

$$i \in S_2 \subseteq S_1 \implies A_i^{S_2} \geq A_i^{S_1} \geq 1. \tag{2.16}$$

Bertsimas and Niño-Mora (1996) also consider an undiscounted version of the above which can be used for problems with a weighted flow-time criterion. Here, consider an expected total cost incurred during the first busy period $[0, T]$ under control u, which may be expressed as

$$\sum_{i \in S} c_i x_i^u, \tag{2.17}$$

where

$$x_i^u = E_u \left\{ \int_0^T t I_i(t) dt \right\}. \tag{2.18}$$

They show that $\underline{x}^u := \{x_i^u\}$, $i \in E$, satisfies GCL(2).

Suppose now that a system satisfies GCL(1), namely that a performance vector $\underline{x}$ satisfies (2.3) and (2.5). Bertsimas and Niño-Mora show that it follows that the *performance space* (the set of all possible $\underline{x}^u$) is a special kind of convex polytope, called (the base of) an *extended polymatroid*, and is given by

$$\mathcal{B}_A(b) = \left\{ \underline{x} \in (R^+)^n \; ; \; \sum_{i \in S} A_i^S x_i \geq b(S), S \subset E \text{ and } \sum_{i \in E} A_i^E x_i = b(E) \right\}. \tag{2.19}$$

That the performance space is contained within $\mathcal{B}_A(b)$ is transparent from (2.3) and (2.5). To obtain the inclusion the other way, we consider the $\mathcal{LP}$

$$\max \left\{ \sum_{i \in E} r_i x_i \; ; \; \underline{x} \in \mathcal{B}_A(b) \right\}. \tag{2.20}$$

By considering the dual program and utilising complementary slackness, Bertsimas and Niño-Mora show that (2.20) is solved by $\underline{x}^{\pi^*}$ where $\pi^* = (\pi_1^*, \pi_2^*, \ldots, \pi_n^*)$ is a Gittins index priority rule. Since every extreme point of a bounded polyhedron is the unique maximiser of some linear objective it follows that the extreme points of $\mathcal{B}_A(b)$ are in the performance space and are performances of fixed priority rules. All other points in $\mathcal{B}_A(b)$ must also be in the performance space, since they may be expressed as convex combinations of extreme points and must therefore (since performances are required to be expectations) be performances of randomisations of fixed priority rules. Hence $\mathcal{B}_A(b)$ is also contained within the performance space and must therefore be equal to it.

From this conclusion it then follows that the $\mathcal{LP}$ in (2.20) has an important relation to the control problem

$$\sup_u \left\{ \sum_{i \in E} r_i x_i^u \; ; \; u \text{ admissible} \right\}. \tag{2.21}$$

Suppose $\underline{x}^*$ solves (2.20). It is clear that any control u^* satisfying $\underline{x}^{u^*} = \underline{x}^*$ must solve (2.21). But it was part of the argument of the preceding paragraph that this was true of any Gittins index priority rule π^* which must therefore be optimal for (2.21).

Gittins index rule $\pi^* = (\pi_1^*, \pi_2^*, \ldots, \pi_n^*)$ is, therefore, optimal for any system satisfying GCL(1) and having a linear objective. Further, Bertsimas and Niño-Mora characterise π^* as follows: suppose $\nu = (\nu_1, \nu_2, \ldots, \nu_n) \in R^n$ satisfies

$$\sum_{i=j}^{n} \nu_i A_{\pi_j^*}^{S_i} = r_{\pi_j^*}, \ 1 \le j \le n, \tag{2.22}$$

where $S_i = \{\pi_1^*, \pi_2^*, \ldots, \pi_i^*\}, 1 \le i \le n$; then $\nu_n \ge 0$ and $\nu_j \le 0, 1 \le j \le n-1$. The quantity

$$\gamma_{\pi_j^*} = \sum_{i=j}^{n} \nu_i \tag{2.23}$$

is the *generalised Gittins index* of job type π_j^* and

$$\gamma_{\pi_1^*} \le \gamma_{\pi_2^*} \le \cdots \le \gamma_{\pi_n^*}, \tag{2.24}$$

i.e., π^* determines a policy which chooses from among the job types present one with maximal Gittins index.

In the same way, if a system satisfies GCL(2) as in (2.4) and (2.6) then

$$\mathcal{B}'_F(f) = \left\{ \underline{x} \in (R^+)^n \ ; \ \sum_{i \in S} F_i^S x_i \le f(S), S \subset E \text{ and } \sum_{i \in E} F_i^E x_i = f(E) \right\} \tag{2.25}$$

is the performance space and the control problem

$$\inf_u \left\{ \sum_{i \in E} c_i x_i^u \ ; \ u \text{ admissible} \right\} \tag{2.26}$$

is solved by the Gittins index priority rule π^*, characterised in (2.22)–(2.24) but with c_i replacing r_i.

3 Suboptimality bounds for policies for branching bandits

The set-up described in Section 2 provides a rich environment within which to analyse effectively a range of controlled stochastic systems. Here we focus initially on the class of discounted branching bandits above which satisfy GCL(1) with performance given by (2.15). Hence the above analysis described around (2.19)–(2.24) applies and Gittins index priority rules are optimal. Previous accounts have frequently been technical and cumbersome. We shall illustrate the power of this new formulation by showing how to develop a range of suboptimality bounds for policies for branching bandits. The results extend earlier ones due to Glazebrook (1982b, 1990). A fuller account may be found in Glazebrook and Garbe (1996).

Let π^* be as in (2.22)–(2.24) but with job types renumbered such that $\pi^* = (1, 2, \ldots, n)$. Suppose that at $t = 0$ a job of type j is present in the system but that a job of type $G > j$ has highest Gittins index then. For clarity, we use the following device: label the job of type j above $j1$ and all other occurrences of job type j in the system by $j2$. Hence we now have two job types replacing the single identifier j.

We now define $(G - j) + 1$ permutations of $[E \setminus \{j\}] \cup \{j1, j2\}$, the set of new identifiers, by

$$\pi(g) = (1, \ldots, j-1, j2, j+1, \ldots, g, j1, g+1, \ldots, n), \ j < g \le G. \tag{3.1}$$

We also write $\pi(j) \equiv \pi^*$, a Gittins index policy, and $\pi(G) \equiv j\pi^*$ policy which chooses $j1$ at time 0 and thereafter operates a Gittins index policy. Suitable application of (2.3) to priority policy $\pi(g)$ yields the conclusion that the corresponding performance $x(g)$ satisfies

$$\sum_{i=1}^{k} A_i^{S_k} x_i = b(\{1, \ldots, k\}), \ 1 \le k \le g,$$

$$\sum_{i=1}^{k} A_i^{S_g} x_i + A_j^{S_g} x_{j1} = b(\{1, \ldots, g, j1\}), \tag{3.2}$$

$$\sum_{i=1}^{k} A_i^{S_k} x_i + A_j^{S_k} x_{j1} = b(\{1, \ldots, k, j1\}), \ g+1 \le k \le n,$$

where $S_k = \{1, \ldots, k\}$. In (3.2) and elsewhere we have written j for $j2$. By considering the corresponding equations for $x(g-1)$ and subtracting we deduce that the vector $\delta(g) = x(g-1) - x(g), \ j+1 \le g \le G$, satisfies

$$\delta_k = 0, \ 1 \le k \le g-1,$$

$$\sum_{i=1}^{k} A_i^{S_k} \delta_i + A_j^{S_k} \delta_{j1} = 0, \ g \le k \le n. \tag{3.3}$$

We now write

$$R(\pi) = \sum_{i=1}^{n} r_i x_i^{\pi}$$

for the total expected reward associated with absolute priority rule π.

From (2.22) and (3.3), simple algebra serves to show that

$$R\{\pi(g-1)\} - R\{\pi(g)\} = \sum_{i=1}^{n} r_i \delta_i(g) + r_j \delta_{j1}(g)$$

$$= \delta_{j1}(g) \left\{ r_j - \sum_{i=g}^{n} \nu_i A_j^{S_i} \right\}$$

$$= \delta_{j1}(g) \left(\sum_{i=j}^{g-1} \nu_i A_j^{S_i} \right), \; j+1 \le g \le G. \tag{3.4}$$

We now conclude that

$$R(\pi^*) - R(j\pi^*) = R\{\pi(j)\} - R\{\pi(G)\}$$

$$\sum_{g=j+1}^{G} [R\{\pi(g-1)\} - R\{\pi(g)\}] = \sum_{g=j+1}^{G} \delta_{j1}(g) \left(\sum_{i=j}^{g-1} \nu_i A_j^{S_i} \right) \tag{3.5}$$

is the reward lost if at time 0 a type j job is chosen instead of one with maximal Gittins index.

Now note that it is trivial to establish that $\delta_{j1}(g) \le 0$, $j+1 \le g \le G$. It was also stated following (2.22) that $\nu_i \le 0$, $1 \le i \le n-1$, and from Definition 2.1 we observe that $A_j^{S_i} \ge 0$, $1 \le i \le n-1$. Equation (3.5) then yields $R(\pi^*) - R(j\pi^*) \ge 0$ from which the optimality of π^* follows via standard dynamic programming arguments. We continue from there to bound $R(\pi^*) - R(j\pi^*)$ by bounding the right hand side of (3.5). For example, we can rewrite that expression as

$$R(\pi^*) - R(j\pi^*) = \sum_{g=j}^{G-1} \left\{ \sum_{i=g+1}^{G} \delta_{j1}(i) \right\} \nu_g A_j^{S_g}$$

$$\sum_{g=j}^{G-1} \{x_{j1}(g) - x_{j1}(G)\} \nu_g A_j^{S_g} \le -x_{j1}(G) \sum_{g=j}^{G-1} \nu_g A_j^{S_g}$$

$$\le -\sum_{g=j}^{G-1} \nu_g/\alpha = (\gamma_G - \gamma_j)/\alpha. \tag{3.6}$$

The last inequality in (3.6) combines an exact calculation for x_{j1}, which is trivial, with an upper bound on $A_j^{S_g}$, $j \le g \le G-1$, derived from a characterisation of these quantities in Bertsimas and Niño-Mora (1996).

An alternative bounding procedure applied to the right hand side of (3.5) yields

$$R(\pi^*) - R(j\pi^*) \le (\gamma_G - r_j)\{1 - E(e^{-\alpha v_j})\}/\alpha, \tag{3.7}$$

where v_j is the duration of the initial period of usage of job type j in policy $j\pi^*$. We summarise these calculations in the following result.

Theorem 3.1 *(Discounted branching bandits – suboptimality bound) If π^* is a Gittins index policy, a job of type j is present at time 0 and job type G has maximal index then,*

$$R(\pi^*) - R(j\pi^*) \le \min[(\gamma_G - \gamma_j)/\alpha; (\gamma_G - r_j)\{1 - E(e^{-\alpha v_j})\}/\alpha].$$

The expressions on the right hand sides of (3.6) and (3.7) may now be used as the basis for the development of suboptimality bounds for general stationary policy π for a discounted branching bandit, i.e., upper bounds for $R(\pi^*) - R(\pi)$. The approach is standard and makes use of conditioning arguments, aggregation and convergence results. In this way, bounds are developed and results obtained which generalise those in Glazebrook (1982a, 1990).

Undiscounted branching bandits satisfy GCL(2) where the performance is given by (2.18) which is assumed always finite. See the comments around (2.25), (2.26) above. When the objective is linear, Gittins index priority rules are optimal for the average cost criterion. We can develop both a dynamic programming proof of the optimality of index policies and suboptimality bounds via an analysis which is very close to that for the discounted case. By following closely the steps which yield (3.5) above we obtain

$$C(j\pi^*) - C(\pi^*) = -\sum_{g=j+1}^{G} \delta_{j1}(g) \left(\sum_{i=1}^{g-1} \nu_i A_j^{S_i} \right). \tag{3.8}$$

The right hand side of (3.8) is easily seen to be nonnegative since $\delta_{j1}(g) \geq 0$, $j+1 \leq g \leq G, \nu_i \leq 0, 1 \leq i \leq n-1$, and $A_j^{S_i} \geq 0, 1 \leq i \leq n-1$. The optimality of Gittins index policy π^* for the average cost case is a straightforward consequence.

As above, we can develop suboptimality bounds from (3.8) analogous to those reported in (3.6) and (3.7) for the discounted case. Here it is straightforward to establish the following result

Theorem 3.2 *(Undiscounted branching bandits – suboptimality bound) If π^* is a Gittins index policy, a job of type j is present at time 0 and job type G has maximal index then,*

$$C(j\pi^*) - C(\pi^*) \leq \min[\{x_{j1}(j) - x_{j1}(G)\} A_j^{S_j} (\gamma_G - \gamma_j); \{x_{j1}(j) - x_{j1}(G)\}(\gamma_G - c_j)].$$

As before it is possible to use Theorem 3.2 to develop reasonably simple bounds on the excess average cost per unit time incurred by using a general policy π instead of optimal policy π^*.

4 Submodular returns and job selection problems

Example 1: Subset selection – maximising net rewards

A set of jobs denoted Δ is available to be processed by a single machine. A discounted reward r_j is realised upon completion of each job $j \in \Delta$. There is also a cost c_j incurred (for tools and materials, say) at time zero upon submission of job j for processing. It may be that it is uneconomic to submit all jobs for processing. The natural optimisation problem involves choosing a subset $S \subseteq \Delta$ to maximise

$$-\sum_{j \in S} c_j + \Psi_1(S), \tag{4.1}$$

where $\Psi_1(S)$ is the return obtained from processing subset S optimally.

Example 2: Choosing a research portfolio of given size

A collection of research projects, Δ, are candidates for inclusion in a company's research portfolio. The chosen portfolio will include only k projects, where $k \leq |\Delta|$. We wish to choose a subset $S \subseteq \Delta$ of size k to maximise $\Psi_2(S)$, the return from the portfolio of projects in S. Once chosen, the projects in S must be scheduled optimally.

Example 3: Splitting problems

A set of customer classes, Δ, can be served at one of two service facilities, which need not be identical. We would like an a priori splitting of Δ into S (to be served at Facility 1) and $\Delta \setminus S$ (Facility 2) to minimise

$$\Phi_1(S) + \Phi_2(\Delta \setminus S), \tag{4.2}$$

where $\Phi_i(S)$ is the expected cost incurred when Facility i serves subset S optimally.

Progress with the above problems is dependent upon understanding the properties of the optimal return functions Ψ_1, Ψ_2, Φ_1 and Φ_2. That these set functions should all be *increasing* is clear. A natural conjecture arising from the discounted reward structure in Example 1 is that $\Psi_1 : 2^\Delta \to R^+$ should be *submodular*, i.e., such that for any $S \subseteq T \subseteq \Delta$ and $j \notin T$

$$\Psi_1(T \cup \{j\}) - \Psi_1(T) \leq \Psi_1(S \cup \{j\}) - \Psi_1(S). \tag{4.3}$$

Inequality (4.3) expresses a notion of diminishing returns as additional jobs are submitted for processing. In Example 3 we might expect Φ_i, $i = 1, 2$, to be *supermodular*. For supermodularity, the direction of the inequality in (4.3) is reversed. Supermodularity would arise from the effect of increasing congestion in a queueing system as additional classes of customers are allowed access to service.

In fact, Federgruen and Groenevelt (1987) elucidated (stochastic) supermodularity of standard performance measures such as work in system, response times etc. in some simple multiclass queueing systems under a FIFO discipline. Weber (1992) showed that the return from a multi-armed bandit with discounted returns is submodular in the set of "arms". In fact, it emerges that the framework of generalised conservation laws is ideal for exploration of the issues raised in the preceding paragraph. We shall show that, under some simply stated additional conditions, systems satisfying GCL(1) have submodular returns when maximising and those satisfying GCL(2) have supermodular returns when minimising a linear objective.

To be specific, consider a system satisfying GCL(1). See (2.3) and (2.5). We shall suppose that E, the set of job types, is expressed as a disjoint union,

$$E = \bigcup_{k \in \Delta} E_k, \tag{4.4}$$

where $\Delta = \{1, 2, \ldots, K\}$ is the set of objects (projects, jobs, customer classes etc.) available for selection and E_k is the set of job types belonging to object k. In multiclass queueing systems we may have $E_k = \{k\}$, representing customer class k. In a branching bandit process, E_k will be the state space of a type k project. Our interest will be in characterising optimal returns when service is restricted to the objects in $\Sigma \subseteq \Delta$, i.e., the job types in $\bigcup_{k \in \Sigma} E_k \equiv S_\Sigma \subseteq E$.

Refer to the full system as $\mathcal{R}(\Delta)$ and a reduced system in which service is restricted to the job types in S_Σ as $\mathcal{R}(\Sigma)$. We shall require that $\mathcal{R}(\Sigma)$ inherits GCL from $\mathcal{R}(\Delta)$ and that performances in $\mathcal{R}(\Sigma)$ can be obtained as performances in $\mathcal{R}(\Delta)$ when job types in S_Σ are given priority. More formally, consider the following definition in which $\mathcal{R}(\Delta)$ is assumed to satisfy GCL(1) as in (2.3) and (2.5) with the matrix A_i^S, $i \in E$, $S \subseteq E$.

Definition 4.1 $\mathcal{R}(\Sigma)$ *is a* <u>reduction</u> *of* $\mathcal{R}(\Delta)$ *if*

(1) *there exists a performance vector* $\underline{y} = \{y(i)\}$, $i \in S_\Sigma$, *satisfying* $GCL(1)$ *with the matrix* A_i^S, $i \in S_\Sigma$, $S \subseteq S_\Sigma$;

(2) *if* π *is a permutation of* S_Σ *then* $y^\pi(i) = x^{\pi'}(i)$, $i \in S_\Sigma$, *where* $\pi' = \{\tilde{\pi}, \pi\}$ *is any permutation of* E *formed from the concatenation of* $\tilde{\pi}$, *a permutation of* S_Σ^c, *with* π.

Obvious adjustments are made to Definition 4.1 when $\mathcal{R}(\Delta)$ satisfies GCL(2).

Definition 4.2 *The system* $\mathcal{R}(\Delta)$ *which satisfies GCL is reducible if there is a reduction to* $\mathcal{R}(\Sigma)$ *for every* $\Sigma \subseteq \Delta$.

Examples

Multiclass queueing systems The examples of multiclass queueing systems cited in Section 2 as instances of the satisfaction of SCL(2) are reducible when job types are identified as customer classes, as indicated above. Plainly the reduced systems must satisfy SCL(2) also. That the requirement of Definition 4.1(2) is met follows immediately from the independence assumptions and the fact that priorities are implemented preemptively.

Branching bandit processes Branching bandit processes in which E_k is the state space of a bandit of type k, $k \in \Delta$, are not reducible in general. The problem concerns the satisfaction of Definition 4.1(1). In general the matrix coefficients A_i^S will be changed upon restriction to project types in Σ, $\Sigma \subseteq \Delta$. If, however, we require that the random variables N_{ij} introduced in Section 2 satisfy

$$N_{ij} = 0 \text{ w.p. 1 whenever } i \in E_k, j \in E_l, k \neq l \tag{4.5}$$

then the coefficients A_i^S (F_i^S) have the property

$$A_i^S = A_i^{S \cap E_k}, \ i \in S \cap E_k, \ S \subseteq E, \tag{4.6}$$

(similarly for F_i^S) and the system is said to be *decomposable with respect to* $\{E_1, E_2, \ldots, E_k\}$. The requirement (4.5) is most obviously met in *multi-armed bandit problems* in which at the end of a period of processing job type $i \in E_k$, i is replaced by $j \in E_k$ where j is obtained from i via a Markovian transition law.

It is trivial to check that both discounted and undiscounted multi-armed bandit problems are reducible. Note also that all systems satisfying SCL are trivially decomposable with respect to $\{1, 2, \ldots, n\}$.

It is easy to obtain the performance space of $\mathcal{R}(\Sigma)$ when $\mathcal{R}(\Delta)$ is a reducible and decomposable system. Definition 4.1(1) implies that the space must be (the base of) an extended polymatroid and fixes the coefficients A_i^S. Definition 4.1(2) then yields the base function (the function b in (2.3) and (2.5)). If $\mathcal{R}(\Delta)$ satisfies GCL(1) and is reducible and decomposable with respect to $\{E_1, E_2, \ldots, E_K\}$ then the performance space of $\mathcal{R}(\Sigma)$ is

$$
\mathcal{B}_A(S_\Sigma, b) = \left\{ \underline{y} \in (R^+)^{|S_\Sigma|} ; \sum_{i \in T} A_i^T y_i \geq b(T \cup S_\Sigma^c) - b(S_\Sigma^c), \ T \subset S_\Sigma, \right.
$$

$$
\left. \text{and} \ \sum_{i \in S_\Sigma} A_i^{S_\Sigma} y_i = b(E) - b(S_\Sigma^c) \right\}. \tag{4.7}
$$

If, however, $\mathcal{R}(\Delta)$ satisfies GCL(2) and is reducible and decomposable with respect to $\{E_1, E_2, \ldots, E_K\}$ then the performance space of $\mathcal{R}(\Sigma)$ is

$$
\mathcal{B}'_F(S_\Sigma, f) = \left\{ \underline{y} \in (R^+)^{|S_\Sigma|} ; \sum_{i \in T} F_i^T y_i \leq f(T \cup S_\Sigma^c) - f(S_\Sigma^c), \ T \subset S_\Sigma, \right.
$$

$$
\left. \text{and} \ \sum_{i \in S_\Sigma} F_i^{S_\Sigma} y_i = f(E) - f(S_\Sigma^c) \right\}. \tag{4.8}
$$

For systems satisfying GCL and the additional requirement of reducibility and decomposability, a natural family of optimisation problems to consider has values given by

$$
\Psi(\Sigma) = \max \left\{ \sum_{i \in S_\Sigma} r_i x_i ; \underline{x} \in \mathcal{B}_A(S_\Sigma, b) \right\} \tag{4.9}
$$

or

$$
\Phi(\Sigma) = \min \left\{ \sum_{i \in S_\Sigma} c_i x_i ; \underline{x} \in \mathcal{B}'_F(S_\Sigma, f) \right\}. \tag{4.10}
$$

From the above discussions, analysis of Ψ will yield properties of the optimal return functions for the example of De, Ghosh and Wells (1993) cited in Section 2 as well as all discounted multi-armed bandit problems. Analysis of Φ would be relevant for the multiclass queueing systems cited in Section 2 with objectives linear in the performances given in (2.7)–(2.12).

For the remainder of the section we shall focus primarily on the set function $\Psi : 2^\Delta \to R^+$ and establish its submodularity under given conditions. The analysis which yields the supermodularity of $\Phi : 2^\Delta \to R^+$ is very similar. The first step from this point is to express Ψ in terms of the base function $b : 2^E \to R^+$. Recall the Gittins index notations π^*, $\gamma_{\pi_j^*}$ and S_j from the conclusions to Section 2. We then have

$$\Psi(\Delta) = \sum_{j=1}^{n-1} b(S_j)(\gamma_{\pi_j^*} - \gamma_{\pi_{j+1}^*}) + b(E)\gamma_{\pi_n^*}. \tag{4.11}$$

To obtain (4.11), we consider the dual of the $\mathcal{LP}$ in (2.20). A solution is easily obtained by means of complementary slackness and (4.11) is the consequential value of the dual. Some of the necessary details may be found in Bertsimas and Niño-Mora (1996). Once the additional conditions are introduced which yield the performance space for $\mathcal{R}(\Sigma)$ in (4.7), namely that the full system $\mathcal{R}(\Delta)$ is reducible and decomposable with respect to $\{E_1, E_2, \ldots, E_K\}$, then an expression for $\Psi(\Sigma)$, $\Sigma \subseteq \Delta$, is easily deduced from (4.11). The new base function is as in (4.7) and decomposability guarantees that the relevant Gittins indices γ_i, $i \in S_\Sigma$, are unchanged upon reduction of the system.

Having expressed Ψ is terms of b, it will now be properties of the base function b which will yield properties of Ψ. The base functions for the systems satisfying SCL(1) (SCL(2)) are well known to be normalised, increasing and supermodular (submodular). That the base functions continue to be normalised in the GCL case is clear, but the other properties do not carry over in general. However, a weakened form of supermodularity (submodularity) is found to survive in the base functions of decomposable systems.

Definition 4.3   *Function* $b : 2^E \to R^+$ *is supermodular with respect to* $\{E_1, E_2, \ldots, E_K\}$ *if, for all* $\Sigma \subset \Delta$, Σ *a proper subset, and subsets* S, T, σ *and* τ *satisfying*

$$T \subseteq S \subseteq \bigcup_{k \in \Sigma} E_k \ \text{and} \ \tau \subseteq \sigma \subseteq \bigcup_{k \notin \Sigma} E_k, \tag{4.12}$$

$$b(T \cup \sigma) - b(T \cup \tau) \le b(S \cup \sigma) - b(S \cup \tau). \tag{4.13}$$

Submodularity with respect to $\{E_1, E_2, \ldots, E_K\}$ has the inequality in (4.13) reversed.

That the base function b of a decomposable system satisfying GCL(1) should have the weakened form of supermodularity described in Definition 4.3 follows by applying (2.3) to carefully chosen priority rules π. Everything is now in place which is needed for the main result of this section.

Theorem 4.4

(1) *If a system satisfying* GCL(1) *is reducible and decomposable with respect to* $\{E_1, E_2, \ldots, E_K\}$ *then*

    (a) $\Psi : 2^\Delta \to R^+$ *is normalised and submodular, and*

    (b) Ψ *is also nondecreasing when* $b : 2^E \to R^+$ *is nondecreasing.*

(2) *If a system satisfying* GCL(2) *is reducible and decomposable with respect to* $\{E_1, E_2, \ldots, E_K\}$ *then*

    (a) $\Phi : 2^\Delta \to R^+$ *is normalised and supermodular, and*

    (b) Φ *is also nondecreasing when* $f : 2^E \to R^+$ *is nondecreasing.*

The proof of Theorem 4.4(1) involves a fair amount of detailed calculation.

It combines

(1) the fact that, under decomposability, Gittins indices do not change upon reduction of the system;

(2) an expression for $\Psi(\Sigma)$, $\Sigma \subseteq \Delta$, deduced from (4.11); and

(3) the supermodularity property of base function b described in Definition 4.3.

See Garbe and Glazebrook (1995) for full details. As a result of Theorem 4.4, we now have established the supermodularity of optimal return functions for the SCL(2) systems described in Section 2 with linear objectives and for the undiscounted version of the multi-armed bandit problem. We also have the submodularity of the optimal return function of a discounted multi-armed bandit as well as for the return function of the De, Ghosh and Wells (1993) model given in Section 2. As a consequence, formulations of Examples 1–3 at the beginning of this section which conform to the requirements of the above theory may be expressed as

Example 1

$$\max\left\{ -\sum_{j \in S} c_j + \Psi_1(S); S \subseteq \Delta\right\}, \ \Psi_1 \text{ nondecreasing and submodular;}$$

Example 2

$$\max\left\{\Psi_2(S); S \subseteq \Delta, |S| = k\right\}, \ \Psi_2 \text{ nondecreasing and submodular;}$$

Example 3

$$\min\{\Phi_1(S) + \Phi_2(\Delta \setminus S); S \subseteq \Delta\}, \ \Phi_1 \text{ and } \Phi_2 \text{ nondecreasing and supermodular.}$$

However, it is trivial to show that

(1) $\Psi : 2^\Delta \to R^+$ is submodular $\Rightarrow \Psi_{\underline{c}} : 2^\Delta \to R^+$ is submodular, where

$$\Psi_{\underline{c}}(S) = -\sum_{j \in S} c_j + \Psi(S), \ S \subseteq \Delta;$$

(2) $\Phi_1, \Phi_2 : 2^\Delta \to R^+$ is supermodular $\Rightarrow \Phi : 2^\Delta \to R^+$ is supermodular,

where $\quad \Phi(S) = \Phi_1(S) + \Phi_2(\Delta \setminus S), \ S \subseteq \Delta;$

(3) $\Phi : 2^\Delta \to R^+$ is supermodular $\Rightarrow -\Phi : 2^\Delta \to R^-$ is submodular.

It then follows that these job selection problems become either

$$\max\{\Psi(S); S \subseteq \Delta\}, \ \Psi \text{ submodular} \tag{4.14}$$

or

$$\max\{\Psi(S); S \subseteq \Delta, |S| = k\}, \ \Psi \text{ nondecreasing and submodular.} \tag{4.15}$$

Such combinatorial optimisation problems as those in (4.14) and (4.15) are known to be difficult in general. There are certainly NP-complete special cases. Hence research has focused on the development and evaluation of approximation schemes. There are strong grounds for believing that greedy algorithms for constructing solutions to the versions of (4.14) and (4.15) arising from the selection/scheduling problems discussed here will perform well. Firstly, Nemhauser and Wolsey (1981) have shown that a 'q-enumeration plus greedy' heuristic has special status in relation to (4.15). This heuristic takes (all) subsets of Δ of size $q < k$, constructs a subset of size k from each one in a greedy fashion and then chooses the best as a solution to the problem. This procedure involves $O(n^{q+1})$ function evaluations. Nemhauser and Wolsey (1981) give a performance bound on the heuristic and demonstrate that any algorithm yielding a tighter performance guarantee will require at least $O(n^{q+2})$ evaluations. It is not difficult to see that the above should map into reasonable properties for greedy heuristics as solutions to (4.14) also. In addition to these general considerations from combinatorial optimisation, we know from work by Stadje (1993) that greedy heuristics are optimal for versions of (4.15) arising from a particularly simple version of Example 2 which schedules the selected projects in a nonpreemptive fashion and earns discounted reward in the process. Aspvall, Flåm and Villanger (1995) propose this as a model for a problem involving the selection and (sequential) extraction of petroleum deposits.

Extensive computational results attest to the quality of performance of greedy procedures for the subset selection problems discussed here. Please note that, of course, the scheduling problems subsequent to selection will always have Gittins index solutions with optimal value given by (4.11) or some suitable variant of it. In 60,000 instances of the job splitting problem involving twelve classes of customers to be served at one of two facilities (with a variety of assumptions about costs and stochastic structure), a greedy solution was never more that 8.9% suboptimal and was optimal in 36.9% of cases. The median level of suboptimality was about 0.3%.

The performance of greedy approaches to instances of Example 2 in which the project scheduling problem subsequent to subset selection was a k-armed bandit with discounted rewards was quite spectacular. Of 80,000 problem instances generated (incorporating a range of assumptions about rewards, stochastic structure and the value of k) a greedy solution was optimal in all but 257 cases.

5 Further work

Further work by the authors and co-workers, either planned or in progress, includes:

(1) A development of the analyses of Section 3 to the case of families of competing Markov decision processes. These systems are multi-armed bandits in which each arm has its own decision structure. Clean approaches to simple cases are now available. More complex problems need to be approached via relaxations which yield mathematical programs with feasible space containing the relevant performance space. Solutions of these programs often

suggest heuristics for the problem of interest.

(2) The notion of reducibility developed in Section 4 may be further exploited in the consideration of scheduling problems in which job types fall into a set of mutually exclusive priority classes. Here we consider reducible systems satisfying GCL with a modest additional condition on the coefficients A_i^S (F_i^S) which is considerably weaker than the requirements for decomposability. Under such conditions, the control problem constrained by the imposition of priorities may be decomposed into a collection of optimisation problems of the form (2.20) for which Gittins indices offer an optimal solution.

(3) The generalised bandits introduced by Nash (1980) will have a suitably defined performance which satisfies GCL(1) when an additional condition is satisfied which guarantees that the indices solving such problems are always positive. Whether some form of GCL holds in general and, if so, whether it can be used to characterise a performance space remain open questions.

(4) The development of numerical and analytical approaches to systems which satisfy GCL in an approximate (or asymptotic) sense.

Bibliography

1. Aspvall, B., Flåm, S. D. and Villanger, K. P. (1995). Selecting among scheduled projects. *Operations Research Letters*, **17**, 37–40.

2. Bertsimas, D. and Niño-Mora, J. (1996). Conservation laws, extended polymatroids and multi-armed bandit problems: a unified approach to indexable systems. *Mathematics of Operations Research*, to appear.

3. De, P., Ghosh, J. P. and Wells, C. E. (1993). Job selection and sequencing on a single machine in a random environment. *European Journal of Operational Research*, **70**, 425–431.

4. Federgruen, A. and Groenevelt, H. (1987). The impact of the composition of the customer base in general queueing models. *Journal of Applied Probability*, **24**, 709–724.

5. Federgruen, A. and Groenevelt, H. (1988a). Characterization and optimization of achievable performance in general queueing systems. *Operations Research*, **36**, 733–741.

6. Federgruen, A. and Groenevelt, H. (1988b). M/G/c queueing systems with multiple customer classes: characterization and control of achievable performance under non-preemptive priority rules. *Management Science*, **34**, 1121–1138.

7. Garbe, R. and Glazebrook, K. D. (1995). Conservation laws, submodular returns and greedy heuristics for job selection and scheduling problems. Technical Report, Newcastle University.

8. Gelenbe, E. and Mitrani, I. (1980). *Analysis and Synthesis of Computer Systems*. Academic Press, New York.

9. Gittins, J. C. (1979). Bandit processes and dynamic allocation indices. *Journal of the Royal Statistical Society* B, **14**, 148–177.

10. Gittins, J. C. (1989). *Bandit Process and Dynamic Allocation Indices*. John Wiley, New York.

11. Gittins, J. C. and Jones, D. M. (1974). A dynamic allocation index for the sequential design of experiments. In *Progress in Statistics* (J. Gani *et al.* eds). North-Holland, Amsterdam, 241–266.

12. Glazebrook, K. D. (1976). Stochastic scheduling with order constraints. *International Journal of Systems Science*, **7**, 657–666.

13. Glazebrook, K. D. (1982a). On the evaluation of suboptimal strategies for families of alternative bandit processes. *Journal of Applied Probability*, **19**, 716–722.

14. Glazebrook, K. D. (1982b). On a sufficient condition for superprocesses due to Whittle. *Journal of Applied Probability*, **19**, 99–110.

15. Glazebrook, K. D. (1990). Procedures for the evaluation of strategies for resource allocation in a stochastic environment. *Journal of Applied Probability*, **27**, 215–220.

16. Glazebrook, K. D. (1993). Indices for families of competing Markov decision processes with influence. *Annals of Applied Probability*, **3**, 1013–1032.

17. Glazebrook, K. D. and Garbe, R. (1996). Reflections on a new approach to Gittins indexation. *Journal of the Operational Research Society*, to appear.

18. Katehakis, M. N. and Veinott, A. F. (1987). The multi-armed bandit problem: decomposition and computation. *Mathematics of Operations Research*, **12**, 262–268.

19. Nash, P. (1980). A generalised bandit problem. *Journal of the Royal Statistical Society* B, **42**, 165–169.

20. Nemhauser, G. L. and Wolsey, L. A. (1981). Maximizing submodular set functions: formulations and analysis of algorithms. *Annals of Discrete Mathematics*, **11**, 279–301.

21. Shanthikumar, J. G. and Yao, D. D. (1992). Multiclass queueing systems: polymatroidal structure and optimal scheduling control. *Operations Research*, **40**, Supplement 2, S293–299.

22. Stadje, W. (1993). Selecting jobs for scheduling on a machine subject to failure (unpublished manuscript).

23. Tsitsiklis, J. N. (1986). A lemma on the multi-armed bandit problem. *IEEE Transactions on Automatic Control*, **31**, 576–577.

24. Tsitsiklis, J. N. (1993). A short proof of the Gittins index theorem. *Annals of Applied Probability*, 4, 194–199.

25. Varaiya, P. P., Walrand, J. C. and Buyukkoc, C. (1985). Extensions of the multi-armed bandit problem: the discounted case. *IEEE Transactions on Automatic Control*, **30**, 426–439.

26. Walrand, J. C. (1988). *An Introduction to Queueing Networks*. Prentice

Hall, Englewood Cliffs, New Jersey.

27. Weber, R. (1992). On the Gittins index for multi-armed bandits. *Annals of Applied Probability*, **2**, 1024–1033.

28. Whittle, P. (1980). Multi-armed bandits and the Gittins index. *Journal of the Royal Statistical Society* B, **42**, 143–149.

29. Whittle, P. (1981). Arm acquiring bandits. *Annals of Probability*, **9**, 284–292.

16

Approximate solutions for open networks with breakdowns and repairs

Ram Chakka

Imperial College

Isi Mitrani

University of Newcastle

Abstract

Network models whose nodes are subject to interruptions of service are notoriously difficult to solve. There are no exact results for systems with more than two nodes, and even the approximations that have been proposed turn out to be rather unsatisfactory. We assess the existing approaches and develop new approximation methods. The latter's accuracy is evaluated by comparisons with simulations and is shown to be acceptable over a large range of parameter values.

1 Introduction

Despite the great importance and widespread use of queueing network models, there have been very few efforts to analyse their behaviour when the servers are not completely reliable. The general problem is difficult because, even under favourable assumptions, the existence of service interruptions destroys the separability of the appropriate multidimensional Markov process and renders its solution intractable.

Most of the existing studies involving breakdowns have concentrated on models with a single job queue served by one or more processors (e.g., see [1, 8, 15, 18]). Some results are available for systems where jobs from a single source are directed to one of several parallel queues, and after completion depart: in [11], the consequences of a server breakdown are the loss of all jobs in the corresponding queue and the re-direction or loss of all arrivals to that queue during the subsequent repair period; [16] treated a similar model but without job losses; [6] examined the case of two queues, only one of which is subject to breakdowns resulting in the transfer of all jobs to the other.

This work was carried out in connection with the Basic Research projects QMIPS and PDCS2, funded by the European Union.

To our knowledge, the only 'genuine network' (in the sense that after completing service at one node jobs may go to another), which has been analysed in the presence of breakdowns and repairs, consists of only two nodes [5]. Even this apparently simple case is not treated in its full generality: a tight coupling is assumed, so that a breakdown interrupts service at both nodes simultaneously, and both nodes complete their repair simultaneously. The analysis in [5] provides ample illustration of the formidable mathematical difficulties posed by this kind of model.

In view of the above, it seems clear that approximate solutions offer the best hope of obtaining performance measures for non-trivial unreliable networks. Indeed, some steps in that direction have been taken already. A simple approach known as the *reduced work-rate approximation* was proposed in the context of both priority interruptions and breakdowns [13, 14, 17]. This is based on replacing an interruptable server with an uninterruptable but slower one, choosing the new service rate so that the overall service capacity remains the same. The evaluation of that approximation has been very limited.

Another possibility is to model the breakdowns and repairs exactly, but reduce the interactions between nodes by considering each of them in isolation. The simplest assumption would be to say that the total arrival stream into a node is Poisson. This *Poisson approximation* was suggested in [7], but its accuracy has not been evaluated.

The object of the present paper is to examine in some detail the problem of approximating the behaviour of open Jackson networks where nodes break down and are repaired independently of each other. The regime of interest is one where both the operative periods and the repair intervals are large compared to the job interarrival and service times. It turns out that, under those conditions, the reduced work-rate approximation is a very poor choice indeed. The Poisson approximation is better, but it too can be inaccurate, especially in networks with cycles.

Two new approximation methods are developed and evaluated. The first, referred to as the *MMPP–approximation*, attempts to capture the non-renewal nature of the arrival stream into a node by modelling it as a Markov-modulated Poisson process; the departure streams are modelled as interrupted Poisson processes (the latter may of course contribute to the former). The second method takes explicit account of the operative/broken states of several (perhaps all) nodes; it will be called the *joint-state approximation*. In both approaches, if the network has cycles, then the parameters of the arrival processes are found by iterations converging to a fixed point. Thus, a considerably higher accuracy is obtained at the expense of increasing the numerical complexity of the solution.

We start by introducing the network model and the two existing approximations, reduced work-rate and Poisson, in section 2. The MMPP and joint-state approximations are described and analysed in sections 3 and 4, respectively. A number of numerical evaluation results and comparisons with simulations are presented in section 5. Section 6 plays the role of an appendix, outlining the spectral expansion method used in the analysis of isolated queues.

2   Jackson networks with breakdowns and repairs

Consider an open Jackson queueing network (e.g., see [4]), with N nodes. The external job arrival streams are Poisson, with arrival rates given by the row vector $\gamma = (\gamma_1, \gamma_2, \ldots, \gamma_N)$. The service times at all nodes are exponentially distributed, with parameters specified by the row vector $\mu = (\mu_1, \mu_2, \ldots, \mu_N)$. The routing decisions are Bernoulli, governed by the matrix $R = [r_{i,j}]$, $(i, j = 1, 2, \ldots, N)$, whose elements are the probabilities that after completing service at node i, a job goes to node j. Every node contains a single server and an unbounded FIFO queue.

The possibility of relaxing some of the above assumptions will be discussed at the end of the paper.

The server at node i goes through alternating operative and inoperative periods, distributed exponentially with parameters ξ_i and η_i, respectively ($i = 1, 2, \ldots, N$). These periods are independent of the states of other nodes and of the number of jobs in the queue. Transitions from the operative to the inoperative state are called *breakdowns*, while those from the inoperative to the operative state are *repairs*. If a breakdown occurs during a service, the latter is resumed from the point of interruption after the repair. Incoming jobs continue to join the queue during inoperative periods.

Because of the breakdowns, the average number of jobs that server i can complete per unit time, ν_i, is given by

$$\nu_i = \mu_i \frac{\eta_i}{\xi_i + \eta_i} \; ; \; i = 1, 2, \ldots, N. \tag{2.1}$$

This is the *service capacity* at node i.

Let $\sigma = (\sigma_1, \sigma_2, \ldots, \sigma_N)$ be the vector of total job arrival rates at the N nodes. These rates are obtained as the unique solution of the set of linear equations

$$\sigma = \gamma + \sigma R. \tag{2.2}$$

The ergodicity condition for this network is simple: the total arrival rates should be smaller than the corresponding service capacities, i.e., $\sigma_i < \nu_i$ for all i. This will be assumed to hold. However, finding the joint or marginal distributions of queue sizes, and hence any performance measures associated with them, is an open problem for $N > 1$.

The reduced work-rate approximation consists of replacing the unreliable servers with reliable ones that have the same service capacity. In other words, the new server at node i has service rate ν_i and breakdown rate 0 ($i = 1, 2, \ldots, N$). The arrival rates and routing probabilities remain the same. The resultant network has a product-form solution and its performance measures are easily calculable. In particular, the reduced work-rate approximation for the average queue size at node i, L_i, is

$$L_i = \frac{\sigma_i}{\nu_i - \sigma_i} \; ; \; i = 1, 2, \ldots, N. \tag{2.3}$$

The major drawback with this approach is that, since ν_i depends only on the ratio of ξ_i and η_i, and not on their separate values, the reduced work-rate approximation depends only on the long-term fraction of time that the server is operative, and is insensitive to the frequency of breakdowns and repairs. Yet the network performance measures depend very strongly on that frequency (see section 5). For example, at a node with Poisson arrivals, if the average lengths of the operative and inoperative periods increase, keeping their ratio and the other parameters fixed, then the average queue size grows without bound while the value given by (2.3) remains constant. An unreliable node may be very lightly loaded, in terms of its service capacity (i.e., it may be operative and idle most of the time), and yet its average queue size may be very large. Thus, the reduced work-rate approximation can be arbitrarily bad when servers break down rarely and take a long time to repair, which is precisely the case of practical interest.

Better results are obtained by treating node i as an isolated, independent M/M/1 queue and modelling its breakdowns and repairs exactly. This approximation consists of assuming that the arrivals into node i form a Poisson stream with rate σ_i; the other assumptions are left intact. The known solution of the isolated queue (e.g., [1, 8]) yields the following Poisson approximation for the average queue sizes:

$$L_i = \frac{\sigma_i}{\nu_i - \sigma_i}\left[1 + \nu_i \frac{\xi_i}{\eta_i(\xi_i + \eta_i)}\right] ; \; i = 1, 2, \ldots, N. \tag{2.4}$$

Whenever breakdowns and repairs occur, the values provided by (2.4) are larger than those in (2.3). Moreover, the difference between the two becomes arbitrarily large when the average operative and inoperative periods grow in a fixed ratio.

The experiments in section 5 show that the Poisson approximation is often quite reasonable. Moreover, it is particularly good in large, strongly connected networks. This is of course due to the fact that, under quite general conditions, the superposition of many input streams tends to be approximately Poisson. However, the quality of the approximation can be unsatisfactory in small to moderately sized networks. There the dependency between nodes and the non-Poisson, non-renewal nature of the traffic ought to be taken into account. This is the object of the following sections.

3   The MMPP approximation

A generalisation of the Poisson approximation is to treat node i as an isolated, independent queue subject to breakdowns and repairs, with jobs arriving according to a Markov-modulated Poisson process (or MMPP, see [3]). Suppose that the arrival rate is controlled by an irreducible Markov chain (environment), $Y(t)$, with T states (phases) numbered $1, 2, \ldots, T$. Here and in the rest of this section we simplify the notation by omitting the node index, i. Let σ_j (rather than $\sigma_{i,j}$) be the instantaneous arrival rate into node i when its environment is in phase j $(j = 1, 2, \ldots, T)$. Denote the generator matrix of the process $Y(t)$ by M, and its steady-state distribution by $\mathbf{a} = (a_1, a_2, \ldots, a_T)$. The latter is obtained by solving the equations

$$\mathbf{a}M = 0; \quad \sum_{j=1}^{T} a_j = 1.$$

The overall arrival rate, σ, is equal to

$$\sigma = \sum_{j=1}^{T} a_j \sigma_j. \tag{3.1}$$

The queue is stable iff $\sigma < \mu\eta/(\xi + \eta)$.

Let the binary variable $\delta(t)$ indicate the state of the server at time t: 0 if broken, 1 if operative. Then the joint state of the server and the MMPP can be represented by a single random integer, $U(t) = T\delta(t) + Y(t)$, taking values in the range $1, 2, \ldots, 2T$. When $U(t) \leq T$, the server is broken; when $U(t) > T$, the server is operative. The state of the queue at time t is described by the pair of integers $[U(t), V(t)]$, where $V(t)$ is the number of jobs present. That pair is a Markov process whose state space is a lattice strip (i.e., finite in one dimension and infinite in the other): $\{1, 2, \ldots, 2T\} \times \{0, 1, \ldots\}$. The corresponding steady-state probabilities, $p_{k,n} = P(U = k, V = n)$, can be determined exactly, for instance by spectral expansion [10, 9], or by the matrix-geometric method [12]. The equations that have to be solved are given in section 6.

The probabilities $p_{k,n}$ can be used to compute various performance measures, such as the average queue size:

$$L = \sum_{n=1}^{\infty} \sum_{k=1}^{2T} n p_{k,n}. \tag{3.2}$$

However, we still have to address the question of how to find the parameters T, M and σ_j, describing the arrival MMPP at each node. This is done by an iterative procedure converging to a fixed point. The first step is to determine the Laplace transform of the interval between consecutive departures from a node, when the arrival MMPP into that node is given and the probabilities $p_{k,n}$ have been calculated.

Let t be an arbitrary point of time in the steady state, and τ be the interval between t and the next departure. Define the following conditional Laplace transforms:

$$\Upsilon(s) = E[e^{-s\tau} \,|\, U(t) \leq T, V(t) > 0]$$

(at time t, the server is broken and the queue is non-empty);

$$\Omega(s) = E[e^{-s\tau} \,|\, U(t) > T, V(t) > 0]$$

(at time t, the server is operative and the queue is non-empty);

$$\Gamma_k(s) = E[e^{-s\tau} \,|\, U(t) = k, V(t) = 0], \quad k = 1, 2, \ldots, 2T$$

(at time t, the joint state of the server and the MMPP is k, and the queue is empty).

By conditioning upon the number of breakdowns that may occur during a service (geometrically distributed with parameter $\xi/(\mu+\xi)$), we obtain

$$\Omega(s) = \frac{\mu(\eta+s)}{(\eta+s)(\mu+\xi+s) - \xi\eta}. \tag{3.3}$$

Also, since a broken server has to be repaired before it resumes service,

$$\Upsilon(s) = \frac{\eta}{\eta+s}\Omega(s) = \frac{\mu\eta}{(\eta+s)(\mu+\xi+s) - \xi\eta}. \tag{3.4}$$

If the server is broken and the queue is empty, then the next event to occur may be an arrival, a repair, or a change of phase in the controlling process Y. Hence, for $j = 1, 2, \ldots, T$,

$$\Gamma_j(s) = \frac{\sigma_j\Upsilon(s)}{\sigma_j + \eta + m_j + s} + \frac{\eta\Gamma_{T+j}(s)}{\sigma_j + \eta + m_j + s} + \sum_{k=1, \neq j}^{T} \frac{m_{j,k}\Gamma_k(s)}{\sigma_j + \eta + m_j + s}, \tag{3.5}$$

where $m_{j,k}$ is the rate at which the process Y jumps from phase j to phase k, and m_j is the total rate at which it leaves state j.

Similarly, if the server is operative and the queue is empty, then the next event to occur may be an arrival, a breakdown, or a change of phase in the controlling process Y. This yields

$$\Gamma_{T+j}(s) = \frac{\sigma_j\Omega(s)}{\sigma_j + \xi + m_j + s} + \frac{\xi\Gamma_j(s)}{\sigma_j + \xi + m_j + s} + \sum_{k=1, \neq j}^{T} \frac{m_{j,k}\Gamma_{T+j}(s)}{\sigma_j + \xi + m_j + s}. \tag{3.6}$$

The linear equations (3.5) and (3.6), together with (3.3) and (3.4), determine all the conditional Laplace transforms. Now consider the queue immediately following a departure instant. Let $\pi_{j,0}$ $(j = 1, 2, \ldots, T)$ be the probability that *at that epoch*, the environment is in phase j and the queue is empty (the server must of course be operative). In the absence of a PASTA property, we express that probability as the fraction of all departures which occur when Y is in phase j, the server is operative and the queue size is 1. Remembering that the overall average number of departures per unit time is σ, given by (3.1), we can write

$$\pi_{j,0} = \frac{p_{T+j,1}\mu}{\sigma}; \quad j = 1, 2, \ldots, T. \tag{3.7}$$

The probability that the queue is empty at a departure epoch is equal to $\pi_0 = \pi_{1,0} + \pi_{2,0} + \cdots + \pi_{T,0}$, while the probability that it is not empty is equal to $1 - \pi_0$. Hence, the unconditional Laplace transform of the interdeparture interval, $\Delta(s)$, is given by

$$\Delta(s) = (1 - \pi_0)\Omega(s) + \sum_{j=1}^{T} \pi_{j,0}\Gamma_{T+j}(s). \tag{3.8}$$

Having obtained $\Delta(s)$, one can determine successive moments of the inter-departure interval by taking derivatives in (3.8), (3.5), (3.6), (3.3) and (3.4), at $s = 0$ (the first moment is of course equal to $1/\sigma$). Those moments can be used to fit an MMPP to the departure process, although the computational task is complex if large numbers of phases are involved. If one decides to approximate the departure process by an MMPP with D phases, there would be D^2 parameters to determine (the off-diagonal elements of the matrix M and the phase-dependent departure rates), and so D^2 moments would be needed.

It is desirable, therefore, to approximate the departure process by an MMPP with a small number of phases. Bearing in mind that there are no departures when the server is broken, we choose a two-phase approximation with departure rate of 0 during one of them. That is, we approximate the departure process from each node by an IPP (Interrupted Poisson Process). Since the latter is determined by just three parameters, only the first three moments of the inter-departure intervals have to be calculated. In order that an IPP can be fitted successfully, the coefficient of variation of the interdeparture intervals must be greater than or equal to 1. This is clearly the case, since service times are distributed exponentially, and random interruptions of service can only increase the variability of the intervals between departures.

Markov-modulated Poisson processes can be split or merged. Bernoulli splitting according to fixed probabilities is simple: the number of phases and the generator matrix remain the same, while the phase-dependent rates are multiplied by those probabilities. Merging is more complicated: if K independent MMPPs with $T_1, T_2, \ldots, T_K$ phases respectively are merged, the resulting process has $T_1 T_2 \ldots T_K$ phases. Expressions for the parameters of the merged process are available [3].

Now, the overall arrival stream into node i is formed by merging its external arrivals (if any) with those departures from node j which are routed to node i, for all j different from i (feedbacks from i to i can be eliminated by suitably changing μ_i). If the arrival processes into those other nodes are known, then solving the corresponding isolated models, fitting an IPP to each of their departure processes and performing the appropriate splitting and merging operations, produces an MMPP arrival process for node i.

If the network is acyclic, i.e., if no job can ever return to a node it has visited already, then one can start with the nodes whose arrivals are entirely external (there must be at least one such node), and by applying the above procedure, construct MMPP arrival streams at all nodes and solve the network in one sweep.

If the network has cycles, then it is necessary to iterate. The initial step is to assume Poisson arrivals into all nodes, with rates obtained from (2.2), and solve the isolated models. From then on, the departure processes computed in the last iteration are used to construct the arrival processes for the next one. The procedure stops when some convergence criterion is satisfied, e.g., the change in the average queue sizes (3.2) is less than some ϵ. We have no proof of the existence or uniqueness of the fixed point; however, during extensive trials with different models, the iterations have never failed to converge.

Numerical experiments show that the MMPP approximation is very reasonable in acyclic networks, but may overestimate the queue sizes considerably when there are cycles (see section 5). To improve the accuracy in the latter case, one should take into account the fact that the arrival rate into a node which is part of a cycle is influenced by whether it, as well as the other nodes, is broken or operative. The following approximation attempts to remedy this deficiency.

4   The joint-state approximation

The joint operational state of all nodes in the network at time t is described by the vector of indicators $\delta_1(t), \delta_2(t), \ldots, \delta_N(t)$, where $\delta_i(t)$ is 0 if node i is broken, 1 if operative. That vector has 2^N possible realisations, which can be numbered $1, 2, \ldots, 2^N$. We can therefore represent the joint operational state of the network nodes at time t as a single integer random variable, $U(t)$, taking values in the range $1, 2, \ldots, 2^N$.

Consider the queue at node i in isolation from the other queues, but controlled by the Markovian environment $U(t)$. That environment changes phase whenever *any* operative node breaks down or a broken one is repaired; the generator matrix associated with those transitions is known, since all ξ and η parameters are given. The instantaneous service rate at node i when $U(t) = k$, $\mu_{i,k}$, is equal to μ_i if node i is operative in phase k and is 0 otherwise.

The joint-state approximation is based on assuming that the arrival process into node i is MMPP and is controlled by $U(t)$. The instantaneous arrival rate is $\sigma_{i,k}$ when $U(t) = k$. Then the pair $[U(t), V_i(t)]$, where $V_i(t)$ is the number of jobs present at node i, is a Markov process on the lattice strip $\{1, 2, \ldots, 2^N\} \times \{0, 1, \ldots\}$. If the rates $\sigma_{i,k}$ are known, the steady-state probabilities, $p_i(k, n) = P(U = k, V_i = n)$, and the desired performance measures, can be determined by solving the equations in section 6.

The departure rate from node i in phase k, $d_{i,k}$, is equal to

$$d_{i,k} = \mu_{i,k}\left[1 - \frac{p_i(k, 0)}{p(k)}\right], \qquad (4.1)$$

where $p(k)$ is the marginal steady-state probability that the environment is in phase k. Hence, the arrival rate into node i in phase k is equal to

$$\sigma_{i,k} = \gamma_i + \sum_{j=1}^{N} d_{j,k} r_{j,i}, \qquad (4.2)$$

where γ_i is the external arrival rate into node i and $r_{j,i}$ is the routing probability from node j to node i.

Now we are in a similar situation to that in the last section. If the network is acyclic, then starting from the nodes whose inputs are entirely external, and using their outputs to calculate the inputs into the downstream nodes, the approximate solution for the whole network is obtained in one sweep. Moreover, the computation can be simplified in this case: any node which does not con-

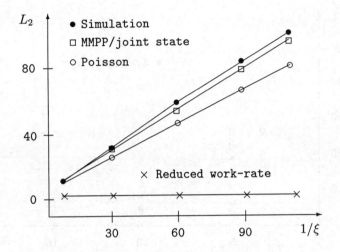

FIG. 1. Approximations for a two-node tandem network
$\sigma = 1; \mu_1 = \mu_2 = 2; \xi_1 = \xi_2 = \xi; \eta_1 = \eta_2 = 5\xi$

tribute, directly or indirectly, to the traffic at node i may be excluded from the latter's Markovian environment.

If the network has cycles, then an iterative procedure is applied: initially, all input rates are assumed to be phase independent and given by (2.2). Subsequently, the departure rates computed in the last iteration are used to find the arrival rates for the next one. The full environment $U(t)$ controls all nodes in every iteration.

5 Numerical results

The first experiment is intended merely to illustrate the inadequacy of the reduced work-rate approximation, so that it can be eliminated from further comparisons. The example involves a simple network of two identical nodes in tandem. In Fig. 1, the approximated and simulated values of the average queue size at node 2 are plotted against the average operative period, $1/\xi$. The ratio η/ξ is kept fixed, so that the service capacities remain constant (each server is operative about 80% of the time). The figure amply confirms the fact that the queue size increases with the lengths of the operative and inoperative periods. Moreover, that increase is almost linear, as suggested by expression (2.4). The reduced work-rate approximation, on the other hand, does not change and therefore becomes progressively worse. In this example, the MMPP and the joint-state approximations are nearly identical and very accurate, while the Poisson approximation is slightly worse.

Next, we examine the quality of the Poisson approximation when the size of the network increases. A number of fully connected networks with 3, 5, 10 and

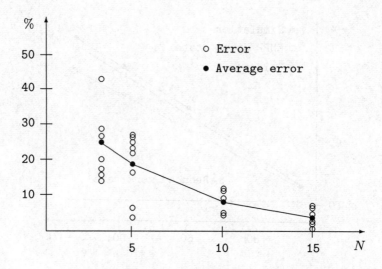

FIG. 2. Poisson approximation: random networks

15 nodes are generated, choosing their parameters at random. The nodes are operative 50%–90% of the time; the average lengths of the operative and repair periods are 2–3 orders of magnitude larger than the interarrival and service times; there are external arrivals into every node; all routing probabilities are non-zero and are picked so that the average number of services required by a job before departing is in the range 2–10; the total effective load at each node is in the range (0.33,0.9).

For each random network, the overall average response time is estimated by means of the Poisson approximation, and also by simulation; the relative error of the former with respect to the latter is evaluated. The results, displayed in Fig. 2, show that both the average and the variability of the error decrease rapidly with N. In the three-node networks, the errors range from 13% to 43%; in the 10-node networks, all observed errors are less than 15%, while in the 15-node ones they are less than 10%.

Thus, the Poisson approximation is adequate in networks where each node has a reasonable number of inputs (10 or more). The nore complex approximations find their applications in small networks, or networks whose structure is such that the number of inputs into each node is small (e.g., nodes in tandem). Further experimentation with that type of network has led us to conclude that when they do not have cycles, both the MMPP and the joint-state approximations are quite accurate.

An acyclic example which was deliberately chosen to be unfavourable to the approximations is shown in Fig. 3. The network has four not very reliable nodes in tandem. Moreover, the first node is faster and has longer operative and repair

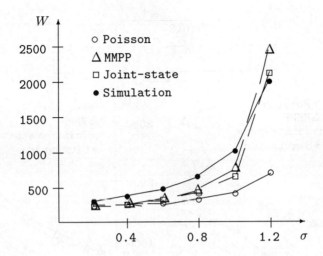

FIG. 3. Average response time in a four-node tandem network
$(\mu_1, \xi_1, \eta_1) = (20, 0.001, 0.002)$; $(\mu_i, \xi_i, \eta_i) = (2, 0.01, 0.02)$, $i = 2, 3, 4$

times than the others.

The performance measure is the total average response time. It is evaluated by the Poisson, MMPP and joint-state approximations, and also by simulation, for different external arrival rates. The results confirm that the MMPP and joint-state approximations are close to each other and are both acceptable, while the Poisson approximation is rather poor.

The situation is different in networks with cycles, especially when the latter contain only a few nodes. Then the joint-state approximation tends to be much preferable to the MMPP one. This is illustrated in Fig. 4, in the context of a three-node network with feedbacks from node 3 to node 1. The performance measure is again the total average response time. In graph 4(a), the servers are operative about 95% of the time, while in 4(b), that fraction is between 60% and 65%. Both graphs show that the Poisson and MMPP approximations strongly overestimate the performance measure. The joint-state approximation is much more accurate.

6 Balance equations

Sections 3 and 4 required the steady-state distribution of Markov processes of the type $[U(t), V(t)]$, where $U(t) = 1, 2, \ldots, K$ represents the environment of a queue and $V(t) = 0, 1, \ldots$ is its size. We had $K = 2T$ in section 3 and $K = 2^N$ in section 4. In both cases, the evolution of the process is governed by the following known parameters:

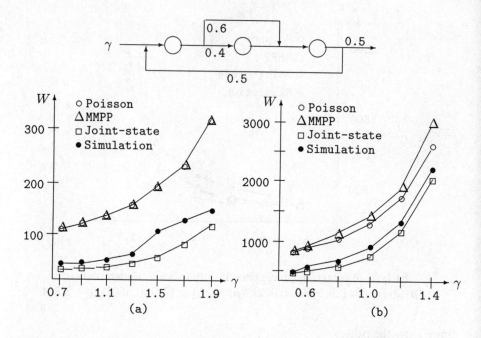

FIG. 4. Average response time in a three-node network with feedbacks
$(\mu_1, \mu_2, \mu_3) = (5, 6, 7)$; $(\eta_1, \eta_2, \eta_3) = (0.002, 0.003, 0.005)$
Graph (a): $\xi_i = 10^{-4}i$; Graph (b): $\xi_i = 10^{-3}i$, $i = 1, 2, 3$

- matrix $A = [a_{j,k}]_{j,k=1}^{K}$, where $a_{j,k}$ is the instantaneous rate at which the environment U jumps from phase j to phase k. The diagonal elements of A are equal to 0.
- row vector $\sigma = (\sigma_1, \sigma_2, \ldots, \sigma_K)$, where σ_k is the arrival rate into the queue when the environment is in phase k.
- row vector $\mu = (\mu_1, \mu_2, \ldots, \mu_K)$, where μ_k is the service rate when the environment is in phase k.

For the MMPP model in section 3, we have

$$
a_{j,k} = \begin{cases}
m_{j,k}; & j, k = 1, 2, \ldots, T; \; j \neq k \\
m_{j-T,k-T}; & j, k = T+1, T+2, \ldots, 2T; \; j \neq k \\
\eta; & j = 1, 2, \ldots, T; \; k = j + T \\
\xi; & j = T+1, T+2, \ldots, 2T; \; k = j - T \\
0; & \text{otherwise,}
\end{cases}
$$

where $m_{j,k}$ are the transition rates of the modulating process Y, and ξ and η are the breakdown and repair rates at node i. The elements of the service rate vector are $\mu_k = 0$ if $k \leq T$, $\mu_k = \mu$ if $k > T$. The arrival rate vector is provided by the iterative procedure, as described in the section.

For the joint-state model in section 4, $a_{j,k} = \xi_\ell$ if the only difference between phase j and phase k is that node ℓ is operative in the former and broken in the latter; $a_{j,k} = \eta_\ell$ if the only difference between phase j and phase k is that node ℓ is broken in the former and operative in the latter; $a_{j,k} = 0$ otherwise. The elements of the service rate vector are of the form $\mu_k = \mu$ if the given node is operative in phase k and $\mu_k = 0$ otherwise. Again, the arrival rate vector is provided by the iterative procedure.

The steady-state probabilities $p_{k,n} = P(U = k, V = n)$, grouped into vectors, $\mathbf{v}_n = (p_{1,n}, p_{2,n}, \ldots, p_{K,n})$, $n = 0, 1, \ldots$, satisfy the following balance equations:

$$\mathbf{v}_0[\sigma I + D^A] = \mathbf{v}_0 A + \mathbf{v}_1 \mu I, \tag{6.1}$$

$$\mathbf{v}_n[\sigma I + D^A + \mu I] = \mathbf{v}_{n-1}\sigma I + \mathbf{v}_n A + \mathbf{v}_{n+1}\mu I \; ; \; n = 1, 2, \ldots, \tag{6.2}$$

where I is the unit matrix of order K and D^A is the diagonal matrix whose kth diagonal element is the kth row-sum of A. The spectral expansion solution of these equations is described in [9].

7 Conclusion

We have made some progress with a difficult problem. The reduced work-rate approximation can generally be dismissed from consideration. The Poisson approximation is acceptable if many arrival streams are merged at all nodes. In small or weakly connected networks, it is better to use non-Poisson approximations. Of the two suggested here, the joint-state approximation is recommended for networks containing cycles. However, its complexity increases rapidly with the number of nodes. The complexity of the MMPP approximation is governed mainly by the number of inputs at each node. Both appear to be adequate in feedforward networks.

The accuracy of the approximations can probably be improved, at the price of increasing their complexity. For example, while a node is broken, its queue tends to accumulate and there may be a large backlog of jobs at the start of the next operative period. Consequently, for an initial fraction of that period the departure rate is higher than average. That phenomenon can be included into the approximation by adding extra phases to the Markovian environment. It is also clear that these approaches can be applied to more general models. For example, one could handle networks with multiple parallel servers at each node, or breakdowns with different consequences (such as the loss of the job in service).

Bibliography

1. Avi-Itzhak, B. and Naor, P. (1963). Some queueing problems with the service station subject to breakdowns. *Operations Research*, **11**, 303–320.

2. Feller, W. (1968). *An Introduction to Probability Theory and its Applications, Vol. 1.* Wiley, New York.

3. Fischer, W. and Meier-Hellstern, K. (1993). The Markov-modulated Poisson

process. *Performance Evaluation*, **18**, 149–171.

4. Kleinrock, L. (1975). *Queueing Systems, Vol. 1*. Wiley, New York.

5. Mikou, N. (1988). A two-node Jackson network subject to breakdowns. *Stochastic Models*, **4**, 523–552.

6. Mikou, N., Idrissi-Kacimi, O. and Saadi, S. (1995). Two processors only interacting during breakdown: the case where the load is not lost. *Queueing Systems*, **19**, 301–317.

7. Mitrani, I. (1974). Networks of unreliable computers. In *Computer Architectures and Networks* (eds E. Gelenbe and R. Mahl). North-Holland, Amsterdam.

8. Mitrani, I. and Avi-Itzhak, B. (1968). A many-server queue with service interruptions. *Operations Research*, **16**, 628–638.

9. Mitrani, I. and Chakka, R. (1995). Spectral expansion solution for a class of Markov models: application and comparison with the matrix-geometric method. *Performance Evaluation*, **23**, 241–260.

10. Mitrani, I. and Mitra, D. (1992). A spectral expansion method for random walks on semi-infinite strips. In *Iterative Methods in Linear Algebra* (eds R. Beauwens and P. de Groen). North-Holland, Amsterdam.

11. Mitrani, I. and Wright, P.E. (1994). Routing in the presence of breakdowns. *Performance Evaluation*, **20**, 151–164.

12. Neuts, M.F. (1981). *Matrix Geometric Solutions in Stochastic Models*. Johns Hopkins University Press, Baltimore, MD.

13. Reiser, M. (1976). Interactive modelling of computer systems. *IBM Systems Journal*, **15**, 309–327.

14. Reiser, M. (1979). A queueing network analysis of computer communication networks with window flow control. *IEEE Transactions on Communications*, **8**, 1199–1209.

15. Sengupta, B. (1990). A queue with service interruptions in an alternating Markovian environment. *Operations Research*, **38**, 308–318.

16. Thomas, N. and Mitrani, I. (1995). Routing among different nodes where servers break down without losing jobs. *Proc. IPDS'95*. Erlangen.

17. Vinod, B. and Altiok, T. (1986). Approximating unreliable queueing networks under the assumption of exponentiality. *Journal of the Operations Research Society*, **37**, 309–316.

18. White, H.C. and Christie, L.S. (1958). Queueing with preemptive priorities or with breakdown. *Operations Research*, **6**, 79–95.

17

Stationary ergodic Jackson networks: results and counter-examples

François Baccelli

INRIA Sophia-Antipolis

Serguei Foss

Novosibirsk State University

Jean Mairesse

BRIMS, Hewlett-Packard Laboratories, Bristol

Abstract

This paper gives a survey of recent results on generalized Jackson networks, where classical exponential or i.i.d. assumptions on services and routings are replaced by stationary and ergodic assumptions. We first show that the most basic features of the network may exhibit unexpected behavior. Several probabilistic properties are then discussed, including a strong law of large numbers for the number of events in the stations, the existence, uniqueness and representation of stationary regimes for queue size and workload.

1 Introduction

Jackson networks provide a very effective mathematical model for packet switching networks. This paper gives a survey of recent results and a preview of ongoing research on *generalized* Jackson networks. Here the classical Markovian assumptions on services and routings, as proposed in Jackson's original model, are replaced by general stationary and ergodic assumptions.

Beyond the natural quest for a better mathematical understanding of such general stochastic networks, the interest in the non-Markovian case stems from two practical observations. First, it enables the incorporation of periodic phenomena, such as the dependence of random variables (services, routings) upon the period of the day or the year. Second, it was recently observed that in several basic communication networks (e.g. Ethernet LANs and the Internet), the point

---

The work of the last author was supported by a post-doctoral grant from INRIA.

processes describing the offered traffic exhibit long range dependence [31], which rules out the classical Markovian representation.

The stability of Jackson networks has been considered in several papers. Without any claim to an exhaustive enumeration, one can cite the works of Jackson [19], Gordon and Newell [18], Borovkov [8, 9], Foss [16, 17], Daduna [14], Sigman [29, 30], Chang [11], Kaspi and Mandelbaum [20, 21] and Meyn and Down [25]. A more complete bibliography on the subject can be found in [17]. All these papers require some sort of independence assumptions or some distributional constraints on services. By constructing counter-examples, the present paper first shows that under more general stationary and ergodic assumptions, all basic quantitative measures of the network may indeed exhibit unexpected behavior. The paper then focuses on positive results: strong laws of large numbers for the daters and counters associated to stations, existence and uniqueness of stationary regimes for the stochastic processes describing the queue sizes and the workloads, with detailed discussions on how these objects may depend on the initial condition. Finally, some special cases are investigated, with the aim of showing more detailed results on the classes of functions which are involved in the stationary regimes of such networks. The results are stated without proofs, the paper serving as a comprehensive review/preview of material to be found in: Baccelli and Foss [2, 3], Baccelli, Foss and Gaujal [4], Baccelli and Mairesse [6], Baccelli, Foss and Mairesse [5]. The counter-examples mentioned above, as well as several constructive results, are original.

The main mathematical tools used are ergodic theory, random graphs, stochastic recurrence equations and stochastic ordering.

2   Definitions

2.1   Model

Definition 2.1 *A Jackson network is a queueing network with K stations, where each station is a single FIFO server with infinite buffer $(./.1/\infty$ FIFO using Kendall's notation). Customers move from station to station in order to receive some service. The data are $(2K)$ sequences*

$$\{\sigma^i(n), n \in I\!N\}, \quad \{\nu^i(n), n \in I\!N\}, \quad i \in \{1, \ldots, K\},$$

where $\sigma^i(n) \in I\!R^+$ and $\nu^i(n) \in \{1, \ldots, K, K+1\}$, $K+1$ being the exit.

The n-th customer to be served by station i after the origin of time requires a service time $\sigma^i(n)$; after completion of its service there, it moves to station $\nu^i(n)$ and is put at the end of the line. We say that $\nu^i(n)$ is the n-th routing variable on station i.

Remark As far as results on throughputs or workload processes are concerned, we can replace FIFO by any non-preemptive, work conserving discipline.

We distinguish between two classes of Jackson networks, open and closed.

- **Open case:** There is an external arrival point process $\{T_n, n \in I\!N\}$, with $0 \leq T_0 \leq T_1 \leq \cdots$. Equivalently, there is an additional saturated station (numbered 0) producing its first customer at time $\sigma^0(0) = T_0$, and further customers with inter-arrival times $\sigma^0(n) = T_n - T_{n-1}$, $n > 0$.

The nth external arrival is routed to station $\nu^0(n) \in \{1, \ldots, K\}$. The description of the arrival process by means of station 0 is systematic throughout the paper. The customers eventually leave the network (see absence of capture below).

- **Closed case:** There are no external arrivals. It is impossible to be routed to the exit, i.e. $\nu^i(n) \in \{1, \ldots, K\}, \forall i \in \{1, \ldots, K\}, \forall n \in I\!N$. The total number of customers in the network is then a constant.

Remark The previous definition of an open Jackson network includes the possibility of bulk arrivals.

It is convenient to denote by $\mathcal{K}$ the set of stations of the network. We have in the open and closed cases respectively:

$$\mathcal{K} = \{0, 1, \ldots, K\} \text{ and } \mathcal{K} = \{1, \ldots, K\}. \tag{2.1}$$

Note that the exit, $K + 1$, is not considered as one of the stations in the open case. We use the notation $*$ with the convention that $* = 0$ in the open case and $* = 1$ in the closed case. For example, the set $\mathcal{K}\backslash*$ has to be interpreted as $\{1, \ldots, K\}$ for an open network and $\{2, \ldots, K\}$ for a closed network.

We use the following compact way of describing a Jackson network J:

$$J = \{(\sigma^k(n), k \in \mathcal{K}), (\nu^k(n), k \in \mathcal{K}), n \in I\!N\}. \tag{2.2}$$

To unify the presentation, the number of customers (in the closed case) is not included in the definition of J but in the initial condition, to be defined below.

Definition 2.2. (Initial condition) *For a Jackson network with K stations, we denote the initial condition by $(Q, R) = \{(Q^k, R^k), k = 1, \ldots, K\}$. The integer Q^k is the number of customers in station k at time 0^-, $M = \sum_{k=1}^{K} Q^k$ is the total number of initial customers and R^k is the residual service time of the customer under service at station k at time 0^-. We adopt the convention $R^k = 0$ when $Q^k = 0$.*

An initial condition is said to be finite if $Q^k < +\infty$ and $R^k < +\infty$, $k = 1, \ldots, K$.

For an open network, the state of station 0 was not included in the above definition as it is always $(Q^0, R^0) = (\infty, T_0) = (\infty, \sigma^0(0))$.

We do not require that $R^k \leq \sigma^k(0)$. If $R^k > \sigma^k(0)$, one can interpret this as the fact that the first customer is frozen in station k until instant $R^k - \sigma^k(0)$ when its service starts.

We assume the initial condition to be deterministic. The case when Q^k and R^k are random variables is further discussed in [5].

Definition 2.3. (Canonical initial condition) *We say that a Jackson network has a canonical initial condition if we have*

- **Open case:** $\{(Q^k, R^k) = (0,0), k = 1, \ldots, K\}$. *In words, the state at the origin of time is that all stations are empty.*
- **Closed case:** $\{(Q^1, R^1) = (M, \sigma^1(0)), (Q^k, R^k) = (0,0), k = 2, \ldots, K\}$. *In words, the state at the origin of time is that all customers are in station 1, and service 0 is just starting on station 1.*

In the following, when nothing is specified, it is always implicit that the Jackson networks have canonical initial conditions. In order to avoid any confusion, we use specific notations indexed by I (e.g. $_IJ$, $_IJ_{[0,n]}$, $_IX_{[0,n]}$, all quantities to be defined later on) for a network with a non-canonical initial condition I.

Let us define the sub-class of closed cyclic Jackson networks.

Definition 2.4 *A cyclic Jackson network (CJN) is a closed Jackson network, where for all n and i, $\nu^i(n) = i + 1$. The numbering of stations has to be understood modulo $[K]$, e.g. station $(K + 2)$ is station 2.*

Remark Jackson networks are a sub-class of free-choice Petri nets. Cyclic Jackson networks are a sub-class of closed event graphs. Free-choice Petri nets and event graphs are sub-classes of Petri nets, which provide an efficient formalism to represent and study discrete event systems with synchronization and/or routing, see for example Murata [26]. In [4], the method presented here for Jackson networks is applied to *open* free-choice Petri nets. Event graphs can be represented as (max,+) linear systems. It yields stronger results than those presented here for the sub-class of Jackson networks which are event graphs (i.e. for CJN), see for example [1], [22].

2.2 Stochastic framework

Let $(\Omega, \mathcal{F}, P)$ be a probability space. We consider a bijective and bi-measurable shift function $\theta : \Omega \to \Omega$. We assume that θ is P-stationary (i.e. $P\{\theta^{-1}(\mathcal{A})\} = P\{\mathcal{A}\}, \forall \mathcal{A} \in \mathcal{F}$) and P-ergodic (i.e. $\mathcal{A} \subset \theta^{-1}(\mathcal{A}) \Rightarrow P\{\mathcal{A}\} = 0$ or 1). The symbols θ^n, $n \geq 0$, denote the iterations of the shift θ (θ^0 is the identity). We use the notation $X \circ \theta$ to denote the r.v. $X \circ \theta(\omega) = X(\theta\omega)$, $\omega \in \Omega$. A sequence of random variables $\{X(n), n \in I\!N\}$ is said to be stationary–ergodic (with respect to θ) if $X(n) = X(0) \circ \theta^n$.

Definition 2.5 *A Jackson network is said to be i.i.d. if the sequences of service times $\{\sigma^k(n), n \in I\!N\}$, $k \in \mathcal{K}$, and routings $\{\nu^k(n), n \in I\!N\}$, $k \in \mathcal{K}$, are i.i.d. and mutually independent.*

As suggested by the title of the article, we are interested in studying more general Jackson networks under stationary–ergodic assumptions. There are several possible definitions of stationary–ergodic Jackson networks, some of which are listed below in increasing order of generality.

SE1 The sequences $\{\sigma^k(n), n \in I\!N\}$, $\{\nu^k(n), n \in I\!N\}$, $k \in \mathcal{K}$, are stationary–ergodic and mutually independent.

SE2 The sequences $\{(\sigma^k(n), k \in \mathcal{K}), n \in I\!N\}$ and $\{\nu^k(n), n \in I\!N\}$, $k \in \mathcal{K}$, are stationary–ergodic and mutually independent.

SE3 The sequence $\{(\sigma^k(n), \nu^k(n), k \in \mathcal{K}), n \in I\!N\}$ is stationary–ergodic.

The stochastic assumptions we are going to work with are different yet again from the previous three. They are defined in eqn (4.3) and denoted **H1**.

Routing matrix   Let us consider a stationary–ergodic (**SE3**) Jackson network. We define its routing matrix as

$$P = (P_{ij}), \quad P_{ij} = P(\nu^i(0) = j), \quad i, j \in \mathcal{K}. \tag{2.3}$$

For an open Jackson network, we identify stations 0 and $K + 1$, setting $P_{i0} = P(\nu^i(0) = K + 1), i = 1, \ldots, K$. Note that this definition of P boils down to the usual one in the case of an i.i.d. network.

In the following, it is always assumed that a Jackson network is at least stationary–ergodic (**SE3**). Furthermore, it is always assumed that a Jackson network has an irreducible routing matrix P (i.e. $\forall i, j, \exists n \in I\!N$ s.t. $P_{ij}^n > 0$). In the open case, the irreducibility of matrix P implies that the system is without capture, i.e. a customer entering the system eventually leaves it.

Remark   When the previous assumptions are not satisfied, one should study separately the maximal irreducible sub-networks. The departure processes from upstream sub-networks provide the arrival processes for downstream ones. Accordingly, to connect together the results obtained for the different sub-networks, one needs to prove that these departure processes are stationary and ergodic. This is partially addressed in §8. For more insights, see also [4] and [5].

Letting α be a left eigenvector associated with the maximal eigenvalue of P, we have

$$\alpha P = \alpha. \tag{2.4}$$

It follows from the Perron–Frobenius theorem that α is unique (up to a constant) and can be chosen to be positive ($\forall i, \alpha_i > 0$). We choose α such that $\alpha_* = 1$. The real α_k can be interpreted as the relative frequency of visits to stations k and $*$.

Periodic networks   By definition, a Jackson network is *periodic* if the sequences of services and routings are periodic. Periodic Jackson networks can be transformed into stationary–ergodic (usually **SE3**) Jackson networks, as shown below.

Example 2.6   Let us consider a closed periodic network with two stations. We have

$$\sigma^1(n) = 0, 3, 0, 3, \ldots, \quad \sigma^2(n) = 5, 5, 5, \ldots, \quad \nu^1(n) = 1, 2, 1, \ldots, \quad \nu^2(n) = 2, 1, 2, \ldots.$$

This network does not satisfy the stationary ergodic assumptions.

Let $(\Omega, \mathcal{F}, P)$ be the probability space with $\Omega = \{\omega_1, \omega_2\}$ and $P(\omega_1) = P(\omega_2) = 1/2$. We consider the P-stationary and P-ergodic shift defined by $\theta(\omega_1) = \omega_2, \theta(\omega_2) = \omega_1$. We define a new network on Ω with services and routings equal to

$$\sigma^1(n, \omega_1) = 0, 3, 0, ..., \sigma^2(n, \omega_1) = 5, \nu^1(n, \omega_1) = 1, 2, 1, ..., \nu^2(n, \omega_1) = 2, 1, 2, ...$$

$$\sigma^1(n, \omega_2) = 3, 0, 3, ..., \sigma^2(n, \omega_2) = 5, \nu^1(n, \omega_2) = 2, 1, 2, ..., \nu^2(n, \omega_2) = 1, 2, 1, ...$$

It is easy to verify that this network is stationary and ergodic (**SE3**).

In general, one builds a stationary–ergodic version of a periodic Jackson network by considering a finite probability space whose cardinal is the *least common multiple* of the periods of the sequences of services and routings.

2.3   First and second order variables

We describe Jackson networks using daters and counters.

Definition 2.7 *We define the* internal dater $X^k(n)$, $k \in \mathcal{K}$, *to be the time at which the n-th service is completed at station k. We define the* internal counter $\mathcal{X}^k(t)$, $k \in \mathcal{K}$, *to be the number of services completed at station k before time t. And last, we define $\mathcal{X}^{k,l}(t)$, $k, l \in \mathcal{K}$, to be the number of customers routed from k to l before time t. The counters are chosen to be right continuous with left-hand limits.*

We are going to study two types of variables, called respectively first and second order variables.

First order variables Counters and daters are called first order variables. Properly scaled, they may converge to throughputs. The throughput at station k (when it exists) is equal to

$$\lambda^k = \lim_{n \to \infty} n/X^k(n) = \lim_{t \to \infty} \mathcal{X}^k(t)/t.$$

The arrival rate in the network is $\lambda^0 = \lim_{t \to \infty} \mathcal{X}^0(t)/t$. The departure rate (when it exists) is $\lambda^{K+1} = \lim_{t \to \infty} \mathcal{X}^{K+1}(t)/t$. For CJN, the *cycle time* is defined as the inverse of the throughput.

Second order variables They include queue length, residual service time, and workload processes. They are called second order variables because they can be defined as differences of counters and daters, as shown below. All processes are chosen to be right continuous with left-hand limits.

(1) Queue length and residual service time process:
$(Q^k(t), R^k(t), k \in \mathcal{K} \backslash 0), t \in I\!R^+$, with $Q^k(t) = Q^k + \sum_{l \in \mathcal{K}} \mathcal{X}^{l,k}(t) - \mathcal{X}^k(t)$ and $R^k(t) = [X^k(\mathcal{X}^k(t) + 1) - t]1_{\{Q^k(t)>0\}}$. We have $(Q(0), R(0)) = (Q, R)$ where (Q, R) is the initial condition.

(2) Workload process:
$(W^k(t), k \in \mathcal{K} \backslash 0), t \in I\!R^+$, with $W^k(t) = [X^k(\mathcal{X}^k(t) + Q^k(t)) - t] \vee 0$. Dually, one can consider the (non-standard) idle time processes $I^k(t) = [t - X^k(\mathcal{X}^k(t) + Q^k(t))] \vee 0$.

Here are different properties that we would like to investigate:

(A) Existence of throughputs $\lambda^k, k \in \mathcal{K}$, for a given initial condition (Def. 2.2)?

(B) Uniqueness of the throughputs $\lambda^k, k \in \mathcal{K}$, for all different initial conditions? For an open network, this uniqueness has to be verified for any finite initial condition, see Def. 2.2. For a closed network, this uniqueness has to be verified for all finite initial conditions such that $\sum_{k=1}^{K} Q^k = M$ for some M (i.e. the number of customers is fixed).

(C) Concavity of the throughputs $\lambda^k(M), k \in \mathcal{K}$, as a function of the number M of customers in the network? This property is of course irrelevant for open networks.

(D) Existence of stationary regimes for second order processes? Uniqueness of the stationary regimes for different initial conditions (see property (B) for a precise statement)?

The answers to these questions are summarized in the following table, where the assumptions which are considered are **H1–2**, see eqn (4.3):

	Open network	Closed network
A	Yes	Yes
B	Yes	No
C	—	No
D	No	No

Table **1.**

3   Counter-examples

We provide counter-examples for all the cases corresponding to an answer "No" in Table 1. The networks considered in these counter-examples are periodic networks. The reader can check that the counter-examples remain valid if we replace these networks by their stationary and ergodic extensions (the initial conditions being kept unchanged), see §2.2. Note also that all these networks (or rather their extensions) satisfy the forthcoming assumption **H1**, see eqn (4.3). It is a direct consequence of the cyclic form for the networks of §3.2, 3.3 and 3.4 and it can be checked directly for the network of §3.1.

3.1   Open network, non-uniqueness of second order limits (D)

We consider an open Jackson network with two stations. The stations can be described as $./D/1/\infty$ FIFO, using Kendall's notation. Here D stands for Deterministic. The service times are $\sigma^1(n) = \sigma^1 = 1/4$ and $\sigma^2(n) = \sigma^2 = 1/2$. The input process is $\{T_n = n+1, n \in I\!N\}$, i.e. one customer arrives each unit of time. The routing sequence of customers leaving station 1 is

$$\nu^1(n) = \{3, 2, 3, 2, 3, \ldots\},$$

where 3 corresponds to the exit. This network is represented in Fig. 1.

Let us show that we obtain several stationary regimes. We fix $1 < c < 1+\sigma^1$. We consider an initial condition of the form $(Q^1, R^1) = (1, c), (Q^2, R^2) = (0, 0)$, i.e. there is one customer in station 1 with residual service time c.

We have represented, in Fig. 2, the Gantt chart corresponding to this network for $c = 1 + 1/8$. The horizontal axis represents time. The blocks correspond to the time spent by the customers in the stations. The colors of the blocks depend

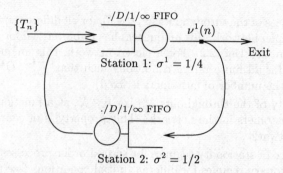

FIG. 1. Open Jackson network with two stations.

on the customer. For example, *black* corresponds to the initial customer and *light gray* to the customer arriving at instant $T_0 = 1$. It follows from the periodicity of $\nu^1(n)$ that each customer (except the initial one) receives exactly two services at station 1 and one at station 2.

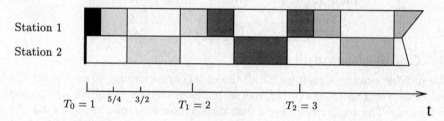

FIG. 2. Initial condition $(Q^1, R^1) = (1, 9/8), (Q^2, R^2) = (0, 0)$.

We recall that $W^1(t)$ is the workload at station 1 at instant t. Using Fig. 2, one can easily see that $\max_t W^1(t) = 1/4 + (c - 1) = W^1(n), n \in \mathbb{N}$. We conclude that there is a continuum of possible stationary regimes for the workload, depending on c.

Remark This counter-example was first mentioned in [4]. The phenomenon of multiplicity of second order limits appears in other types of open networks. A folk example consists of a multiserver queue of the form $D/P/2/\infty$ where arriving customers are allocated to the server with the smallest workload. For more details, see for example Brandt, Franken and Lisek [10], Example 5.5.2.

3.2 Closed network, non-uniqueness of second order limits (D)

We consider a cyclic Jackson network (CJN) with three stations and two customers. The stations are of type $./D/1/\infty$ FIFO. The service times are $\sigma^1(n) = \sigma^2(n) = \sigma^3(n) = 1$. This network is represented in Fig. 3.

In Figs 4 and 5, the Gantt charts corresponding to two different initial conditions are given. The color of the blocks differs according to the customer served: *light gray* is for the customer originally in station 1 and *dark gray* for the one originally in station 3, see Fig. 3.

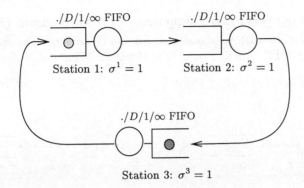

FIG. 3. Cyclic Jackson network, three stations and two customers.

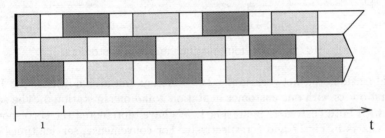

FIG. 4. Initial condition $(Q^1, R^1) = (1, 1/2), (Q^2, R^2) = (0, 0), (Q^3, R^3) = (1, 1)$.

Let us consider for example the idle time $\{I^1(t)\}$ at station 1. For the initial conditions corresponding to Figs 4 and 5, we obtain $\{I^1(n) = 1/2, 1/2, 1/2, \ldots\}$ and $\{I^1(n) = 0, 1, 0, 1, \ldots\}, n \in I\!N$, respectively. It is easy to see that there is a continuum of possible limiting regimes for $I^1(t)$ depending on the initial condition.

In such a deterministic model, initial delays between customers never vanish. In fact, even first order limits may depend on the initial condition, see next section.

Remark Such counter-examples are well-known in the literature. In fact, a complete classification of CJN having multiple second order stationary regimes can be made using the (max,+) theory, see [22, 23].

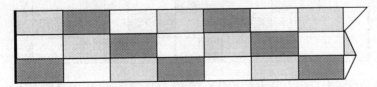

FIG. 5. Initial condition $(Q^1, R^1) = (1, 1), (Q^2, R^2) = (0, 0), (Q^3, R^3) = (1, 1)$.

3.3   Closed network, non-uniqueness of the throughput (B)

We consider a CJN with four stations and two customers. The stations are of type $./P/1/\infty$ FIFO, where P stands for Periodic. The service times are

$$\sigma^1(n) = 2,0,2,0,\ldots, \quad \sigma^2(n) = 1, \quad \sigma^3(n) = 0,2,0,2,\ldots, \quad \sigma^4(n) = 1.$$

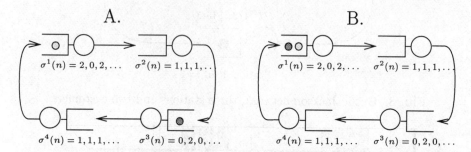

FIG. 6.   CJN, four stations and two customers.

We consider the network under two different initial conditions, see Fig. 6. The first one is with one customer in station 1 and one in station 3. The second one is with both customers in station 1. We have represented the corresponding Gantt charts in Figs 7 and 8 respectively. For convenience, service times equal to 0 have been materialized and represented by slim bars.

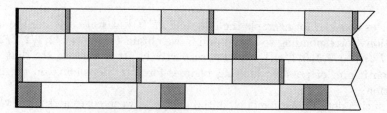

FIG. 7. $(Q^1, R^1) = (1,2), (Q^2, R^2) = (0,0), (Q^3, R^3) = (1,0), (Q^4, R^4) = (0,0).$

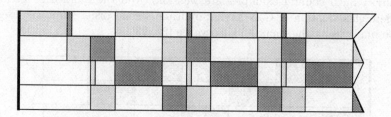

FIG. 8. $(Q^1, R^1) = (2,2), (Q^2, R^2) = (0,0), (Q^3, R^3) = (0,0), (Q^4, R^4) = (0,0).$

In Fig. 7, the throughput is $1/3$. In Fig. 8, the throughput is $1/2$. We conclude that the throughput depends on the initial position of customers.

In Fig. 7, the *light gray* customer always receives long services and the *dark gray* customer always waits before getting served. In Fig. 8, services and waiting times are more equally shared between the two customers. This increases the efficiency of the network.

Remark This example was first introduced in [23], Chap. 8. A closely related counter-example is displayed by Bambos in [7]. His model is a CJN with distinguishable customers. It means that the service times depend on the station *and* on the customer. For each station–customer couple, the sequence of service times is periodic. As customers do not overtake, the cyclic ordering of customers in the network is an invariant. It is shown in [7] that the throughput may depend on the cyclic ordering but also on the initial positioning of customers given a cyclic ordering. This last result is close but slightly different from the one illustrated in Figs 7 and 8. Let us explain why.

We fix a cyclic ordering of customers. For a given initial positioning of customers, we can define the sequence $\{s^i(n), n \in I\!N\}$ of services received at station i. It is easy to see that $\{s^i(n)\}$ is periodic. The difference with our model is that the sequences $\{s^i(n)\}$ depend on the initial position of customers. For another initial position, we will obtain sequences of the form $\{s^i(n + k_i), n \in I\!N\}$.

3.4   Closed network, non-concavity of the throughput (C)

We investigate the behavior of the network with respect to the number of customers. We consider the network of Fig. 6.A and we add a second customer in station 1. The new Gantt chart is represented in Fig. 9.

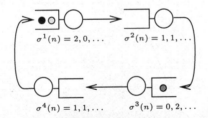

FIG. 9. CJN, four stations and three customers. Initial condition $(Q^1, R^1) = (2,2), (Q^2, R^2) = (0,0), (Q^3, R^3) = (1,0), (Q^4, R^4) = (0,0)$.

The average cycle time (see Def. 2.4) of a customer is $21/4 = 5.25$. For example, the cycle time of the *light gray* customer computed from station 2 to station 2 is $\{4, 5, 6, 6, 4, 5, 6, 6, \ldots\}$.

Comparing Figs 7 and 9, one checks that the average cycle time of the original two customers has decreased from 6 to 5.25. The addition of one customer has increased the speed of the original customers!

For the purpose of this example, let us introduce some notation. We consider a network with M customers. We denote by $\gamma(M)$ the cycle time of a customer and by $\lambda(M)$ the throughput (which is the same at each station). We have

$$\lambda(M) = M/\gamma(M).$$

In the previous example, we have obtained $\gamma(3) < \gamma(2)$. It implies $\lambda(3)/3 > \lambda(2)/2$. We conclude that there is no concavity of the throughput.

We have represented in Fig. 10 the throughput $\lambda(M)$ for the network of Fig. 9. For $M > 2$, each new customer is added in the buffer of station 1. When M becomes large, we obtain the expected behavior, i.e. the throughput becomes constant and is imposed by the bottleneck (slowest) station(s).

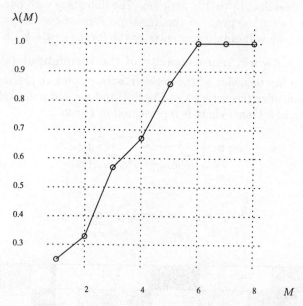

FIG. 10. Throughput $\lambda(M)$ as a function of the number of customers M.

In contrast, we can consider the same network with all the customers in station 1. In this case, we obtain the expected behavior, i.e. the cycle time $\gamma(M)$ is an increasing function of M.

Remark To the best of our knowledge, this kind of counter-example is original. A similar paradox is provided by the network of Braess, studied in Cohen and Kelly [13]. It is a transportation network where the withdrawal of one of the existing routes increases the speed of all the customers. However, the two models are completely different. Braess' model is that of an open congested network with optimal routing. The paradox comes from the non-compatibility between customer optima and global optima, in a game theoretic sense. Our model is closed, uncongested and with a predetermined routing. The paradox comes from the (in)compatibility between the number of customers and the periods of the sequences of service times.

Summary   The best we can expect to prove is the complement of the previous counter-examples. More precisely, we propose in Table 2 a new detailed version of Table 1, with the sections where the positive results are stated.

		Open network			*Closed network*		
		i.i.d.		*stat. erg.*		*i.i.d.*		*stat. erg.*
Table 2.	A	Yes §5.2		Yes §5.2		Yes §5.1		Yes §5.1
	B	Yes §6.3		Yes §6.3		Yes §6.2		No §3.3
	C	—		—		??? §7.1		No §3.4
	D	Yes §8.1		No §3.1		No §3.2		No §3.2

4   Restriction of a Jackson network

4.1   Euler property

Let J be a Jackson network. We associate with J a random graph G, called the routing graph, which is defined from the information carried by the routing sequences only. The set of nodes of G is $\mathcal{K}$ and a routing $\{\nu^i(n) = j\}$ is interpreted as an arc labeled n from node i to node j. In the open case, a routing $\{\nu^i(n) = K + 1\}$ is interpreted as an arc from node i to node 0. If $(i, j) \in \mathcal{K}^2$ is such that $\mathbb{P}_{ij} > 0$ (see (2.3)), then there is an infinite number of arcs from i into j. The arcs originating from a given node are totally ordered by the labels. An initial condition for the graph is defined as a finite number of tokens placed on the nodes. Each node might contain several tokens.

Definition 4.1   (Game $\mathcal{G}$)   *Given a routing graph G and an initial condition, we move the tokens according to the following rules*

- *Step 1: Select one of the tokens, say one on node i, and move it to node $\nu^i(0)$. Remove the arc $\{\nu^i(0)\}$.*
- *Step $n > 1$: Select one of the tokens, say one on node i, and move it to node $\nu^i(p)$ where $\{\nu^i(p)\}$ is the arc originating from i with the lowest label. Remove the arc $\nu^i(p)$.*
- *Termination rule: Each time a token **returns** to station $*$, it is frozen there and cannot be considered for further moves. Equivalently, there is a step 0 for the game which consists of removing the arcs $\nu^*(I^*), \nu^*(I^* + 1), \ldots,$ where I^* is the initial number of tokens in station $*$.*

A sequence of moves following the previous rules is called an execution *of game $\mathcal{G}$. The game ends when there are no unfrozen tokens left in the network. We denote by T the step at which the game ends.*

This game can in some sense be interpreted as an untimed version of the evolution of the Jackson network. There are of course several possible executions of the game depending on which token is selected at each step. In the closed case, given an execution of the game, it is easy to build sequences $\{(\tilde{\sigma}^k(n), k \in \mathcal{K}), n \in I\!N\}$ of service times such that the Jackson network $\{(\tilde{\sigma}^k(n), \nu^k(n))\}$ evolves exactly as $\mathcal{G}$. In the open case, it would be necessary to allow the addition or removal of tokens to obtain the same interpretation.

Proposition 4.2 (Euler property) *Let J be a stationary and ergodic* **(SE3)** *Jackson network. Let G be the corresponding routing graph. Let us consider a finite initial condition for G. For all executions of the game $\mathcal{G}$, the game ends in finite time, i.e. $T < \infty$. Furthermore both T and the set of arcs of G which are not removed at time T do not depend on the execution.*

Proof The key ingredient is the irreducibility of the routing matrix $I\!\!P$. A proof was proposed in [2] under (H_1) type conditions. For a proof under **(SE3)**, see [5]. □

4.2   Restriction of a network

Let J be a Jackson network. For all integers $l \geq 0$, we define the Jackson network

$$J_{[0,l]} = \left\{ (\sigma^k_{[0,l]}(n), k \in \mathcal{K}), (\nu^k_{[0,l]}(n), k \in \mathcal{K}) \right\}, \tag{4.1}$$

with canonical initial conditions, where

$$\sigma^*_{[0,l]}(n) = \left\{ \begin{array}{ll} \sigma^*(n) & \text{for } 0 \leq n \leq l \\ \infty & \text{otherwise,} \end{array} \right.$$

and $\sigma^k_{[0,l]}(n) = \sigma^k(n), \forall n \geq 0, \forall k \in (\mathcal{K}\backslash *)$ and $\nu^k_{[0,l]}(n) = \nu^k(n), \forall n \geq 0, \forall k \in \mathcal{K}$.

We say that $J_{[0,l]}$ is a *restriction* of J. The intuitive interpretation is that we block station $*$ after l services there.

In a consistent way with the notations of Def. 2.7, we denote the daters and counters associated with the network $J_{[0,l]}$ by $X^k_{[0,l]}(n)$ and $\mathcal{X}^k_{[0,l]}(t)$. We define the *maximal daters* associated with the network $J_{[0,l]}$ as

$$X^k_{[0,l]} = \max_{n \geq 0} X^k_{[0,l]}(n), \quad X_{[0,l]} = \max_{k \in \mathcal{K}} X^k_{[0,l]}, \tag{4.2}$$

where the maxima are taken over the finite terms only. It follows from Prop. 4.2 that $X_{[0,l]}$ is finite P–a.s. The interpretation is that $X_{[0,l]}$ is the date of the last event to take place in $J_{[0,l]}$. Let us define

$$l^k_{[0,l]} = \lim_{t \to \infty} \mathcal{X}^k(t) = \mathcal{X}^k(X_{[0,l]}), \quad l_{[0,l]} = \sum_{k \in \mathcal{K}} l^k_{[0,l]}.$$

The integer $l^k_{[0,i]}$ is the total number of services completed at station k until instant $X_{[0,l]}$ (or equivalently until ∞). By definition, we have $l^*_{[0,l]} = l + 1$. It follows from Prop. 4.2 that $l^k_{[0,l]}$ is finite and depends only on the routings, not on the services.

We define the following notations.

- Services used up to time $X_{[0,l]}$:

$$\sigma^k_{[0,l]} = (\sigma^k(0), \cdots, \sigma^k(l^k_{[0,l]} - 1)), \quad \sigma_{[0,l]} = (\sigma^k_{[0,l]}, k \in \mathcal{K}).$$

- Routings used up to time $X_{[0,l]}$:

$$\nu^k_{[0,l]} = (\nu^k(0), \cdots, \nu^k(l^k_{[0,l]} - 1)), \quad \nu_{[0,l]} = (\nu^k_{[0,l]}, k \in \mathcal{K}).$$

- Total service time received up to time $X_{[0,l]}$:

$$\|\sigma^k_{[0,l]}\| = \sum_{i=0}^{l^k_{[0,l]}-1} \sigma^k(i), \quad \|\sigma_{[0,l]}\| = \sum_{k \in \mathcal{K}} \|\sigma^k_{[0,l]}\|.$$

For all $l \geq 0$, we define the Jackson network

$$J_{[l,\infty]} = \left\{ (\sigma^k_{[l,\infty]}(n), k \in \mathcal{K}), (\nu^k_{[l,\infty]}(n), k \in \mathcal{K}), n \in I\!N \right\},$$

where $\sigma^k_{[l,\infty]}(n) = \sigma^k(n + l^k_{[0,l-1]})$ and $\nu^k_{[l,\infty]}(n) = \nu^k(n + l^k_{[0,l-1]})$, $k \in \mathcal{K}$ (with the convention $l^k_{[0,-1]} = 0$).

The interpretation is that $J_{[l,\infty]}$ is the network obtained by unblocking $J_{[0,l-1]}$ after instant $X_{[0,l-1]}$.

The definition of the restricted networks $J_{[l,p]}, l \leq p$, follows naturally. The notations used for quantities associated with $J_{[l,p]}$ are consistent with the previous ones. For example, we denote the internal daters of $J_{[l,p]}$ by $X^k_{[l,p]}(n)$.

For convenience, we use the following abridged notation $J_{[p]} = J_{[p,p]}$ and accordingly $X_{[p]} = X_{[p,p]}, X^k_{[p]}(n) = X^k_{[p,p]}(n), \ldots$.

Remark The restrictions $J_{[l,p]}$ are defined as Jackson networks, and the conventions of §2.1 will be used without further mention. It implies that $J_{[l,p]}$ starts its evolution at time 0 and not at time T_l or $X_{[0,l-1]}$.

Example 4.3 Let us consider the closed Jackson network of Fig. 6.B. We have

$$l_{[i]} = 4, \quad \nu_{[i]} = \{\nu^1_{[i]}, \ldots, \nu^4_{[i]}\} = \{2,3,4,1\},$$

$$\sigma_{[i]} = \{\sigma^1_{[i]}, \ldots, \sigma^4_{[i]}\} = \{2,1,0,1\}, i \text{ even}, \quad \sigma_{[i]} = \{0,1,2,1\}, i \text{ odd}.$$

Note that in this case, the stations visited and the services received by a customer in the networks J and $J_{[i]}$ are the same. One can easily convince oneself that it is not the case as soon as the routings allow customers to overtake.

4.3   Stochastic assumptions

We consider a probability space $(\Omega, \mathcal{F}, P, \theta)$ as defined in §2.2. The main forthcoming results (§5, 6, 7) are to be given under the following stochastic assumptions:

- **H1** The sequence $\{(\sigma_{[i]}, \nu_{[i]}) = (\sigma^k_{[i]}, \nu^k_{[i]}, k \in \mathcal{K}), i \in I\!N\}$ is stationary and ergodic, i.e. we have

$$(\sigma_{[i]}, \nu_{[i]}) = (\sigma_{[0]}, \nu_{[0]}) \circ \tilde{\theta}^i, \tag{4.3}$$

for some P-stationary and P-ergodic shift $\tilde{\theta}$.

- **H2** The total service time received in $J_{[0]}$ is integrable, i.e. $E(\|\sigma_{[0]}\|) < +\infty$.

Note that these assumptions are made on the quantities associated with the restricted networks $J_{[i]}$. They may not seem natural at first glance. All the results to come will show that they are the right ones, see Remark 6.2. The following lemma gives simple sufficient assumptions on J under which the above assumptions are satisfied (see [5]).

Lemma 4.4 *Let us consider a Jackson network J such that the sequences $\{(\sigma^k(n)), n \in I\!N\}, k \in \mathcal{K}$, are stationary–ergodic and mutually independent and the sequences $\{\nu^k(n), n \in I\!N\}, k \in \mathcal{K}$, are i.i.d., mutually independent and independent of the services. Then $\{(\sigma_{[p]}, \nu_{[p]}), p \in I\!N\}$ is a stationary and ergodic sequence.*

The previous lemma fails to be true if we only assume that $\{(\sigma^k(n), k \in \mathcal{K}), n \in I\!N\}$ is stationary–ergodic.

5   First order limits for canonical initial conditions (A)

It is easy to prove, see [5], that α_k, defined in (2.4), is also the expected number of visits to station k in $J_{[0]}$:

$$\alpha_k = E(l_{[0]}^k) \, . \tag{5.1}$$

5.1   Closed network

Theorem 5.1 *Let J be a closed Jackson network. Under the assumptions $\mathbf{H1}$–$\mathbf{2}$, we have for all $k \in \mathcal{K}$:*

$$\lim_{n \to \infty} \frac{n}{X^k(n)} = \lim_{t \to \infty} \frac{\mathcal{X}^k(t)}{t} = \alpha_k \lambda, \ P - a.s.$$

$$\lim_{n \to \infty} \frac{n}{EX^k(n)} = \lim_{t \to \infty} E\left(\frac{\mathcal{X}^k(t)}{t}\right) = \alpha_k \lambda,$$

for some constant λ and where α_k is defined in (2.4) or (5.1).

Proof The proof is based on the sub-additivity of the maximal daters $X_{[m,n]}$. It is given in [5]. $\square$

5.2   Open network

Let J be a Jackson network satisfying assumptions $\mathbf{H1}$–$\mathbf{2}$. We define $J\{0\}$ to be the Jackson network obtained from J by modifying only the arrival process and setting $T_n = 0, \forall n \geq 0$. In words, $J\{0\}$ is the saturated network associated with J. We define, with obvious notations, the quantities $X^k(n)\{0\}, \mathcal{X}^k(t)\{0\}, \ldots$.

Theorem 5.2 *Under the previous assumptions, there exists a constant $\lambda(0)$ such that*

$$\lim_{t \to \infty} \frac{\mathcal{X}^{K+1}(t)\{0\}}{t} = \lambda(0), \ P - a.s. \ and \ in \ L_1. \tag{5.2}$$

The constant $\lambda(0)$ is the asymptotic throughput of departures from the network $J\{0\}$. Furthermore, we have:

$$1/\lambda(0) = \max_{k=1,\ldots,K} E\left(\|\sigma_{[0]}^k\|\right).$$

We recall that $\|\sigma_{[0]}^k\|$ is the total service time received at station k in $J_{[0]}$.

Proof The first proofs were given in [2] and [4]. A shorter proof will appear in [5]. □

Theorem 5.3 *Let J be an open Jackson network satisfying assumptions H1–2. Let $\lambda(0)$ be the constant defined in (5.2). We assume that $\lambda^0 < \lambda(0)$. We have for all $k \in \mathcal{K} \cup \{K+1\}$:*

$$\lim_{n\to\infty} \frac{n}{X^k(n)} = \lim_{t\to\infty} \frac{X^k(t)}{t} = \alpha_k \lambda^0, \; P-a.s. \text{ and in } L_1,$$

where $\alpha_k, k \in \mathcal{K}$, is defined in (2.4) and where $\alpha^{K+1} = 1$. If we assume that $\lambda^0 > \lambda(0)$, then we have:

$$\lim_{n\to\infty} \frac{n}{X^{K+1}(n)} = \lim_{t\to\infty} \frac{X^{K+1}(t)}{t} = \lambda(0), \; P-a.s. \text{ and in } L_1.$$

Proof A weaker version of the result was proved in [2]. The proof of this version is given in [5]. □

It follows from Theorems 5.2 and 5.3 that

$$1/\lambda^{K+1} = \max(1/\lambda^0, 1/\lambda(0)) = \max_{k=0,1,\ldots,K} E\left(\|\sigma_{[0]}^k\|\right).$$

The constant $\lambda(0)$ is the throughput of the network when we saturate the input. The interpretation of Theorem 5.3 is that $\lambda(0)$ is the maximal possible throughput for the network. This follows from the fact that the saturation rule of [3] can be applied. Theorem 5.2 provides a simple practical way to compute the maximal throughput of an open network.

Example 5.4 We consider the network of §3.1. We obtain $\sigma_{[0]}^1 = (1/4, 1/4)$ (two services, each of length $1/4$) and $\sigma_{[0]}^2 = (1/2)$. It follows that $\|\sigma_{[0]}^1\| = \|\sigma_{[0]}^2\| = 1/2$. We conclude that the maximal asymptotic throughput of the network is 2.

6 First order limits for different initial conditions (B)

We want to obtain extensions of the results of §5 for networks with arbitrary initial conditions. The results will be completely different for open and closed networks. The throughputs do not depend on the initial condition in the open case and they do in the closed case.

6.1   Compatibility

Let $_IJ = \{(\sigma^k(n), \nu^k(n)), k \in \mathcal{K}, n \in I\!N\}$ be a Jackson network with an arbitrary finite initial condition $I = \{(Q^k, R^k), k \in \mathcal{K}\}$. We consider a modification $_IJ_{[\emptyset]}$ of $_IJ$ obtained by setting $\sigma^*(0) = +\infty$. All other quantities (services, routings and initial condition) are the same in $_IJ$ and $_IJ_{[\emptyset]}$. The interpretation is that $_IJ_{[\emptyset]}$ is obtained by immediately blocking station $*$. Associated quantities are denoted accordingly, e.g. $_IX_{[\emptyset]}$ for the maximal dater.

It follows from Prop. 4.2 that $_IX_{[\emptyset]}$ is P-a.s. finite. Furthermore at instant $_IX_{[\emptyset]}^+$, the network is empty (open case) or all the customers are in station 1 (closed case).

We define a new network $_I\widehat{J}$ with service times $\{_I\hat{\sigma}^k(n) = \sigma^k(n + {}_Il_{[\emptyset]}^k), n \in I\!N\}$, $k \in \mathcal{K}$, and routings $\{_I\hat{\nu}^k(n) = \nu^k(n + {}_Il_{[\emptyset]}^k), n \in I\!N\}$, $k \in \mathcal{K}$. The initial condition of $_I\widehat{J}$ is canonical. The interpretation is that $_I\widehat{J}$ is the network obtained by unblocking $_IJ_{[\emptyset]}$ after instant $_IX_{[\emptyset]}$.

We are now ready to define *compatibility*. Let $_1J$ and $_2J$ be two Jackson networks differing only by their initial conditions I_1 and I_2. Let $_1J_{[\emptyset]}$ and $_2J_{[\emptyset]}$ be the blocked networks defined as above. Let $_1\hat{J}$ and $_2\hat{J}$ be the unblocked networks defined as above. Let $\{(_1\hat{\sigma}_{[i]}, {}_1\hat{\nu}_{[i]}), i \geq 0\}$ and $\{(_2\hat{\sigma}_{[i]}, {}_2\hat{\nu}_{[i]}), i \geq 0\}$ be the services and routings used in networks $_1\hat{J}_{[i]}$ and $_2\hat{J}_{[i]}$ respectively.

Definition 6.1 *The initial conditions of $_1J$ and $_2J$ are said to be compatible if*

(1) $\{(_1\hat{\sigma}_{[i]}, {}_1\hat{\nu}_{[i]}), i \geq 0\}$ *is stationary and ergodic.*

(2) $\{(_1\hat{\sigma}_{[i]}, {}_1\hat{\nu}_{[i]}), i \geq 0\}$ *has the same distribution as* $\{(_2\hat{\sigma}_{[i]}, {}_2\hat{\nu}_{[i]}), i \geq 0\}$.

Lemma 4.4 provides assumptions under which the first condition is always satisfied. The next lemma is straightforward.

Lemma 6.2 *Let us consider an i.i.d. Jackson network, see Def. 2.5. Then all initial conditions are compatible.*

Without the i.i.d. assumption, compatibility is not always satisfied.

Example 6.3 Let us consider the model of §3.3. The networks of Figs 6.A and 6.B, say $_AJ$ and $_BJ$, differ only by their initial condition.

• $_AJ$. Let us consider the network $_AJ_{[\emptyset]}$ as defined above. It gets blocked after the *dark gray* customer has received exactly one service at stations 3 and 4. It implies that the network $_A\hat{J}$ has the following sequences of service times:

$$\hat{\sigma}^1(n) = 2,0,2,0,\ldots \quad \hat{\sigma}^2(n) = 1,1,1,\ldots \quad \hat{\sigma}^3(n) = 2,0,2,0,\ldots \quad \hat{\sigma}^4(n) = 1,1,1,\ldots.$$

We have $\hat{\nu}_{[i]}^k = k + 1 \,[\mathrm{mod}\ 4]$ for all $i \geq 0$ and $\hat{\sigma}_{[i]} = (\hat{\sigma}_{[i]}^1, \ldots, \hat{\sigma}_{[i]}^4) = (2, 1, 2, 1)$, i odd, $\hat{\sigma}_{[i]} = (0, 1, 0, 1)$, i even.

• $_BJ$. The quantities associated with the restricted network $_BJ_{[i]}$ are $\hat{\nu}_{[i]}^k = k + 1 \,[\mathrm{mod}\ 4]$ and $\hat{\sigma}_{[i]} = (2, 1, 0, 1)$, i odd, $\hat{\sigma}_{[i]} = (0, 1, 2, 1)$, i even.

We conclude that the initial conditions of $_AJ$ and $_BJ$ are not compatible.

6.2   Closed network

The main theorem is the following one.

Theorem 6.4 *Let $_1J$ and $_2J$ be two closed Jackson networks with finite initial conditions I_1 and I_2. We assume that I_1 and I_2 are compatible. The networks are assumed to satisfy assumption* **H2**. *Let $_1X^k(n),_1\mathcal{X}^k(t)$ and $_2X^k(n),_2\mathcal{X}^k(t)$ denote the daters and counters associated with $_1J$ and $_2J$ respectively. There exists a constant λ such that for all $k \in \mathcal{K}$:*

$$\lim_{n\to\infty} \frac{n}{_1X^k(n)} = \lim_{t\to\infty} \frac{_1\mathcal{X}^k(t)}{t} = \lim_{n\to\infty} \frac{n}{_2X^k(n)} = \lim_{t\to\infty} \frac{_2\mathcal{X}^k(t)}{t} = \alpha_k\lambda, \ P-a.s.,$$

$$\lim_{n\to\infty} \frac{n}{E_1X^k(n)} = \lim_{t\to\infty} E\left(\frac{_1\mathcal{X}^k(t)}{t}\right) = \lim_{n\to\infty} \frac{n}{E_2X^k(n)}$$

$$= \lim_{t\to\infty} E\left(\frac{_2\mathcal{X}^k(t)}{t}\right) = \alpha_k\lambda,$$

where α_k is the expected number of visits to station k in $_1\hat{J}_{[0]}$ or $_2\hat{J}_{[0]}$, see (5.1).

Proof The networks $_1\hat{J}$ and $_2\hat{J}$ are equivalent in distribution. They also have a canonical initial condition. It implies that we can apply Theorem 5.1. The last step is to prove that the first order limits of $_iJ$ and $_i\hat{J}, i = 1, 2$, are identical. For a more detailed proof, see [5]. □

For non-compatible initial conditions, the first order limits need not be the same.

Example 6.5 Let us consider the counter-example of §3.3. The throughputs of the networks $_AJ$ and $_BJ$ are different, equal to $1/3$ and $1/2$ respectively. We know from Example 6.3 that the initial conditions of $_AJ$ and $_BJ$ are not compatible.

However, we obtain the next theorem as a corollary of Lemmas 4.4 and 6.2.

Theorem 6.6 *In an i.i.d. closed Jackson network satisfying assumption* **H2**, *the first order limits $(n/X^k(n), \mathcal{X}^k(t)/t, \ldots)$ exist and do not depend on the finite initial condition.*

Remark The results of this section show that first order limits depend only on the distribution of $\{(\sigma_{[p]}, \nu_{[p]}), p \in I\!N\}$. It implies that the "unusual" stochastic assumption that we make on our Jackson networks (**H1**, see (4.3)) is the "natural" one. It is not an artefact of our method of proof.

6.3   Open network

Let $_IJ$ be an open Jackson network with an arbitrary finite initial condition $I = \{(Q^k, R^k), k \in \{1, \ldots, K\}\}$.

Let J be the associated network with exactly the same sequences of services and routings but with a canonical initial condition (see Def. 2.3). We assume that J satisfies assumptions **H1–2**.

Theorem 6.7 *We assume that $\lambda^0 < \lambda(0)$, see eqn (5.2). For all finite initial conditions I, we have for all $k \in \mathcal{K} \cup \{K + 1\}$:*

$$\lim_{n \to \infty} \frac{n}{_I X^k(n)} = \lim_{t \to \infty} \frac{_I \mathcal{X}^k(t)}{t} = \alpha_k \lambda^0, \ P - a.s. \ and \ in \ L_1,$$

where α_k is defined in eqn (2.4). If we assume that $\lambda^0 > \lambda(0)$, then we have:

$$\lim_{n \to \infty} \frac{n}{_I X^{K+1}(n)} = \lim_{t \to \infty} \frac{_I \mathcal{X}^{K+1}(t)}{t} = \lambda(0), \ P - a.s. \ and \ in \ L_1.$$

Proof The first proof was given in [2], for initial conditions which are compatible with the canonical one. The general proof is given in [5]. □

7   First order limits: concavity of throughput (C)

7.1   Results and conjecture

The throughput at a station is an increasing function of the vector of initial queue lengths.

Proposition 7.1 *Let $_U J$ and $_V J$ be two closed Jackson networks differing only in their initial condition. We denote by $_U \lambda$ and $_V \lambda$ the throughputs at station 1 in $_U J$ and $_V J$ respectively. Let us assume that $_U Q^k \geq \ _V Q^k, k \in \mathcal{K}$. Then we have $_U \lambda \geq \ _V \lambda$.*

Proof The result follows from a sample path argument. It was originally proved by Shanthikumar and Yao [28]. □

As far as concavity is involved, the only result which is known is for exponential networks.

Proposition 7.2 *Let J be an i.i.d. closed Jackson network. We assume furthermore that all the service times are exponentially distributed, i.e. there exist constants $\beta_k, k \in \mathcal{K}$, such that $P\{\sigma^k(0) \geq u\} = \exp(-\beta_k u)$. Let M be the number of customers in the network and $\lambda^k(M)$ be the throughput at station $k, k \in \mathcal{K}$. Then $\lambda^k(M)$ is a concave function of M.*

Proof The proof depends heavily on the product form solution for the stationary distribution of the vector of queue lengths. For details, see Shanthikumar and Yao [27] and also Dowdy *et al.* [15]. □

Proposition 7.2 fails to be true for a stationary–ergodic (**H1–2**) Jackson network. Note first that it is now necessary to record the exact initial position of the customers because of the non-uniqueness of the throughput. However, the concavity fails even for an increasing (in the sense of the partial ordering on $(Q^1, \ldots, Q^K)$) sequence of initial conditions. A counter-example is provided in §3.4, see Fig. 10 in particular.

To bridge the gap between this counter-example and Prop. 7.2, the next step would be to (dis)prove the following result.

Conjecture Let J be an i.i.d. closed Jackson network. Then $\lambda^k(M), k \in \mathcal{K}$, is a concave function of M, the number of customers in the network.

8 Second order limits (D)

We obtain different types of results for open and closed Jackson networks. In the open case, we prove the existence of minimal stationary regimes for second order processes. In the closed case, there are no general ways of constructing the stationary regime. In both cases, there is no uniqueness of the stationary solutions in general.

8.1 Open network

Let us consider an open network J with a canonical initial condition and satisfying assumptions **H1–2**. The second order processes of J are denoted by $(Q(t), R(t)) = (Q^k(t), R^k(t), k \in \mathcal{K} \backslash 0)$ and $W(t) = (W^k(t), k \in \mathcal{K} \backslash 0)$, see §2.3.

Theorem 8.1 *If $\lambda^0 > \lambda(0)$, then there exists $k \in \mathcal{K}$ such that we have P – a.s.:*

$$Q^k(t) \overset{t}{\to} +\infty, \quad W^k(t) \overset{t}{\to} +\infty.$$

If $\lambda^0 < \lambda(0)$, then the processes $(Q(t), R(t))$ and $W(t), t \geq 0$, converge weakly to finite and stationary–ergodic limit processes $(Q_\infty(t), R_\infty(t))$ and $W_\infty(t), t \geq 0$.

Proof The proof is given in [2], §6. □

The stationarity of the processes $(Q_\infty(t), R_\infty(t))$ and $W_\infty(t)$ has to be interpreted as a Palm stationarity (i.e. with respect to the instants $\{T_n\}$). These processes are called the minimal stationary regimes for reasons explained in §9. They are explicitly computed under special assumptions in §9. In many cases, there will be multiple stationary regimes depending on the initial condition, see Example 8.2.

Example 8.2 Let us consider the network of §3.1. It was shown in Example 5.4 that $\lambda(0) = 2$. In the example of §3.1, the arrival rate is 1; hence we are in the case $\lambda^0 < \lambda(0)$. However, we have shown that there are several limits for second order processes depending on the initial condition. The minimal stationary regimes are the ones corresponding to $c = 0$. For example, the minimal queue length process is the randomized version of

$$Q_\infty(t) = \begin{cases} (1,0) & \text{if} \quad t \in [0,1] \bigcup_{n>0}[n-1/4, n+1/4] \\ (0,1) & \text{if} \quad t \in \bigcup_{n>0}[n+1/4, n+3/4]. \end{cases}$$

Under stronger assumptions, we obtain the following refined result.

Theorem 8.3 *We consider a Jackson network $_IJ$ with a finite initial condition I. We assume that the sequence of service times $\{(\sigma^k(n), k \in \mathcal{K}), n \in I\!N\}$ is stationary and ergodic and the sequences of routings $\{\nu^k(n), n \in I\!N\}, k \in \mathcal{K}$, are i.i.d., mutually independent and independent of the services. Then the results of Theorem 8.1 apply independently of the initial condition. In particular, when $\lambda^0 < \lambda(0)$, there is a unique stationary regime for the second order processes $(Q(t), R(t))$ and $W(t)$. Furthermore, they converge in total variation to their stationary distribution.*

Proof The proof of this result is given in [2], §7. □

Stability region   Saying that $[0, \lambda(0)]$ is the *stability region* of an open Jackson network is a way to summarize the results of Theorems 5.2, 6.7 and 8.1. For an input rate $\lambda^0 \in [0, \lambda(0)]$, the output rate (λ^{K+1}) is λ^0 and second order processes are finite. For an input rate $\lambda^0 \notin [0, \lambda(0)]$, the output rate is $\lambda(0)$ and second order processes are asymptotically infinite.

8.2   Closed network

Let J be a closed Jackson network satisfying assumptions **H1–2**. In general there is no uniqueness of the stationary regimes for second order processes. This is illustrated by the counter-example of §3.2. Furthermore, we have not been able to obtain a counterpart of Theorem 8.1, i.e. to construct a stationary regime in general. Some insights into the difficulties that arise are given in [6].

Even for i.i.d. closed Jackson networks, no general necessary and sufficient conditions for stability are known. There exist, however, some good sufficient conditions, see [8, 9], [17], [29] or [20]. The most recent ones are provided in [24].

9   Fluid open Jackson networks

The object of this section is to give more detailed results on the class of functions involved in the weak limits for second order variables, like those mentioned in Theorem 8.1. We will concentrate on a rather special model with fluid routing and with deterministic service times, but which still allows one to handle general stationary ergodic external arrival processes. More general cases can be considered along these lines (see [5]).

9.1   Evolution equations for counters

In this section, we consider an open Jackson network J with canonical initial condition, and such that $\sigma^k(n) = \sigma^k = \text{Const.}$ for all $k \neq 0$. We will make use of the following counters:

- $\mathcal{Y}^k(t)$, the total number of services *initiated* in station k in $[0, t]$;
- $\mathcal{A}^k(t)$, the total number of *external* arrivals in station k in $[0, t]$.

In addition, let

$$\Pi_{jk}(p) = \sum_{l=1}^{p} 1_{\{\nu^j(l)=k\}}, \quad p \in \mathbb{N}, j, k \in \mathcal{K} \setminus 0. \tag{9.1}$$

Using the FIFO hypothesis and the assumption that the network is initially empty, we obtain the following set of recurrence equations, holding for all $t > 0$ and all $k \neq 0$ (see [4] for more details on these equations):

$$\mathcal{Y}^k(t) = \left(\mathcal{Y}^k(t - \sigma^k) + 1\right) \wedge \left(\sum_{j=1}^{K} \Pi_{jk}\left(\mathcal{Y}^j(t - \sigma^j)\right) + \mathcal{A}^k(t)\right), \tag{9.2}$$

with initial condition $\mathcal{Y}^k(t) = \mathcal{A}^k(t) = 0$, for $t < 0$.

Fluid networks By definition, the *fluid* network $\overline{J}$ associated with (9.2) is that with evolution equation

$$\overline{\mathcal{Y}}^k(t) \;=\; \left(\overline{\mathcal{Y}}^k(t - \sigma^k) + 1\right) \wedge \left(\sum_{j=1}^{K} \mathbb{P}_{jk}\overline{\mathcal{Y}}^j(t - \sigma^j) + \mathcal{A}^k(t)\right), \qquad (9.3)$$

where $\mathbb{P}$ is the routing matrix. Such fluid models, which differ completely from the usual fluid models of queueing theory (here only the routing is fluid, whereas the services remain unchanged), were introduced in [12] for a class of Petri nets. Although the state variables stop being integer valued, we will go on speaking of numbers of customers etc.

When each of the K sequences $\{\nu^k(p)\}$ is stationary, then for all (deterministic) $X \in \mathbb{N}$, $E(\Pi_{jk}(X)) = \mathbb{P}_{jk}E(X)$. Assume that the sequences $\{\nu^k(p)\}$ and the arrival process are independent and that each of the sequences $\{\nu^k(p)\}$ is i.i.d. Then, when X is a random vector of $\mathbb{N}^K$, the relation $\sum_j \Pi_{jk}(X^j) = \sum_j \mathbb{P}_{jk}E(X^j)$ also holds true whenever X is a stopping time of the sequences $\{\nu^k(p)\}$, $k = 1,\ldots,K$. This fact and the concavity of the mappings involved in eqn (9.3) are the key ingredients to prove the following relation between the fluid and the non-fluid equations (see [5] for more details).

Lemma 9.1 *If the sequences $\{\nu^k(p)\}$ and the arrival process are independent and if each of the sequences $\{\nu^k(p)\}$ is i.i.d., then for all $f : \mathbb{R}^K \to \mathbb{R}$, nondecreasing and concave, and for all $t \in \mathbb{R}^+$, $E\left(f(\mathcal{Y}(t))\right) \le E\left(f(\overline{\mathcal{Y}}(t))\right)$.*

We will from now on concentrate on the fluid model.

9.2 Evolution equations for second order variables

When letting t go to infinity in the above evolution equations, we obtain that the total number of events $\overline{\mathcal{Y}}^k = \overline{\mathcal{Y}}^k(\infty)$, $\mathcal{A}^k = \mathcal{A}^k(\infty)$, satisfy the equation $\overline{\mathcal{Y}} = \overline{\mathcal{Y}}\mathbb{P}^* + \mathcal{A}$ where $\mathbb{P}^*$ denotes the restriction of $\mathbb{P}$ to the coordinates $\{1,\ldots,K\}$. The total numbers of events do not depend on the values of the service times. This property is a special case of Prop. 4.2 (for fluid).

When $\overline{\mathcal{Y}}$ is finite, let

$$\overline{\mathcal{M}}^k(t) = \overline{\mathcal{Y}}^k - \overline{\mathcal{Y}}^k(t), \quad \overline{\mathcal{B}}^k(t) = \mathcal{A}^k - \mathcal{A}^k(t), \qquad (9.4)$$

be the processes which count the *residual* number of events to take place in station k after time t. These second order variables satisfy the following system of equations:

$$\overline{\mathcal{M}}^k(t) \;=\; \left(\overline{\mathcal{M}}^k(t - \sigma^k) - 1\right) \vee \left(\sum_j \mathbb{P}_{jk}\overline{\mathcal{M}}^j(t - \sigma^j) + \overline{\mathcal{B}}^k(t)\right). \qquad (9.5)$$

One can reconstruct the total number $\overline{Q}^k(t)$ of customers present in queue k at time t, from $\overline{\mathcal{M}}^k(.)$, via the formula

$$\overline{Q}^k(t) \;\; = \;\; \overline{M}^k(t) - \sum_j \mathbb{P}_{jk} \overline{M}^j(t - \sigma^j) - \overline{B}^k(t). \tag{9.6}$$

Remark Similar evolution equations can be derived for the initial non-fluid network (see [4]), including the case of random services (see the last section therein).

9.3 Stationary regime

In this section, we consider a slotted model where all service times and inter-arrival times are equal to 1. The only randomness comes from the number of external arrivals at time n in station $k = 1, \ldots, K$, described by the random sequence $\{\mathcal{A}_{[n]}^k\}$, $n \in \mathbb{Z}$, which is assumed to be such that $\mathcal{A}_{[n]}^k = \mathcal{A}_{[0]}^k \circ \theta^n$, for all k and n. These stochastic assumptions are slightly different from the previous ones in that θ is not necessarily the Palm shift of the arrival process anymore.

Let $M_n^k \equiv \overline{\mathcal{M}}_{[-n,0]}^k(n)$, $n \geq 0$, be the residual process in the system which starts empty at time 0, has arrivals at time $1, 2, \ldots, n{+}1$, with respective numbers of arrivals $\mathcal{A}_{[-n]}$, $\mathcal{A}_{[-n+1]}$, $\ldots$, $\mathcal{A}_{[0]}$. We have $M_0^k = \overline{\mathcal{Y}}_{[0]}^k$, and from the relation $\overline{\mathcal{Y}}_{[-n-1,0]}^k(n+1) \circ \theta = \overline{\mathcal{Y}}_{[-n,0]}^k(n+1)$, we obtain that

$$M_{n+1}^k \circ \theta \;\; = \;\; \overline{\mathcal{Y}}_{[1]}^k + (M_n^k - 1) \vee \left(\sum_j \mathbb{P}_{jk} M_n^j \right). \tag{9.7}$$

It is easy to check that the sequences M_n^k are non-decreasing in n, and that the limit $M = \lim_{n \to \infty} M_n$ is either a.s. finite or a.s. infinite. When the limit $M = \lim_{n \to \infty} M_n$ is a.s. finite, it is the minimal solution of the functional equation

$$M^k \circ \theta = \overline{\mathcal{Y}}_{[1]}^k + (M^k - 1) \vee \left(\sum_j \mathbb{P}_{jk} M^j \right). \tag{9.8}$$

These minimal stationary variables allow one to construct an associated "minimal" stationary queueing process (giving the queueing process just before arrival instants), via relation (9.6). More generally, it is in this sense that the limit variables mentioned in Theorem 8.1 are said to be minimal.

Remark In the case where the transpose of matrix $\mathbb{P}$ is sub-stochastic, (9.7) can be rewritten in vector form as $M_{n+1} = \phi(M_n)$, where the random map ϕ is monotone, sub-homogeneous and non-expansive. The existence of the a.s. limit $\max_k M_n^k/n$ follows then immediately from general theorems on such maps given in [6].

Theorem 9.2 below, which is proved using (9.7) and (9.8), gives a representation of the transient and the stationary variables M_n and M.

The following notations will be needed: for $p \leq q$, let

$$a_{[p,q]}^k = \overline{\mathcal{Y}}_{[p,q]}^k - (q - p). \tag{9.9}$$

For $2 \leq h \leq \infty$, let $l^1, l^2(j_1), l^3(j_1, j_2), \ldots, l^h(j_1, \ldots, j_{h-1})$ be a family of integers indexed by $j = (j_1, \ldots, j_{h-1})$ in $\{1, \ldots, K\}^{h-1}$. Let $\Delta_n(h)$ denote the set of all families $L = \{l^q(j)\}_{q=1,\ldots,h,\ j \in \{1,\ldots,K\}^{h-1}}$ such that for all j, $0 < l^1 < l^2(j) < \cdots < l^h(j) \leq n$. Finally, let $\Delta = \Delta_\infty(\infty)$.

Theorem 9.2 *Under the foregoing assumptions, the variable M_n^k, $n < \infty$, $k = 1, \ldots, K$, admits the following representation:*

$$M_n^k = \max_{h=0,\ldots,n} A_n^k(h), \qquad (9.10)$$

with $A_n^k(0) = a_{[-n,0]}^k$, $A_n^k(1) = \max_{0 < l_1 \leq n} a_{[-l^1+1,0]}^k + \sum_j \mathbb{P}_{jk} a_{[-n,-l^1]}^j$, and more generally

$$A_n^k(h) = \max_{\substack{L \in \Delta_n(h)}} \sum_{\substack{p=0,\ldots,h \\ 1 \leq k_1,\ldots,k_p \leq K}} a_{[-l^{p+1}(k_1,\ldots,k_p)+1,\,-l^p(k_1,\ldots,k_{p-1})]}^{k_p} \prod_{q=1}^{p} \mathbb{P}_{k_q,k_{q-1}}, \qquad (9.11)$$

with the conventions $l^0 = 0$, $l^{h+1} = n+1$, $k_0 = k$ and $\prod_1^0 = 1$.

In the case where the stability condition $E(\mathcal{Y}_{[0]}^k) < 1$, for all k, is satisfied, *the stationary variable M^k admits the representation*

$$M^k = \sup_{\substack{L \in \Delta}} \sum_{\substack{p \geq 0 \\ 1 \leq k_1, k_2, \ldots \leq K}} a_{[-l^{p+1}(k_1,\ldots,k_p)+1,\,-l^p(k_1,\ldots,k_{p-1})]}^{k_p} \prod_{q=1}^{p} \mathbb{P}_{k_q,k_{q-1}}. \qquad (9.12)$$

Example 9.3 Consider a feedback queue with unit service times, where feedback takes place with probability $p < 1$. Equation (9.7) reads

$$M_{n+1} \circ \theta = \frac{A_{[1]}}{1-p} + (M_n - 1) \vee (pM_n). \qquad (9.13)$$

The solution of this equation is

$$
\begin{aligned}
M_n \; = \; & \tfrac{1}{1-p}(A_{[-n,0]} - n) \\
\vee \; & \max_{0 < l^1 \leq n} \tfrac{1}{1-p}(A_{[-l^1+1,0]} - (l^1-1)) + \tfrac{p}{1-p}(A_{[-n,-l^1]} - (n-l^1)) \vee \cdots \\
\vee \; & \max_{0 < l_1 < \cdots < l_q \leq n} \tfrac{1}{1-p}(A_{[-l^1+1,0]} - (l^1-1)) \\
& + \cdots + \tfrac{p^q}{1-p}(A_{[-n,-l^q]} - (n-l^q)) \vee \cdots \\
\vee \; & \sum_{q=0}^{n} \tfrac{p^q}{1-p}(A_{[-q]}-1).
\end{aligned}
$$

The sequence M_n is non-decreasing, and it tends to the limit

$$M \; = \; \sup_{0 < l_1 < l_2 < \cdots} \sum_{q \geq 0} \frac{p^q}{1-p} A_{[-l^{q+1}+1,\,-l^q]},$$

as n goes to ∞, which is finite when $E(A_{[0]}) < 1-p$.

Bibliography

1. Baccelli, F., Cohen, G., Olsder, G.J. and Quadrat, J.P. *Synchronization and Linearity*. Wiley, New York, 1992.

2. Baccelli, F. and Foss, S. Ergodicity of Jackson-type queueing networks. *Queueing Syst.*, **17**, 5–72, 1994.

3. Baccelli, F. and Foss, S. On the saturation rule for the stability of queues. *J. Appl. Probab.*, **32(2)**, 494–507, 1995.

4. Baccelli, F., Foss, S. and Gaujal, B. Structural, temporal and stochastic properties of unbounded free-choice Petri nets. INRIA Report **2411**, 1994. To appear in *IEEE Trans. Autom. Control*, 1996.

5. Baccelli, F., Foss, S. and Mairesse, J. On the stability of Jackson networks under stationary and ergodic assumptions. In preparation, 1996.

6. Baccelli, F. and Mairesse, J. Ergodic theory of stochastic operators and discrete event networks. In J. Gunawardena, editor, *Idempotency*. Cambridge University Press, 1995.

7. Bambos, N. On closed ring queueing networks. *J. Appl. Probab.*, **29**, 979–995, 1992.

8. Borovkov, A. Limit theorems for queueing networks. I. *Theor. Probab. Appl.*, **31**, 413–427, 1986.

9. Borovkov, A. Limit theorems for queueing networks. II. *Theor. Probab. Appl.*, **32**, 257–272, 1988.

10. Brandt, A., Franken, P. and Lisek, B. *Stationary Stochastic Models*. Probab. and Math. Stat. Wiley, New York, 1990.

11. Chang, C.S. Stability, queue length and delay of deterministic and stochastic queueing networks. *IEEE Trans. Autom. Control*, **39**, 913–931, 1994.

12. Cohen, G., Gaubert, S. and Quadrat, J.P. Polynomial discrete event systems. In J. Gunawardena, editor, *Idempotency*. Cambridge University Press, 1995.

13. Cohen, J. and Kelly, F. A paradox of congestion in a queueing network. *J. Appl. Probab.*, **27**, 730–734, 1990.

14. Daduna, H. Note on the ergodicity of closed queuing network. Technical Report **88-3**, Hamburg University, 1988.

15. Dowdy, L., Eager, D., Gordon, K. and Saxton, L. Throughput concavity and response time convexity. *Inf. Process. Lett.*, **19**, 209–212, 1984.

16. Foss, S. Some properties of open queueing networks. *Probl. Inf. Transm.*, **25(3)**, 90–97, 1989.

17. Foss, S. Ergodicity of queueing networks. *Siberian Math. J.*, **32**, 184–203, 1991.

18. Gordon, W. and Newell, G. Closed queuing systems with exponential servers. *Oper. Res.*, **15**, 254–265, 1967.

19. Jackson, J.R. Jobshop-like queueing systems. *Manage. Sci.*, **10**, 131–142, 1963.

20. Kaspi, H. and Mandelbaum, A. Regenerative closed queueing networks. *Stochastics Stochastic Rep.*, **39**, 239–258, 1992.

21. Kaspi, H. and Mandelbaum, A. On Harris recurrence in continuous time. *Math. Oper. Res.*, **19(1)**, 211–222, 1994.

22. Mairesse, J. Products of irreducible random matrices in the (max,+) algebra. Technical Report RR-1939, INRIA, Sophia Antipolis, France, 1993. To appear in *Adv. Appl. Probab.*, June 1997.

23. Mairesse, J. *Stabilité des systèmes à événements discrets stochastiques. Approche algébrique.* PhD thesis, Ecole Polytechnique, Paris, 1995. In English.

24. Mairesse, J. New conditions for the stability of closed i.i.d. Jackson networks. In preparation, 1996.

25. Meyn, S. and Down, D. Stability of generalized Jackson networks. *Ann. Appl. Probab.*, **4(1)**, 124–149, 1994.

26. Murata, T. Petri nets: properties, analysis and applications. *Proc. IEEE*, **77(4)**, 541–580, 1989.

27. Shanthikumar, G. and Yao, D. Second-order properties of the throughput of a closed queueing network. *Math. Oper. Res.*, **13(3)**, 524–534, 1988.

28. Shanthikumar, G. and Yao, D. Stochastic monotonicity in general queueing networks. *J. Appl. Probab.*, **26**, 413–417, 1989.

29. Sigman, K. Notes on the stability of closed queueing networks. *J. Appl. Probab.*, **26**, 678–682, 1989.

30. Sigman, K. The stability of open queueing networks. *Stochastic Proc. Appl.*, **35**, 11–25, 1990.

31. Willinger, W. Traffic modeling for high-speed networks: theory versus practice. In F. Kelly and R. Williams, editors, *Stochastic Networks*, volume **71** of *IMA*. Springer-Verlag, New York, 1995.

18

The Cesaro limit of departures from certain ·/GI/1 queueing tandems

T. Mountford

University of California at Los Angeles

B. Prabhakar

BRIMS, Hewlett-Packard Laboratories, Bristol

Abstract

We consider an infinite tandem of independent, identical ·/GI/1 queues with mean service rate equal to 1 subjected to stationary and ergodic inputs of rate $\alpha < 1$. Of some interest in the study of such queueing tandems are the following three inter-related questions: (1) For each $\alpha < 1$, does there exist a rate α stationary and ergodic process, $\mathbf{I}_\alpha$, which is an invariant distribution for the queue in the sense that $\mathbf{I}_\alpha \overset{d}{=} T(\mathbf{I}_\alpha)$? (Here $T(\mathbf{I}_\alpha)$ is the equilibrium departure process corresponding to an input of $\mathbf{I}_\alpha$.) (2) For a fixed α, is this invariant distribution unique? (3) When a stationary and ergodic arrival process of rate $\alpha < 1$ is input to the first queue, do the successive departure processes converge in distribution to the invariant distribution $\mathbf{I}_\alpha$ (assuming it exists)?

For general non-exponential server queues, it is not yet known if invariant distributions exist. However, for each $\alpha < 1$, should one exist, it is known to be unique. This paper contributes to the third question when the service time distribution of each queue in the tandem has an *increasing hazard rate*. It is shown that when a stationary and ergodic arrival process of rate $\alpha < 1$ is passed through a tandem of such queues, the Cesaro averages of the successive departure processes converge weakly to a limit which is an invariant distribution for the queue.

1 Introduction

In this paper we study the effect of passing a subcritical stationary point process through a series of ·/GI/1 queues. We are interested in establishing the distributional convergence of successive departure processes to a limit, assuming that such a limit exists. In [9] coupling arguments were used to establish the Poisson convergence of successive departures from a series of ·/M/1 queues. Our goal here is to adapt these coupling arguments for non-memoryless services. As

This work was supported in part by NSF grant DMS9157461, a grant from the Sloan Foundation, and the NSERC.

opposed to the case of $\cdot/M/1$ queues which are known to possess the Poisson process as the (only) invariant distribution [3, 1], a major problem in the case of non-exponential server queues is that it is not yet known if they possess an invariant distribution. We do not address this question, but rather establish some convergence results which take place when requisite invariant distributions exist. In this sense, this work is somewhat limited. In the remainder of this section we introduce some relevant terminology and describe our result.

Let A^0 be an ergodic stationary point process of rate α strictly less than one. Suppose that it is passed through a $\cdot/GI/1$ queue whose service times have mean one to obtain another stationary ergodic output process A^1 of rate α. That A^1 is stationary and ergodic and of rate α is guaranteed by Loynes' construction (for details of Loynes' construction see, for example, [2]). The output process A^1 can in turn be regarded as an input process for another $\cdot/GI/1$ queue, independent of and identical to the first. As before, we may then obtain a stationary and ergodic process A^2 as output from the second queue. Proceeding thus, we may obtain a series of output processes $A^n, n \in \mathcal{Z}^+$, each stationary and ergodic of rate α. The following is the question we wish to address: what can be said about the point process A^n as $n \longrightarrow \infty$? In the case where the services are i.i.d. exponentials of rate one it was recently shown (see [9]) that as n tends to infinity, A^n converges in distribution to a Poisson process of rate α. We would like to show an analogous result for general service distributions. Basically the method of [9] consisted of inputting a rate α Poisson process P^0 into the "same" queue as A^0 and obtaining a sequence of output processes P^i. By the well known Burke's Theorem, the processes P^i will all be Poisson of rate α. The proof then consisted of showing that as n becomes large, the processes A^n and P^n come closer and closer, and that A^n converged in distribution to a Poisson process.

Say that an input point process A, with law μ, is *invariant* for a $\cdot/GI/1$ queue if the corresponding departure process D is also distributed according to μ. As mentioned above, it is not known whether a general $\cdot/GI/1$ queue with mean service rate equal to one admits an invariant distribution of rate α ($\alpha < 1$). However, it is known that should such an invariant distribution exist, then it must necessarily be unique [4, 10]. This uniqueness of invariant distributions of a given rate is obtained by showing that, as a mapping on the space of stationary point processes, the queueing operator is contractive under a metric $d(.,.)$. That is, for distinct stationary arrival processes, A and B, of rate α, there is a metric $d(.,.)$ such that $d(A, B) > d(T(A), T(B))$, where $T(A)$ (respectively, $T(B)$) is the departure process resulting from an input of A (respectively, B) [4, 10]. The fact that the queueing operator is contractive under this metric also implies that any stationary invariant arrival process of *pathwise rate* α (i.e. a stationary invariant process whose ergodic components are all of rate α) must necessarily be ergodic (see [10]). We are careful to mention this here because later on we establish the convergence of Cesaro means of successive departure processes to a limit which is an invariant distribution for the queue. By the above discussion, this invariant distribution (assuming it exists) is unique and ergodic.

Attention, in this paper, is restricted to the family of service distributions of *increasing hazard rate.* For a random variable X, the hazard rate f (if it exists) is defined by

$$P[X \in (x, x + dx)|X > x] = f(x)dx + o(dx).$$

A random variable (or distribution) has an increasing hazard rate if f is increasing. If a random variable has increasing hazard rate then some exponential moments exist. The family of distributions with increasing hazard rate includes the exponential distribution whose hazard rate is constant on the positive half line. The following is the main result of this paper.

Theorem 1.1 *Let A^0 be a stationary and ergodic point process of rate α which is passed through a sequence of i.i.d. ·/GI/1 queues whose service times have increasing hazard rate and have unbounded distributional support. Let the output from the n-th queue be A^n with corresponding law μ_n. If the ·/GI/1 queue admits a rate α invariant distribution, μ, then*

$$\bar{\mu}_n = \frac{1}{n} \sum_{i=1}^{n} \mu_i$$

converges in distribution to μ.

Two assumptions are made for the service distribution. The first one, increasing hazard rate, is demanded by our whole approach; the second, unbounded support, is merely a simplification. At the end of the paper we sketch how this extra assumption may be removed.

We can only prove convergence of the Cesaro means because in this case limit points can be shown to be invariant. However, we believe that the original sequence of processes, $\{A^n, n \in \mathcal{Z}^+\}$, converges in distribution to an invariant limit when n tends to infinity.

The proof of the theorem requires a modification of Loynes' construction. We proceed to develop this in the next section.

2   A modification of Loynes' construction

In the following, for notational convenience only, we consider simple arrival processes. The standard development of Loynes' construction for a ·/GI/1 queue can be briefly summarized as follows. Arriving customers are associated with service times chosen in an i.i.d. fashion from some general service time distribution. Without loss of generality, these customers can be assumed to belong to an arrival process A^0 of rate $\alpha < 1$ and the service times can be assumed to have a mean value equal to one. Letting $A^{0,n}$ denote the (non-stationary) process obtained by suppressing all arrivals before time $-n$, it is then shown that the queue size at time t, X_t^n, resulting from an input of $A^{0,n}$ grows with n along each realization ω of the arrival and service processes. Therefore, $\lim_{n \to \infty} X_t^n = X_t^\infty$ exists. The condition $\alpha < 1$ (arrival rate < service rate) is then used to show that X_t^∞ is an a.s. finite stationary and ergodic process. If $A^{1,n}$ is the output corresponding to an input of $A^{0,n}$, it then easily follows that a stationary limit,

A^1, of the non-stationary outputs $A^{1,n}$ exists. See [2] for details.

For the special case of exponential servers, a different method of construction of A^1 from A^0 is used in [7], [1] and [9]. Here, the queue is associated with a rate one Poisson process N^0 representing virtual service times. For $t \in N^0$, a customer is served at time t if and only if the queue is non-empty at time t^-. Defining the service in this way, for an arrival process A^0 of rate $\alpha < 1$ one can construct the output process A^1 and queue process X^∞ as before by considering $A^{0,n}$ and taking the limit as $n \to \infty$. Further details of this procedure may be found in [9].

To study queueing systems with increasing hazard rate, we introduce a further modification to Loynes' construction for memoryless servers. Instead of a simple Poisson process of virtual service times we will derive our queue from a Poisson process N on $R^1 \times R^+$ of Lebesgue intensity. Given a queueing process $X(s)$, $-\infty < s < \infty$, a service will occur at time t if $(t,r) \in N$ for some r with $r \le f(t)$, where $f(t)$ is the hazard rate of the customer in the queue currently being served. It follows from Markov properties of the Poisson process that for a queueing process $X(t)$, the point process N defines a server with appropriate i.i.d. service times.

The following is a simple large deviations lemma stated without proof.

Lemma 2.1 *Fix $\beta < 1$. Given an arrival process $\cdots t_{-1} < 0 \le t_0 < t_1 \cdots$, consider the queueing process X_t^n starting at t_{-n} with a single customer in the queue (the $-n$-th customer), and with service derived from the Poisson process N defined above. If $t_{-n} < -n/\beta$, then the chance that $X_t^n > 0$ for each $t \in [t_{-n}, 0]$ is $\le Ce^{-nc}$ for some finite, positive C, c.*

It is easy to see that for each t the size of the queue, X_t^n, increases with n. We show that this limit must a.s. be finite.

Lemma 2.2 *The increasing limit of the processes $X^n, n \in Z^+$, is a.s. finite as $n \to \infty$.*

Proof Fix $\alpha < \beta < 1$. By the fact that the rate of A is $\alpha < \beta$, there exists M so that for all $n \ge M$, $t_{-n} < -n/\beta$. By Lemma 2.1, for $n \ge M$, the chance that the queue starting with a single customer at time t_{-n} is never empty in $(t_{-n}, 0]$ is $\le Ce^{-nc}$.

Thus, by the Borel–Cantelli Lemma, there exists M' so that for all $n \ge M'$ the queue started with one customer at time t_{-n} is empty at some time in $(t_{-n}, 0]$. Now, if $X_s^{n+1} = 0$ for $s \in (t_{-n-1}, 0]$, then $X_r^n \ge X_r^{n+1}$ for all $r \in [s, \infty)$. However, by monotonicity in n, $X_r^n \le X_r^{n+1}$ for $r \in [s, \infty)$. In particular, $X_r^n = X_r^{n+1}$ for all $r \in [0, \infty)$ and, therefore, for all $n \ge M'$, $X_s^n = X_s^{M'}$ for $s \in [0, \infty)$ implying the finiteness of $\lim_{n \to \infty} X_t^n$. $\qquad\square$

Lemma 2.3 *The limit of the output processes corresponding to the increasing limit of the X^n exists a.s.*

Proof First suppose that A is ergodic. From the usual Loynes construction it is clear that all we have to do is show that with probability one there exist times tending to $-\infty$ at which $X_t^\infty = \lim_{n\to\infty} X_t^n$ is equal to zero. By ergodicity of (A, N), this will be the case if we can show that the chance that X_t^∞ is zero at a fixed point is strictly positive. By stationarity, this in turn will follow if we can show that with positive probability, on some interval of positive length the limiting queue size is zero. By Lemma 2.2, we can choose L so large that $P[X_0^n < L] > 1/2$ for each n. However, in the notation of Lemma 2.1, for m large enough $t_m > m/\beta$. Thus we can apply this lemma to the time interval $[0, t_m]$ (for m large enough), to conclude that the limit of X^n ($= X^\infty$) is zero on an interval in $(0, m)$ with positive probability. This completes the proof for ergodic A.

If A is of rate α but not ergodic then it can be expressed as a mixture of rate α ergodic point processes. The above argument applied to each of these ergodic processes gives the result for this A. $\square$

3   A coupling of point processes

In this section we consider two arrival processes A^0 and B^0. A^0 is stationary and ergodic of rate $\alpha < 1$. B^0, also stationary and ergodic of rate α, is the (unique) invariant distribution for our queue (assumed to exist). We suppose that they are served by the same Poisson process N^0 on $R^1 \times R_+$, producing, respectively, A^1 and B^1 as output. These are then fed into a second queue served by an independent Poisson process N^1 (identically distributed to N^0). Proceeding thus, a series of departure processes $\{A^n\}$ and $\{B^n\}$, $n \in Z^+$, are obtained. We wish to examine how close A^n and B^n become as n tends to infinity. Initially all the customers in the A and B processes are labelled individually. If x is a customer in the A process, then $s_x(n)$ will denote the arrival time of x in the process A^n. Similarly if y is a customer in the B process then $t_y(n)$ will denote the arrival time of y in B^n. Note that in the following, we may switch the ordering of customers in A^i (or B^i) so that if x_1 and x_2 are customers in the A queues, it may be that x_1 arrives at queue i before x_2, but leaves the queue after x_2.

We wish to match up the A and B customers in such a way that for all large n the processes A^n and B^n are "close" to each other in distribution. The aim is that once customers are matched they remain so for ever; equivalently, if $x \in A^n$ is matched with $y \in B^n$, then x and y will be matched in (A^m, B^m) for all m greater than n. We now describe how customers are matched up at a queue.

For ease of exposition we introduce a colouring scheme similar to that employed in [9].

If customer $x \in A^n$ is matched with customer $y \in B^n$, then they are both coloured yellow. If customer $x \in A^n$ is unmatched, it is coloured blue. If customer $y \in B^n$ is unmatched, it is coloured red. So red or blue customers may become yellow subsequently, but we will prove in Proposition 3.1 that once a customer becomes yellow, it will remain so for ever. Say that x and y are matched in the pair (A^n, B^n) (or, equivalently, after queue n) if $s_x(n), t_y(n)$ satisfy

(∗) $s_x(n) = \inf\{s_z(n) \in [t_y(n), t_y^+(n)] : z$ is not matched with $y' \neq y\}$

where $t_y^+(n) = \inf\{t > t_y(n) : t \in B^n\}$. In other words, x and y are matched after queue n if x is the first unmatched A customer to depart from stage n after y, but before the next B customer who departs at time $t_y^+(n)$. This definition is ambiguous only in the trivial case where A^n is identical to B^n.

Observe that x and y may be matched after queue M but that $s_x(M) = t_j(\in B^M) \neq t_y(M)$ or $t_y(M) = s_i(\in A^M) \neq s_x(M)$. That is, matching doesn't imply coincidence of points and vice versa. This is because in our definition of matching we consider the interval $[t_y(n), t_y^+(n)]$ and not the interval $[t_y(n), t_y^+(n))$.

In order that the coupling argument be successful, it is crucial that once x and y are matched they remain so ever after. It is easy to see that a simple service policy like FIFO does not preserve matchings. For, consider two matched customers x and y belonging respectively to processes A and B at queue n. In particular, this implies that y departs from queue n before x. However, at queue $n + 1$, it is quite possible that y is delayed long enough by the presence of other B customers (who will not affect x) so that x departs from this queue before y, thus annulling the matching between x and y. In order to preserve matchings, we will therefore be required to modify the service policy for the A and B customers, sometimes giving priority to yellow customers and sometimes to non-yellow customers. This in itself will not be sufficient to preserve matchings and we will be forced to relabel customers from time to time.

Before describing the priority scheme and the relabelling procedure, we would like to motivate them by giving a small list of things we would want them to achieve.

(a) If $x \in A$ and $y \in B$ are matched after stage n then to preserve (∗) we would like to serve y before x at all subsequent stages and once y has been served, we would like to serve x as soon as possible. In this case, we might have to bump the priority of x (a yellow customer) over that of some other unmatched (blue) A customer.

(b) If x and y are matched after stage n and if y is at the $(n+1)$th stage while x has yet to arrive, we would like to avoid serving y, if possible.

We now proceed to introduce the priority scheme. Consider customers arriving according to (A^n, B^n) at the $(n + 1)$th queue. These customers are coloured yellow, red or blue depending on whether they are matched when departing from the nth stage or not. We begin by maintaining two imaginary internal queues for each of the A and B customer types. That is, at every stage n, each of the A and B queues will have separate FIFO buffers for yellow and non-yellow customers. Associated with queue A and present time t will be a random variable $\tau_A(t)$ which represents the amount of time that queue A has been non-empty since the last service was rendered. Similarly $\tau_B(t)$ is associated with queue B and time t. If a point (t, r) occurs in the service generating Poisson process N, then a customer will be served at time r at queue A (respectively, B) if and only if $r \leq f(\tau_A(t))$ (respectively, $f(\tau_B(t))$), where f is the hazard rate of the service distribution. The priority scheme and relabelling procedure amount to a rule for deciding when to release a yellow customer over a non-yellow one.

There are essentially three distinct possibilities to address for $(t, r) \in N$:

(1) $f(\tau_B(t)) < r \leq f(\tau_A(t))$: *i.e. there is a service for an A type customer but not for a B type customer.*

A decision as to which A type customer is served needs to be made only if both the internal queues are non-empty. (Equivalently, if there are yellow and non-yellow customers in the A queue.) If both queues are non-empty, then the first yellow customer is released if and only if its matched B partner has already been served. If not, the first blue customer is released.   ◇

(2) $f(\tau_A(t)) < r \leq f(\tau_B(t))$: *i.e. there is a service for a B customer but not for an A customer.*

Again a decision about which type of B customer is to be released needs to be made only if both internal queues are non-empty. If we have a choice between a red and a yellow customer, then the first yellow customer is released if and only if its matched A partner is presently in the A queue. Otherwise, the first red customer is released.   ◇

(3) $r \leq f(\tau_A(t)), f(\tau_B(t))$: *i.e. there is a service for both an A customer and a B customer.*

Consider the B queue first. If there are both yellow and red customers, then release the first yellow customer only if its partner is also in the A queue. If the partner has not yet arrived to the A queue, then release the first red customer. If there are only yellow (red) customers present, then the first yellow (respectively, red) customer has to be released, since service is non-idling.

Consider the A queue next. If there are any yellow customers present, then release the first of these. If there are no yellow customers, then release the first blue customer.   ◇

These are the rules for reprioritizing customers and we will now describe the relabelling procedure. But first, we briefly discuss the need to do this.

Consider Rule (2) above. It is quite possible that the only B customer present is a yellow customer, y, while queue A has only blue customers. This means that the partner of y, x, is yet to arrive at queue A. Now, it is possible that between the departure of y and the arrival of x there is service at queue A. Should this happen one of the blue customers, z say, is let go and this leads to an annulment of the matching between x and y, since x will no longer be the first A customer departing from stage $n + 1$ after y.

A similar situation is encountered under Rule (3) when we are forced to release a yellow B customer, say y, and a blue A customer, say z, simultaneously while x, the partner of y, has yet to arrive.

In either of the above cases, we simply swap the labels of customers x and z to ensure that the first A customer to leave queue $n + 1$ in time $[t_y(n + 1), \infty)$ is the partner of y, i.e. x.

We are now ready to prove the following proposition which establishes that under the above stated rules for customer reprioritizing and relabelling, the matchings of customers are preserved for ever.

Proposition 3.1 *For each* n, *if customers* x *and* y *belonging, respectively, to processes* A *and* B *are matched after queue* n, *then they will continue to be matched after queue* $n + 1$.

Proof We wish to show that all matched pairs of customers in (A^n, B^n) are also matched in (A^{n+1}, B^{n+1}). By the construction in Section 2, it is sufficient to show this for processes begun at time $-N$, with empty queues. We may disregard the arrival of the first yellow A customer if its partner arrived before time $-N$. Look at the first time that there is a departure so that the relation (∗) fails for some previously matched customers x and y. This could occur in three ways: $s_x(n) < t_y(n)$, $s_x(n) > t_y^+(n)$ or there are intervening blue customers between y and x. The last cannot occur because of relabelling so we concentrate on the first two.

Suppose that it is the first and, hence, that x leaves queue n before y. Now x arrived at the A queue after y arrived at the B queue. So y must be queueing while x departs. By Rules (1) and (3), when customer x is served there must be no blue customers present and x must be the longest waiting yellow customer in the A queue. By Rule (3), there cannot be a B customer served at the same time as x is served since this would require that either y be this B customer, or that (x, y) not be the first matched pair who fail to satisfy (∗). Therefore, the three facts (a) x is served alone, (b) when x is served there are no blues, and (c) y arrived before x, together imply that the last service of any kind before x is served must be that of a B customer but not of an A customer (recall that service has increasing hazard rates).

Now, if this last service occurred before the arrival of x, then at this point the A queue must have been empty and, since y arrives before x, increasing hazard rate properties of the server imply that a service must take place at queue B before or with the next service at the queue A. On the other hand if it occurred after the arrival of x, then by Rule (2) y must be served since it is the first yellow in the B queue (using the fact that (x, y) is the first pair who become unmatched). Either way there is a contradiction.

Next suppose that $s_x(n) > t_y^+(n)$. Let w be the B customer who departs after y at time $t_y^+(n)$. Now, customer y is served either alone or simultaneously with an A customer. Since the argument is similar in both cases, we only treat the case where y is served alone. Suppose first that x is present in the A queue when y is served. Because of increasing hazard rate properties of the server, the next customer served will be of type A (possibly simultaneously with a B customer). Since we assume that (x, y) is the first pair for whom things go wrong, by Rule (1), this next A customer must be x. This contradicts $s_x(n) > t_y^+(n)$. On the other hand, if x is not present when y is served then, by Rule (2), at time $t_y(n)$ there must be no red customers in the B queue. By property (∗) there are also no other yellow customers in the B queue. Thus when x arrives (possibly with a

B customer) the B queue is empty. Thus the next service after $t_y(n)$ must be of an A customer (possibly simultaneously with a B customer). By Rules (1) and (3) and the assumption that (x,y) is the first pair to go wrong, this A customer must be x. Again, this contradicts $s_x(n) > t_y^+(n)$. □

Similar arguments establish

Proposition 3.2 *Suppose that after queue n customer x is coloured blue, customer y is coloured red and $t_y(n) < s_x(n)$. If $s_x(n+1) < t_y(n+1)$ (i.e. x overtakes y at queue n), then x must be coloured yellow at all queues after $n+1$.*

Sketch of Proof Suppose that blue customer x arrives at queue $n+1$ after red customer y, but leaves strictly before y. Let $s_x^-(n+1)$ be the time of the A-departure from queue $n+1$ immediately prior to $s_x(n+1)$. Then either queue B is non-empty throughout the interval $(s_x^-(n+1), s_x(n+1)]$ or $s_x^-(n+1) < t_y(n) < s_x(n)$ and queue A is empty in the interval $(s_x^-(n+1), s_x(n))$. Either way a (necessarily red) customer must be released from queue B in the interval $(s_x^-(n+1), s_x(n+1)]$.

4   Proof of eventual matching of customers

For ease of exposition, we will assume throughout this section that the support of the service times is unbounded. With appropriate non-trivial modifications, the methods can be extended to the case of bounded service times using techniques similar to those in [10]. The remark at the end of the paper provides an idea of some of the steps involved in implementing the changes. We are now ready to prove the following proposition.

Proposition 4.1 *If B^0 is an invariant distribution of rate α and A^0 is stationary and ergodic of rate α, then the densities of matched customers in A^n and B^n must increase to α as n increases to ∞.*

Proof The argument is essentially an adaptation of that given in [6]. We argue by contradiction and suppose that the density of matched customers increases to a value strictly less than α. Then we must have customers who remain red or blue for ever. Let us call such customers ever-reds and ever-blues.

By the preservation of order among blues and reds established in Proposition 3.2, if customer x is an ever-blue, customer y is an ever-red, and $s_x(0) < t_y(0)$, then, for each n, it must be the case that $s_x(n) < t_y(n)$. This enables us to talk of intervals of ever-reds and ever-blues. We say customers $y_1, y_2, \ldots, y_r$ with $t_{y_1}(0) < t_{y_2}(0) < \ldots < t_{y_r}(0)$ form an interval of ever-reds if
  (i) For all i, y_i is an ever-red.
  (ii) There is no ever-blue x with $t_{y_i}(0) \leq s_x(0) \leq t_{y_j}(0)$.
  (iii) The interval (y_1, y_r) is not contained in a strictly larger subset with properties (i) and (ii).
We similarly define intervals of ever-blues.

Given such an interval (y_1, y_r) of ever-reds with $t_{y_1}(0) < t_{y_r}(0)$, call y_1 a left ever-red and y_r a right ever-red. A similar labelling procedure applies for ever-blues.

Now, consider the process of ever-reds and ever-blues. By the stationarity of (A^0, B^0), and the translation invariant nature of the queueing operation, the process of ever-reds and ever-blues must be stationary. However, a priori these need not be ergodic and hence the density of the ever-reds (and similarly of the ever-blues) must be considered a random quantity R where $E(R) < \alpha$. However, the following fact follows from ergodicity of B^n. The (non-random) density of red customers in B^n must decrease to $E(R)$ as n goes to infinity. Similarly for blue customers in A^n. Thus R is non-random.

Therefore initially there coexist, with probability one, ever-blue and ever-red customers, both of density $R > 0$. Consider the points of $A^0 \cup B^0$ that represent customers that never become yellow as an alternation of intervals of ever-blue customers with intervals of ever-red customers. By the order preservation property shown in Proposition 3.2, these intervals will have their "orderings" preserved. This and the fact that ever-reds and ever-blues coexist a.s. imply that the processes of left ever-red and left ever-blue customers as defined above exist. Now the process of left ever-reds is stationary (for the same reason that the ever-reds are stationary) and thus possesses a (possibly random) density. Denote this density by the random variable L_R and note that it must be true that $E(L_R) > 0$. Therefore for some $\epsilon > 0$, $P(L_R > \epsilon) > \epsilon$. Now, $L_R = L_R^l + L_R^g$, where L_R^l is the density of left ever-reds y_1 such that there is another left ever-red $y_2 \in (y_1, y_1 + 2/\epsilon)$ and L_R^g is the density of the remaining left ever-reds for whom the next left ever-red is at a distance greater than $2/\epsilon$. Since, by definition, $L_R^g \leq \epsilon/2$ P a.s., this means $L_R^l \geq \epsilon/2$ whenever $L_R > \epsilon$.

A vital observation originating in Ekhaus and Gray [6] is that between two left ever-reds there must be a left ever-blue, by definition of intervals of ever-reds/ever-blues. We deduce from this observation and the previous paragraph that the density of ever-reds y in B^n such that there is an ever-blue x with $s_x(n) \in (t_y(n), t_y(n) + 2/\epsilon]$ must be at least $\epsilon/2$ with probability ϵ. Now ever-blues and ever-reds are distinguished customers, identified by looking into the future; however, they must, respectively, be blue and red in A^n and B^n for any n. It now follows from the above that the density of reds y in B^n such that there is a blue x with $s_x(n) \in (t_y(n), t_y(n) + 2/\epsilon]$ must be at least $\epsilon/2$ with probability ϵ. Appealing now to the ergodicity of the pair (A^n, B^n) it is easy to see that this density must, in fact, be $\geq \epsilon/2$ with probability one.

By the stability of the queues and the strong law of large numbers every customer y in B^n has a strictly positive chance of arriving at an empty queue. Denote this probability by $p_n(y)$. Let us fix $\delta > 0$ so that (for all n by the invariance of B^0), the density of customers y in B^n with $p_n(y) < \delta$ is less than $\epsilon/2$. Therefore it follows (for every n) that with density at least $\epsilon/2 - \epsilon/4 = \epsilon/4$ there are customers $y \in B^n$ so that

(i) there exists a blue customer in $[t_y(n), t_y(n) + 2/\epsilon]$

(ii) $p_y(n) > \delta$.

Suppose now that after such a customer y arrives, it is not served within time $2/\epsilon$ (this event has strictly positive probability). Then it is easy to see that when y leaves, it must be matched to a blue customer.

Thus the density of matched customers in A^{n+1}— the density of matched customers in $A^n \geq (\epsilon/4)\,\delta\,(1 - F(2/\epsilon))$, where F is the c.d.f. of the service distribution. By our unboundedness assumption for the support of service times, this is a positive number. Therefore the density of matched customers in A^n (and B^n) must grow at least linearly with n. This contradicts the fact that the density of A^n, and hence that of the matched customers of A^n, is bounded by α for each n. □

The proof of Theorem 1.1 is now completed in the next section.

5   Proof of Theorem 1.1

We use a simple uniformity property for Loynes' construction.

Lemma 5.1 *Consider a stationary and ergodic arrival process V^0 of rate $\alpha < 1$ for a ·/GI/1 queue whose service times have an increasing hazard rate and are of mean 1. Suppose the stationary and ergodic equilibrium output, V^1, has the property that for all $x \geq x_0$,*

$$P\left[\frac{1}{x}V^1(-x,0) > \alpha + \epsilon\right] < \epsilon$$

where $\alpha + 2\epsilon < 1$; then

$$P\left[V^{n,1} = V^1 \text{ on } [0,\infty)\right] \geq 1 - \epsilon - e^{-nc(\epsilon)}$$

for some $c(\epsilon)$ strictly positive. Here $V^{n,1}$ is the output obtained from $V^{n,0}$ and $V^{n,0}$ is as in Section 2.

Proof As has already been noted, if X^∞ is the limiting queueing process for V^0 and X^n is the queueing process for $V^{n,0}$, then $X^\infty \geq X^n$. From this and the Markov property for our Poisson process, it follows that if $X^\infty_t = 0$ for some $t \in (-n, 0)$, then $V^{n,1} = V^1$ on $[0, \infty)$. But because the hazard rate of the service times is increasing, if $X^\infty > 0$ in $(-n, 0)$, then it must be the case that the distribution of the number of points of V^1 in $(-n, 0)$ is stochastically greater than the distribution of the number of service renewals in $(-n, 0)$. Thus by standard large deviations results (recall service times have some exponential moments) $P\left[V^1(-n, 0) \leq n\,(\alpha + \epsilon), \; X^\infty > 0 \text{ on } (-n, 0)\right] < e^{-c(\epsilon)n}$. □

Corollary 5.2 *Let $A^{r+1,k}$ be the point process obtained by suppressing all points in A^r that occur before time $-k$ and then passing the resulting point process through the ·/GI/1 queue generated by Poisson process N^r. For each $\epsilon > 0$ sufficiently small, there exists k sufficiently large so that for all r large enough*

$$P\left[A^{r+1,k} = A^{r+1} \text{ on } (0,\infty)\right] > 1 - \epsilon.$$

Proof Fix $\epsilon > 0$. By ergodicity of the B point processes, there exists x_0 so that for all $x > x_0$ (and for all r, since the B processes are identically distributed for all r), $P[B^r(-x, 0) > x(\alpha + \epsilon/2)] < \epsilon/2$. On any interval, the number of matched points in B^r can differ from the number of matched points in A^r by at most one. Therefore

$$\frac{A^{r+1}(-x,0)}{x} \leq \frac{B^{r+1}(-x,0)}{x} + \frac{C^{r+1}(-x,0)}{x} + \frac{1}{x}$$

where C^{r+1} denotes the unmatched points in A^{r+1}. Now by Proposition 4.1, the density of unmatched points in A^n tends to zero as n tends to infinity. Therefore

$$P\left[\frac{A^{r+1}(-x,0)}{x} > \alpha + 2\epsilon/3\right] < 2\epsilon/3$$

for r sufficiently large if x is greater than x_0. The statement of the corollary now follows from Lemma 5.1.   □

Proof of Theorem 1.1 Proposition 4.1 implies that the sequence of measures $\bar{\mu}_n = \frac{1}{n}\sum_{i=1}^{n}\mu_i$ is tight and that any limit point must have density α. It remains to show that these measures converge to μ, the invariant measure of density α. By the uniqueness result of [4], it is sufficient to show that any limit point is an invariant distribution for the $\cdot/GI/1$ queue. In the following T stands for the queueing operator, i.e. $T(\mu_i) = \mu_{i+1}$.

So we must show that if $\bar{\mu}_{n_i} \to \nu$ in distribution, then $\langle \nu, f \rangle = \langle T\nu, f \rangle$ for every bounded continuous function f of compact support. See, for example, [5]. Or equivalently, $\langle \nu, f \rangle = \langle \nu, T^*f \rangle$ for every bounded continuous function f of compact support, where T^* is the adjoint of T.

We suppose without loss of generality that f has support in $(0,\infty)$. $\bar{\mu}_{n_i} \to \nu$ weakly, implies that $\langle \bar{\mu}_{n_i}, f \rangle \to \langle \nu, f \rangle$. Also

$$\langle \bar{\mu}_{n_i}, T^*f \rangle = \frac{1}{n_i}\sum_{j=1}^{n_i}\langle \mu_j, T^*f \rangle = \frac{1}{n_i}\sum_{j=1}^{n_i}\langle T\mu_j, f \rangle = \frac{1}{n_i}\sum_{j=1}^{n_i}\langle \mu_{j+1}, f \rangle$$

and the last equation implies that

$$\langle \bar{\mu}_{n_i}, T^*f \rangle = \langle \bar{\mu}_{n_i}, f \rangle + \frac{1}{n_i}\left(\langle \mu_{n_i+1}, f \rangle - \langle \mu_1, f \rangle\right) \longrightarrow \langle \nu, f \rangle.$$

However, it is not immediate that as i tends to infinity, $\langle \bar{\mu}_{n_i}, T^*f \rangle \to \langle \nu, T^*f \rangle$ since the function T^*f is not necessarily continuous on the set of discrete measures. We can address this technical point by introducing continuous functions $T^{*,k}f(N) = T^*f(N^{-k})$ where N^{-k} is obtained from N as follows. All arrivals of N before $-(k+1)$ are removed with probability one, customers who arrive at $s \in [-(k+1), -k]$ are removed with probability $k+1+s$, and all arrivals after time $-k$ are kept. This yields a continuous function as the influence of points of N near the cutoff goes to zero. By Corollary 5.2, for every $\epsilon > 0$ we can find a k so that

$$\limsup \left|\langle \bar{\mu}_{n_i}, T^*f \rangle - \langle \bar{\mu}_{n_i}, T^{*,k}f \rangle\right| < \epsilon$$

and

$$\left|\langle \nu, T^*f \rangle - \langle \nu, T^{*,k}f \rangle\right| < \epsilon.$$

As convergence in distribution of μ_{n_i} to ν implies that for each k

$$\langle \bar{\mu}_{n_i}, T^{*,k} f \rangle \to \langle \nu, T^{*,k} f \rangle,$$

we now have that $\langle \bar{\mu}_{n_i}, T^* f \rangle \to \langle \nu, T^* f \rangle$ and, consequently, that $\langle \nu, f \rangle = \langle T\nu, f \rangle$. This proves Theorem 1.1. □

Remark. Extension to the case of bounded service times: The main point at which the unboundedness assumption on the service time distributions is invoked is in the last paragraph of Section 4. Note that by order preservation properties of ever-reds and ever-blues, we are guaranteed a minimum density (at least $\epsilon/4$) of blue customers who are within a distance of $2/\epsilon$ of red customers at every stage n. Now, given that the service times are unbounded, this means that with probability $1 - F(2/\epsilon)$ the reds may be held at a stage for at least $2/\epsilon$ time units, causing a minimum density of customer matchings at each stage and generating the desired contradiction.

Obviously, this argument fails when the service times are bounded and modifications are necessary. The key idea is to realize that although red and blue customers are not guaranteed to match at one stage (because of bounded service times), they can be forced to match over several stages. This is done by giving a red customer higher service times than a blue customer behind it over several stages, thus causing the blue to "catch up" with the red and couple with it. Since the blues in question are within $2/\epsilon$, and the service times are i.i.d., the chance of accomplishing this within a certain fixed number, say S, of stages is strictly positive. This, then, is a description (*sans* details) of how one might extend the result of the paper to the case of ·/*GI*/1 queues with bounded service times.

Acknowledgements The authors thank the anonymous referee for a thorough review of the paper which has led to an improvement in the presentation.

Bibliography

1. Anantharam, V. (1993). Uniqueness of stationary ergodic fixed point for a ·/M/k node. *Annals of Applied Probability*, **3**, 1, 154–173.

2. Baccelli, F. and Brémaud, P. (1987). *Palm Probabilities and Stationary Queues*. Lecture Notes in Statistics **41**. Springer, New York.

3. Burke, P.J. (1956). The output of a queueing system. *Operations Research*, 4, 699–704.

4. Chang, C.S. (1994). On the input-output map of a G/G/1 queue. *Journal of Applied Probability*, **31**, 4, 1128–1133.

5. Daley, D.J. and Vere-Jones, D. (1988). *An Introduction to the Theory of Point Processes*. Springer Series in Statistics. Springer, New York.

6. Ekhaus, M. and Gray, L. (1993). A strong law for the motion of interfaces in particle systems. *In preparation*.

7. Liggett, T.M. and Shiga, T. A note on the departure process of an infinite series of ·/M/1 nodes. *Unpublished manuscript*.

8. Loynes, R.M. (1962). The stability of a queue with non-independent inter-arrival and service times. *Proceedings of the Cambridge Philosophical Society*, **58**, 497–520.

9. Mountford, T.S. and Prabhakar, B. (1995). On the weak convergence of departures from an infinite sequence of ·/M/1 queues. *Annals of Applied Probability*, **5**, 1, 121–127.

10. Prabhakar, B. (1995). Uniqueness of fixed points for the ·/GI/1 queueing operator. *In preparation.*

19

The Poisson-independence hypothesis for infinitely-growing fully-connected packet-switching networks

Yu. M. Suhov and D. M. Rose

University of Cambridge

Abstract

We analyse a packet-switching network modelled by a complete graph with nodes $0, 1, \ldots, N$. For each pair of nodes, (i, j), there is a rate-λ Poisson flow $\alpha_{i,j}$ of messages generated at i and destined for j. The numbers of packets in the message (the message lengths) are IID variables with distribution μ. With probability p, a packet is directly transmitted from i to j on link (i, j), and with probability $q_N = (1 - p)/(N - 1)$, on any of the two-link paths joining i and j. The queueing discipline at each node is FCFS, and the end-to-end delay is the time of delivery of the last packet of a message. Under certain conditions on λ, μ and p, we prove that as $N \to \infty$, the probability distribution of the end-to-end delay is deduced from the so-called Poisson-independence hypothesis, which is shown to be valid. The main requirement is that most packets should be routed directly to their destinations, i.e., the network is a small perturbation of the 'ideal' one in which the links act as independent entities.

1 Introduction

We consider a fully-connected packet-switching network with nodes $0, 1, \ldots, N$, and $(N + 1)N$ links, (i, j), $i, j = 0, 1, \ldots, N$, $i \neq j$. Messages of IID lengths L are generated at the nodes of the network, and messages with different (source, destination) pairs are generated as IID Poisson processes, $\alpha_{i,j}$, of rate λ. The distribution of L is μ, which is supported by a finite number of positive integers.

At the time of its generation, a message is divided into unit-length packets. With probability p, a packet directed from i to j chooses the one-link path (in short, 1-path), (i, j), and with probability $q_N = (1 - p)/(N - 1)$ a two-link path (in short, 2-path), (i, k, j), $k \neq i, j$. The *end-to-end delay*, T, is defined as the time from the message generation at the source to the moment of delivery of the last of its packets at the destination.

Although a number of simplifying assumptions (e.g., those relating to symmetry) have been made, the analysis of the finite network under consideration is difficult. This paper deals with the situation when N becomes large $(N \to \infty)$. Under certain conditions on λ, μ and p (see (1.3)–(1.5) below), the packets of a given message sent through 2-paths join asymptotically independent queues which, in turn, are independent of the queue on the 1-path. This is the essence of the *Poisson-independence* hypothesis for the class of networks under consideration. Although there appears to be no standard terminology for it in the literature, the concept of the Poisson-independence hypothesis is popular amongst network practitioners and theoreticians. See, e.g., Dobrushin *et al.* 1990; Graham and Méléard 1993, 1994a, 1994b; Kel'bert and Sukhov 1990; Kelly 1986; Kleinrock 1964 (and the references therein). We focus on the existence of a stationary regime, which implies the need to consider the whole time axis $\mathbb{R}$; this complicates the proofs.

The Poisson-independence hypothesis enables us to express the limiting distribution function (d.f.) $F_T(x)$ for $x \geq 2$ as

$$
\begin{aligned}
F_T(x) \;=\; & \sum_{l \geq 1} \mu(l) \left[(1-p)^l \left[(F_w \circ F_w)(x-2) \right]^l \right. \\
& \left. + \sum_{l^0=1}^{l'} \binom{l}{l^0} p^{l^0} (1-p)^{l-l^0} F_w(x-l^0) \left[(F_w \circ F_w)(x-2) \right]^{l-l^0} \right] \quad (1.1)
\end{aligned}
$$

where F_w is the d.f. of waiting time w in the stationary $M/GI/1$ queue described below, and $F_w \circ F_w$ denotes the d.f. of the self-convolution of the law of w. l^0 indicates the number of packets of the message which have been sent along the 1-path, and $l' = \min(l, \lfloor x \rfloor)$. If $1 \leq x < 2$, $F_T(x) = \mu(1) p F_w(x-1)$, and for $x < 1$, the distribution function of the end-to-end delay is 0, regardless of the value of N.

The aforementioned $M/GI/1$ queue describes the stationary regime at the input port of a fixed link, say (i,j), as $N \to \infty$. It is formed by a Poisson arrival flow, $\alpha_{i,j}^*$, which is the superposition of three independent Poisson subflows, $\alpha_{i,j}^0$, $\beta_{i,j}^*$ and $\gamma_{i,j}^*$, each with its own sequence of IID service times. Symbolically,

$$
\alpha_{i,j}^* = \alpha_{i,j}^0 \vee \beta_{i,j}^* \vee \gamma_{i,j}^*
$$

where $\vee$ denotes the superposition of independent flows.

More precisely,

(1) $\alpha^0 \ (= \alpha_{i,j}^0)$ is the flow of packet sequences which follow the 1-path (i,j). (These sequences shall be called 1-*quasimessages*: their constituent packets must be served without interruption.) The distribution of the length of a 1-quasimessage is $\mu^0(l) = (\sigma^0)^{-1} \sum_{\ell : \ell \geq l} \mu(\ell) \binom{\ell}{l} p^l (1-p)^{\ell-l}$, where $\sigma^0 = 1 - \mathbb{E}[(1-p)^L]$, and the rate of $\alpha_{i,j}^0$ equals

$$
\lambda^0 = \lambda \sigma^0.
$$

(2) $\beta_{i,j}^*$ is the limiting flow of (2,1)-*packets*, i.e., the packets routed from node

i to some destination on a 2-path having (i, j) as its first link. The rate of flow $\beta_{i,j}^*$ equals

$$\nu^* = \lambda(1 - p)\mathbb{E}(L)$$

and the service times are precisely one.

(3) $\gamma_{i,j}^*$ is the limiting flow of $(2, 2)$-*packets*, routed from some source to node j on a 2-path having (i, j) as its second link. Their incorporation into $\gamma_{i,j}^*$ occurs once they are at the input port of (i, j). It is easy to see that $\gamma_{i,j}^*$ has the same distribution as $\beta_{i,j}^*$.

Flow $\beta_{i,j}^*$ is obtained, as $N \to \infty$, from $\beta_{i,j}$, the Poisson flow of $(2, 1)$-*quasimessages*, i.e., sequences of packets directed from i to some destination on a 2-path having (i, j) as its first link. Both the rate of $\beta_{i,j}$ and the distribution of the length of the $(2, 1)$-quasimessage depend on N, but the load created is the same as that of $\beta_{i,j}^*$. As $N \to \infty$, the rate of $\beta_{i,j}$ tends to ν^* and the length distribution to the Dirac measure concentrated at 1 (i.e., the $(2, 1)$-quasimessages are reduced in the limit to unit packets). Flow $\beta_{i,j}^*$ is independent of $\alpha_{i,j}^0$, as $\alpha_{i,j}^0$ and $\beta_{i,j}$ are independent of each other.

The situation is far less trivial with $\gamma_{i,j}^*$, which is the limit of a flow $\gamma_{i,j}$ formed by packets using (i, j) as the second link of their 2-path to destination j. A formal proof of the fact that $\gamma_{i,j}$ converges to a Poisson flow, and becomes asymptotically independent of $\alpha_{i,j}^0$ and $\beta_{i,j}$, is required. This is a necessary step in justifying the Poisson-independence hypothesis.

In other words, the flow $\bar{\alpha}_{i,j}$ forming the queue for link (i, j) and defined as the superposition of $\alpha_{i,j}^0$, $\beta_{i,j}$ and $\gamma_{i,j}$ converges to $\alpha_{i,j}^*$. In addition, the Poisson-independence hypothesis asserts that flows $\bar{\alpha}_{i,j}$ for different links (i, j) become asymptotically independent. The proof of these statements is the subject of the present paper.

The rate of flow $\alpha_{i,j}^*$ is $\lambda^* = \lambda\sigma^*$ where $\sigma^* = 1 - \mathbb{E}[(1 - p)^L] + 2(1 - p)\mathbb{E}(L)$, and the distribution of service times is given by

$$\mu^*(l) = \begin{cases} (\sigma^*)^{-1} \sum\limits_{\ell:\ell \geq l} \mu(\ell)\binom{\ell}{l}p^l(1 - p)^{\ell-l}, & l \geq 2 \\ (\sigma^*)^{-1}\left[\sum\limits_{\ell \geq 1} \mu(\ell)\ell p(1 - p)^{\ell-1} + 2(1 - p)\mathbb{E}(L)\right], & l = 1 \end{cases}$$

so that the workload created equals $\rho^* = \lambda(2 - p)\mathbb{E}(L)$. Hence the subcriticality condition (terminology adopted from Loynes (1962)) in the limiting queue should read

$$\lambda(2 - p)\mathbb{E}(L) < 1 \ \text{ or } \ [2 - 1/(\lambda\mathbb{E}(L))]^+ < p \leq 1 \tag{1.2}$$

where $x^+ = \max(0, x)$. The waiting time d.f., F_w, may be calculated in a standard way (see Kleinrock (1975) or Prabhu (1965)).

We expect eqn (1.1) and other assertions related to the Poisson-independence hypothesis to be valid under the bound (1.2) (which would be of some practical value), but our present methods force us to use a much more restrictive condition determined by the behaviour of *majorising* flows (see below). The main result of the paper is that for each λ, μ such that

$$\lambda \mathbb{E}(L) < 1 \tag{1.3}$$

there exists a $p_0 \in (0, 1)$, more precisely,

$$1 > p_0 > 1/2\,[1 + \lambda \mathbb{E}(L)] > [2 - 1/(\lambda \mathbb{E}(L))]^+, \tag{1.4}$$

which does not depend on N, and is such that eqn (1.1) holds when

$$p_0 \le p \le 1. \tag{1.5}$$

Under conditions (1.3) and (1.5), the network is a small perturbation of the analytically ideal network corresponding to $p = 1$. In the ideal network, all packets choose a 1-path, and the network is decomposed into a collection of isolated queues.

Unfortunately it is difficult to give a simple bound for p_0 which is the outcome of a lengthy construction that involves stochastic majorisation procedures (see Section 3) and the extinction of a certain Galton–Watson process (see Section 4).

2   Statement of the results

A convenient theoretical basis is provided by random marked point processes (RMPPs): in this paper a random flow and an RMPP are synonymous. Various aspects of RMPP theory are considered in Baccelli and Brémaud (1987), Brandt *et al.* (1990) and Franken *et al.* (1982).

A mark in $\alpha_{i,j}$ is a collection $x = (l^0, \{l_k,\ k = 0, 1, \ldots, N,\ k \ne i, j\})$. l^0 (as in Section 1) and l_k are nonnegative integers giving the number of packets that have chosen the 1-path (i, j) (i.e., the length of the 1-quasimessage created from the given message), and the 2-path (i, k, j) (i.e., the length of the corresponding $(2,1)$-quasimessage) respectively. Write $l^1 = \sum_k l_k$, so that $l = l^0 + l^1$. A trajectory X $(= X_{i,j})$ of flow $\alpha_{i,j}$ is a sequence of pairs $\mathbf{x}_s = [t_s; x_s]$, $s \in \mathbb{Z}$, where $\{t_s\} \subset \mathbb{R}$ is a strictly increasing sequence of the times of generation of the messages, and $\{x_s\}$ is the sequence of their marks. For the enumeration in X, we adopt the convention that $t_0(X) < 0 < t_1(X)$. $\mathbf{x}_s(X) = [t_s(X); x_s(X)]$ is called the sth message from X (or, when X is random, from flow $\alpha_{i,j}$). We also write $\mathbf{x} = [t; x] \in X$ and say that $\mathbf{x}$ is a message from X if $\mathbf{x} = \mathbf{x}_s(X)$ for some $s \in \mathbb{Z}$. Similar notation and terminology is used for other random flows we consider in the sequel.

The distribution of $\alpha_{i,j}$ is determined by the conditions that the generation times $t_s(X)$ form a Poisson process of rate λ, and the marks $x_s(X)$ form an IID sequence; the marginal distribution of these marks is defined in a straightforward way.

It is convenient to associate the family $\{\alpha_{i,j}\}$ with the total generation flow $\alpha = \bigvee \alpha_{i,j}$. A trajectory of α is denoted by $\mathbf{X}$ $(= \mathbf{X}^{(N)})$ which is a sequence of pairs $\bar{\mathbf{x}}_s = [t_s; \bar{x}_s]$, $s \in \mathbb{Z}$, where $\{t_s\}$ is again a strictly increasing sequence on $\mathbb{R}$, representing the message generation times, and $\{\bar{x}_s\}$ is a sequence of marks of the form $\bar{x}_s = (i_s, j_s; x_s)$, where i_s and j_s $(i_s \ne j_s)$ are the source and destination respectively, and x_s is as before. α is again a Poisson flow, of rate $\lambda(N+1)N$, with IID marks. The pair (i_s, j_s) has the discrete uniform distribution (subject to the condition $i_s \neq j_s$) and the conditional distribution of $\bar{x}_s$ given that $i_s = i$, $j_s = j$ is $\Lambda_{i,j}$.

Conversely, if we extract the subsequences $\bar{\mathbf{x}}_s^{i,j} = [t_s^{i,j}; \bar{x}_s^{i,j}]$ with $\bar{x}_s^{i,j} = (i, j; x_s)$ from $\mathbf{X}$, then this leads (after appropriate resequencing) to trajectories $X_{i,j}$ of flows $\alpha_{i,j}$.

Our aim is to construct and analyse *extended* flows that describe the stable regime in the network, first on the half-infinite time axis $[\tau_0, \infty)$ (with the system empty at τ_0) and then, by letting $\tau_0 \to -\infty$, on the whole real line. These extended flows are denoted by $\phi^{(\tau_0)}$ and ϕ respectively. A mark in flows $\phi^{(\tau_0)}$ and ϕ is a pair $(\bar{x}, y)$ where $\bar{x} = (i, j; x)$ as before, and y is a vector $(w^0, \{(w_k, \mathbf{w}_k'), k \neq i, j\})$, with components $w^0 \geq 0$, $w_k \geq 0$ and $\mathbf{w}_k' = (w_{k1}', \ldots, w_{kl_k}') \geq \mathbf{0}$, describing the waiting times assigned to the packets of the message. Specifically,

(1) w^0 is the waiting time of the 1-quasimessage,

(2) w_k is the time spent waiting for the link (i, k) by the $(2, 1)$-quasimessage routed on the 2-path (i, k, j), and

(3) $\mathbf{w}_k'$ is the vector of the waiting times in the queue for (k, j) of the collection of $(2, 2)$-packets (into which the $(2, 1)$-quasimessage fragments while being processed on (i, k)).

The waiting times will be defined as functions of trajectory $\mathbf{X}$ of flow α (and τ_0 for the network with initial time τ_0).

To determine the waiting times, we use a system of Lindley-type equations (Lindley 1952) that reflects the FCFS character of the queues in the network. More precisely, fix a trajectory $\mathbf{X}$ and assume that waiting times are assigned to all messages from $\mathbf{X}$ (all messages generated after time τ_0 in the case of the initial condition). Take a message $\bar{\mathbf{x}} = [t; \bar{x}] \in \mathbf{X}$, where $\bar{x} = ((i, j), x)$, with a waiting-time vector $y = (w^0, \{w_k, \mathbf{w}_k'\})$. Then find (1) a *1-predecessor*, (2) a family of $(2, 1)_k$-*predecessors* and (3) a family of $(2, 2)_{k,m}$-*predecessors* of message $\bar{\mathbf{x}}$. The message $\bar{\mathbf{x}}^-$ is a *predecessor* of $\bar{\mathbf{x}}$ if one of its packet sequences is the last to be transmitted on a link before the corresponding packet of $\bar{\mathbf{x}}$ is. The components of y must be related to the waiting times assigned to the predecessors via the relevant Lindley equations. (When the network runs from a finite initial point τ_0, then predecessors of certain types will not exist for some messages. In such cases, the corresponding waiting times should take the value 0.)

For example, consider the mth $(2, 2)$-packet, $x_{k,m}$, of the message $\bar{\mathbf{x}} \in X_{i,j}$ which is to be transmitted on link (k, j). Then the $(2, 2)_{k,m}$-predecessor of $\bar{\mathbf{x}}$ is the message that contains a packet sequence directly preceding $x_{k,m}$ on the link. If such a sequence turns out to be the 1-quasimessage created by a message $\bar{\mathbf{x}}^- = [t^-; \bar{x}^-] \in X_{k,j}$, with $\bar{x}^- = ((k, j), x^-)$ and a waiting-time vector y^-, then the corresponding Lindley equation is

$$w_{km}' = \left[w^{0-} + l^{0-} - [(t + w_k + m) - t^-] \right]^+ \tag{2.1}$$

since $t + w_k + m$ is the time when the mth $(2,2)$-packet of $\bar{\mathbf{x}}$ joins the queue for link (k, j). Similar equations may be specified in the other possible cases: taken together they constitute the *system* of *network equations*.

It is not difficult to check that for each $\tau_0 \in \mathbb{R}$ there always exists a unique extended flow $\phi^{(\tau_0)}$ for which the above conditions, including the system of network equations, are satisfied. Under the appropriate subcriticality condition, a stationary extended flow, $\phi = \lim_{\tau_0 \to -\infty} \phi^{(\tau_0)}$, which again satisfies the system of network equations, exists. We are able to show that ϕ gives the *unique* weak solution (Brandt *et al.* 1990) to this system.

Theorem 2.1 below is related to the asymptotic analysis (as $N \to \infty$) of flow $\phi_{i,j}$ which describes the waiting times assigned to the messages from $\alpha_{i,j}$. $\phi_{i,j}$ is obtained from ϕ by extracting the 'extended' messages $[t_s^{i,j}; \bar{x}_s^{i,j}, y^{i,j}]$ with source i and destination j. The analysis performed in Section 1 suggests that, as $N \to \infty$, each length $(l_k)_s$ associated with the flow $\phi_{i,j}$ assumes the values 0 or 1 (which means that no two packets of the same message choose the same 2-path (i, k, j), i.e., the $(2,1)$-quasimessages are reduced to unit packets). Consequently, the vectors $(\mathbf{w}_k')_s$ will be reduced to (nonnegative) scalars $(w_k')_s$. For the purpose of Theorem 2.1 it is convenient to suppress the indices k for which $l_k \geq 1$ and consider an (unordered) collection of 2-dimensional vectors $\left(w_s^{(1)}, w_s'^{(1)}\right), \ldots, \left(w_s^{(l_s^1)}, w_s'^{(l_s^1)}\right)$. The first and second components correspond respectively to the first and second link waiting times of the packets created by the sth message, without reference to the intermediate node k of the path.

Theorem 2.1 *Let $N \to \infty$ and assume that conditions (1.3)–(1.5) are fulfilled, where p_0 is close enough to 1. Then the limiting distribution of the waiting times w_s^0 (of the 1-quasimessages), $w_s^{(m)}$ and $w_s'^{(m)}$ (the waiting times of the $(2,1)$- and $(2,2)$-packets) in flow $\phi_{i,j}$ is defined by the following properties:*

(1) *The $\{w_s^0\}$ are related by the stochastic equation*

$$w_s^0 = \max\left[w_{s-1}^0 + l_{s-1}^0 + S\left(t_{s-1}, t_s\right) - (t_s - t_{s-1}), \\ \sup_{\tilde{t}: \, t_{s-1} < \tilde{t} < t_s} \left(S(\tilde{t}, t_s) - (t_s - \tilde{t})\right) \right]^+, \qquad (2.2)$$

where $S(t, t')$ denotes the workload that has been contributed during the interval (t, t') by a Poisson flow having intensity $2\lambda(1 - p)\mathbb{E}(L) \ (= 2\nu^*)$, unit service times, and which is independent of t_s and x_s. (This Poisson flow is merely the superposition of $\beta_{i,j}^*$ and $\gamma_{i,j}^*$.)

(2) *Given a trajectory X of flow $\alpha_{i,j}$ and a sequence w_s^0, the waiting times $w_s^{(m)}$ and $w_s'^{(m)}$, $m = 1, \ldots, l_s^1(X)$, $s \in \mathbb{Z}$, correspond with IID variables each having the d.f. F_w (see (1.1)).*

Moreover, for different pairs (i, j), the flows $\phi_{i,j}$ are asymptotically independent.

Corollary Under the conditions of Theorem 2.1, the distribution of the end-to-end delay in any of the flows $\phi_{i,j}$,

$$T_s = \max\left[w_s^0 + l_s^0, \max_{1 \le m \le l_s^1}\left[w_s^{(m)} + w_s'^{(m)} + 2\right]\right],$$

tends to the limit (1.1).

The following theorem deals with flow $\bar{\alpha}_{i,j}$, defined in Section 1 as the superposition of $\alpha_{i,j}^0$, $\beta_{i,j}$ and $\gamma_{i,j}$, that forms the queue for link (i, j). All three flows here are considered as RMPPs with positive integer marks (the lengths of the 1- and (2, 1)-quasimessages in flows $\alpha_{i,j}^0$ and $\beta_{i,j}$ and the unit lengths of the (2, 2)-packets in flow $\gamma_{i,j}$). To define flow $\gamma_{i,j}$, pick the extended messages $[t^{k,j}; \bar{x}_{k,j}, y_{k,j}]$ from flows $\phi_{k,j}$, $k \ne i, j$, with $l_i \ge 1$. Recall that l_i is the length of the (2, 1)-quasimessage travelling along (k, i, j) and w_i is its waiting time for link (k, i). Then $\bar{t}^{k,j}(n) = t^{k,j} + w_i + n$ gives the time when the nth (2, 2)-packet joins the queue for link (i, j), $n = 1, \dots, l_i$ (cf. eqn (2.1)). Flow $\gamma_{i,j}$ is formed by the pairs $[\bar{t}^{k,j}(n); 1]$.

Theorem 2.2 *Under the conditions of Theorem 2.1, flow $\gamma_{i,j}$ converges to Poisson flow $\gamma_{i,j}^*$ and flow $\bar{\alpha}_{i,j}$ to $\alpha_{i,j}^*$ (see Section 1). Furthermore, for different links (i, j), flows $\bar{\alpha}_{i,j}$ are asymptotically independent.*

3 A family of majorising flows

3.1 The majorisation procedure: step one

Throughout this subsection we fix a link of our graph, say $(0, 1)$, and concentrate on constructing majorants for flow $\phi_{0,1}$.

Given a trajectory $Y = \{[t; l]\}$, of a random flow φ (on the whole time axis $\mathbb{R}$ or on an interval $[\tau_0, \infty)$), with marks (lengths) $l > 0$, the trajectory of the *virtual waiting time* (VWT) process $V(\tau)$, $\tau \in \mathbb{R}$, is given by

$$V(\tau) = \left[\sup_{\tilde{\tau} \le \tau} \left(R[\tilde{\tau}, \tau] - (\tau - \tilde{\tau})\right)\right]^+ \tag{3.1}$$

where $R[a^1, a^2]$ stands for the workload created during a (possibly random) time interval $[a^1, a^2]$:

$$R[a^1, a^2] = \sum_{[t;l] \in Y:\ a^1 \le t \le a^2} l. \tag{3.2}$$

When we are concerned with the distribution of $R[a^1, a^2]$, or $V(\tau)$, we refer to the underlying flow by using the notation $R(a^1, a^2; \varphi)$ or $V(\tau; \varphi)$. In particular, we work with processes $V(\tau; \bar{\alpha}_{0,1})$ and quantities $R(a^1, a^2; \bar{\alpha}_{0,1})$ related to the flow $\bar{\alpha}_{0,1}$.

As was mentioned in Section 1, the main difficulty in dealing with the flow $\bar{\alpha}_{0,1}$ (specifically, with $\gamma_{0,1}$) is that it is determined in a complicated way from α. The aim of this subsection is to construct a Poisson flow $\psi_{0,1}^>$ with IID positive integer marks, which creates a dominant VWT in the sense that, for N

large enough, there exists a joint distribution of flows α and $\psi_{0,1}^>$ such that with probability one

$$V(\tau; \psi_{0,1}^>) \geq V(\tau; \bar{\alpha}_{0,1}), \quad \tau \in \mathbb{R}. \tag{3.3}$$

Formally, $\psi_{0,1}^>$ is defined as the superposition $\alpha_{0,1}^0 \vee \beta_{0,1}^> \vee \gamma_{0,1}^>$ of three independent Poisson flows, $\alpha_{0,1}^0$ (the original flow of 1-quasimessages), $\beta_{0,1}^>$ and $\gamma_{0,1}^>$, all with IID marks. $\beta_{0,1}^>$ is supposed to majorise flow $\beta_{0,1}$, and $\gamma_{0,1}^>$, flow $\gamma_{0,1}$.

To simplify the notation, we omit the subscript $_{0,1}$ whenever possible. The rate of flow $\beta^>$ is $\nu^* + \epsilon$ (where $\nu^* = \lambda(1-p)\mathbb{E}(L)$ as before), and the marks take two values, 1 and $\bar{l}$. $\epsilon > 0$ is specified below and $\bar{l}$ is the maximum message length (which is finite due to our assumptions): $\bar{l} = \max\,[l \geq 1 : \mu(l) > 0]$. The mark distribution in $\beta^>$ assigns probability $1 - \epsilon$ to length 1 and ϵ to $\bar{l}$.

Similarly, the rate of flow $\gamma^>$ equals $1 - p$, and the marks take two values, 2 and $\bar{l} + 1$. The mark distribution analogously assigns probability $1 - \epsilon$ to length 2 and ϵ to $\bar{l} + 1$. The value of ϵ is chosen so that the flow $\psi^>$ is subcritical:

$$\lambda p\mathbb{E}(L) + (\nu^* + \epsilon)(1 + \epsilon(\bar{l} - 1)) + (1 - p)(2 + \epsilon(\bar{l} - 1)) < 1, \tag{3.4}$$

which is possible due to (1.3)–(1.5).

Theorem 3.1 *Under the conditions* (1.3)–(1.4) *and* (3.4), *flow* $\psi^>$ *satisfies the majorisation property* (3.3).

Proof We have to guarantee that there exists a joint distribution of flows $\beta^>$, $\gamma^>$ and flow α such that

(1) $\beta^>$ and $\gamma^>$ are independent of each other, and

(2) when N is large enough, with probability one,

$$R\left(a^1, a^2; \beta_{0,1}\right) \leq R\left(a^1, a^2; \beta^>\right), \tag{3.5}$$

and

$$R\left(a^1, a^2; \gamma_{0,1}\right) \leq R\left(a^1, a^2; \gamma^>\right) \tag{3.6}$$

for any bounded interval $[a^1, a^2] \subset \mathbb{R}$.

It is easy to couple flows $\beta^>$ and α. Flow $\beta^>$ is constructed on the basis of the Poisson flow $\beta_{0,1}$: each time a $(2,1)$-quasimessage from $\beta_{0,1}$ appears, it is incorporated into $\beta^>$. Furthermore, if the length of this quasimessage is greater than 1, its counterpart in $\beta^>$ is assigned length $\bar{l}$. It is not difficult to check that bound (3.5) is satisfied when N is large enough for ϵ to exceed the probability that a $(2,1)$-quasimessage consists of more than one packet.

A more elaborate argument is needed to perform the coupling of flow $\gamma^>$ and α. We essentially use the fact that $\gamma_{0,1}$ is formed by the $(2,2)$-packets that have already been processed (as $(2,1)$-packets) on the previous links, $\{(k,0), k \neq 0, 1\}$, of their respective 2-paths. As before, these packets are incorporated into $\gamma_{0,1}$ at the times they join the queue for $(0,1)$, which correspond to the times they are at the output ports of the $\{(k,0)\}$. The subflow of $\gamma_{0,1}$ associated with those packets emerging from the output port of $(k,0)$ will be denoted by $\gamma(k)$. Naturally the joint distribution of the $\gamma(k)$ is determined by the (extended) network flow ϕ.

We consider the output port of link $(k,0)$, $k \neq 0,1$, and focus on the joint distribution of the processes $\{\gamma(k)\}$. Our method for the construction of a majorant, $\gamma^>$, of $\gamma_{0,1}$ is motivated by the following remarks.

(1) Packets from the same 1- or $(2,1)$-quasimessage arrive successively, one time unit apart, at this output port. More generally, it is clear that the arrival times of any two packets at the output port of $(k,0)$ differ by at least one unit. That is, at most one packet appears during any unit interval, $[n, n+1)$, $n \in \mathbb{Z}$.

(2) The first packet of a particular quasimessage to emerge at the output port (during the 'slot' $[n, n+1)$, say) will choose to continue along $(0,1)$ with probability at most q_N, independently of all packets which arrived previously at the output ports of $(k,0)$ as well as $(k^*,0)$ $(k^* \neq 0)$.

(3) The decision by the rth $(r > 1)$ packet from a fixed quasimessage (arriving at the output port of link $(k,0)$) on whether to use $(0,1)$ is dependent on the choices made by the previous $r-1$ packets (which occupy the previous unit slots), especially by the first (in part, because they have the same source and destination nodes). In summary, the fate of a quasimessage is determined by the fate of its first packet.

(4) Packets exiting (different) links $(k_1,0)$ and $(k_2,0)$ independently make their respective decisions to join the queue at the input port of link $(0,1)$.

The development of $\gamma^>$ is based on how the original flows, $\{\gamma(k)\}$, function, but the dependence structure described in item (3) needs to be eliminated when constructing this majorant. To achieve this goal, and the majorising one, we begin as follows.

Consider an integer lattice $\mathbb{Z}$ on the time axis $\mathbb{R}$ and define a discrete-time random process $U = \{U_n, n \in \mathbb{Z}\}$, as the sequence of IID random variables, taking three values, 0, 2 and $\bar{l}+1$, with probabilities $1 - q_N$, $(1-\epsilon)q_N$ and ϵq_N, respectively. Take $N-1$ independent copies, $U^k = \{U_n^k, n \in \mathbb{Z}\}$, of the sequence $\{U_n\}$ labelled by $k \neq 0,1$.

We next construct a joint distribution of the flow α and the array of random variables, $\{U_n^k, k = 2, \ldots, N, n \in \mathbb{Z}\}$, such that, with probability one, for any $n \in \mathbb{Z}$ and k, if a packet appears in flow $\gamma(k)$ (or $\gamma^{(\tau_0)}(k)$) during the unit interval $[n, n+1)$ then either $U_n^k = 2$ or $U_{n-r}^k = \bar{l}+1$ for some $r = 0, 1, \ldots, \bar{l}-1$. This last assertion means that each packet arriving in a flow $\gamma(k)$ is 'overshadowed' (dominated) by a portion of the corresponding process U_n^k.

The construction of the joint distribution proceeds as follows. If the first packet (of some message) appears at the output port of a link $(k,0)$ in a time slot $[n, n+1)$ and decides to join the queue at the start of link $(0,1)$ (i.e., is part of $\gamma(k)$), then we incorporate it into the process U^k by assigning it one of the values $2, \bar{l}+1$. If this packet is followed by other packets of the same message, this value is $\bar{l}+1$, otherwise it is taken to be 2. The reason for our not taking the values $\bar{l}$ and 1 results from the uncertainty as to where within slot $[n, n+1)$ the packet arrives.

The above procedure and remarks guarantee the properties required. In particular, (1) the workload in U^k is no less than that associated with the original process $\gamma(k)$, and (2) the dependence structure between packets in $\gamma(k)$ (see item (3) above) is not present in U^k.

For our purposes, it is preferable to work with majorising flows in continuous time (more especially with Poisson processes), instead of the discrete-time $\gamma(k)$. So we make a refinement to our majorant by constructing $\gamma^>$ as the superposition of $N-1$ IID Poisson flows. The kth such Poisson flow is denoted by $\pi(k)$, and has rate q_N and IID marks taking values 2 and $\bar{l}+1$ with probabilities $1-\epsilon$ and ϵ, respectively. A joint distribution of U^k and $\pi(k)$ exists such that U^k is majorised by the Poisson flow, as required. $\qquad\qquad\square$

3.2   The majorisation procedure: step two

Unfortunately, the majorising flows $\psi^>_{i,j}$ constructed in the preceding subsection are correlated for different links (i,j). We have to take this into account because each message generated in α eventually involves up to $2\bar{l}$ links (two for each packet that has chosen a 2-path). The $(2,1)$- and $(2,2)$-packets are responsible for the interaction in the network. To limit the effect of this mutual dependence, it is necessary to perform another step in the majorisation procedure.

We will require some definitions connected with *busy periods* in a random flow φ with marks $l > 0$. A busy period $B(a;\varphi) = (b^-_a, b^+_a)$ covering a (possibly random) time point a in φ is defined by

$$b^-_a = \sup[b \le a : V(b;\varphi) = 0], \quad b^+_a = \inf[b \ge a : V(b;\varphi) = 0]. \qquad (3.7)$$

The length of $B(a;\varphi)$ is $b^+_a - b^-_a$.

More generally, for a fixed $c > 0$, $b^+_{a,c}$ is defined by

$$b^+_{a,c} = \inf[b \ge a+c : V(\tilde{\tau};\varphi) = 0 \quad \forall \tilde{\tau} \in [b-c,b]]. \qquad (3.8)$$

$B_c(a;\varphi) = (b^-_a, b^+_{a,c})$ shall be called a c-augmented busy period covering a in φ. In practical terms, $B_c(a;\varphi)$ is the busy period still covering a in φ when a service time of c is added at point a. $B(a;\varphi)$ is extended by c, which may mean that it merges with subsequent busy periods.

Finally, given a bounded interval, $I \subset \mathbb{R}$, we denote by $B(I;\varphi)$ the minimal interval containing all $B(a;\varphi)$ and by $B_c(I;\varphi)$ the minimal interval containing all $B_c(a;\varphi)$, $a \in I$. Interval $B_c(I;\varphi)$ is called the c-augmented busy period covering I.

Now again fix a link, $(0,1)$, and omit subscript $_{0,1}$ when it does not cause any ambiguity. Consider a flow β^+ which is the superposition of $4\bar{l}$ independent copies of flow $\beta^>$ (see the preceding subsection). In addition, let γ^+ be the superposition of $2\bar{l}$ independent copies of flow $\gamma^>$. Finally, take $2\bar{l}$ independent copies, $\alpha^0(1), \ldots, \alpha^0(2\bar{l})$, of flow α^0 ($= \alpha^0_{0,1}$), independent of β^+ and γ^+. For each $n = 1, \ldots, 2\bar{l}$ consider a flow $\psi^+(n)$ defined as the superposition $\alpha^0(n) \vee \beta^+ \vee \gamma^+$. Value p should be close enough to 1 and value ϵ small enough for flow $\psi^+(n)$ to be subcritical:

$$\lambda p \mathbb{E}(L) + 4\bar{l}(\nu^* + \epsilon)(1 + \epsilon(\bar{l}-1)) + 2\bar{l}(1-p)(2 + \epsilon(\bar{l}-1)) < 1. \qquad (3.9)$$

Next divide the family $\{\psi^+(1),\ldots,\psi^+(2\bar l)\}$ in two and take

(1) the $\bar l$-augmented busy periods $B_{\bar l}(0;\psi^+(m))$ covering time zero in flows $\psi^+(m)$, $m = 1,\ldots,\bar l$, and

(2) the $\bar l$-augmented busy periods

$$\bar B_{\bar l}(m) = B_{\bar l}\left(B_{\bar l}(0;\psi^+(m));\ \psi^+(m+\bar l)\right)$$

covering $B_{\bar l}(0;\psi^+(m))$ in flows $\psi^+(m+\bar l)$, $m = 1,\ldots,\bar l$.

Let **B** denote the shortest (random) interval containing $\bar B_{\bar l}(1),\ldots,\bar B_{\bar l}(\bar l)$, and let **L** be the length of **B**.

Suppose conditions (1.3)–(1.5) hold, p_0 is close to 1, $\epsilon > 0$ is sufficiently small for (3.9) to be fulfilled and N is large enough. Then the gist of the second step of the majorisation procedure (which is strengthened and stated more accurately in Theorem 3.2 below) is that for each message $\mathbf{x}_s$ from flow $\alpha_{0,1}$, there exists a joint distribution of random variable **L** and flow α such that the following two properties hold.

(1) The busy periods in which the 1- and $(2,1)$-quasimessages, and the $(2,2)$-packets, created by $\mathbf{x}_s$ fall, are within **B**, and thus have respective lengths at most **L**.

(2) In the sequel, we will often use the term discordant element (d.e.) to refer to a $(2,1)$-quasimessage or $(2,2)$-packet. Packets using a 2-path are responsible for the interaction between flows in the network: without them, we would have the analytically ideal situation corresponding to $p = 1$. Consider the number of d.e.s that appeared in the queues for the links of their paths during the busy periods in the corresponding flows $\bar\alpha_{.,.}$ where the packets of message $\mathbf{x}_s$ fall as they progress through the network. Then that number does not exceed the number of points in $\beta^+ \vee \gamma^+$ on an interval of length **L**. That interval may be (and is) chosen in such a way that it contains all the points from the original flows, and usually several others as well, because it is developed from the majorisation construction.

The number, $4\bar l = 2 \times 2\bar l$, of the summand flows $\beta^>$ in the superposition leading to β^+ reflects the bidirectional character of the influence produced by flows $\beta_{i,j}$ in our network. Specifically, $2\bar l$ flows are needed for majorising the $(2,1)$-quasimessages that are generated in the flows $\alpha_{j,k}$, where (j,k) is one of the links engaged by the packets of $\mathbf{x}_s$ (i.e., $(l_k)_s \geq 1$). The other $2\bar l$ copies of the flow $\beta^>$ are used for majorising the $(2,1)$-quasimessages from flows $\beta_{j,k}$.

The above procedure for determining **L** can be repeated for each d.e. which appeared during busy periods $B_{\bar l}(0;\psi^>(1)),\ldots,B_{\bar l}(0;\psi^>(\bar l))$ and $\bar B(1),\ldots,\bar B(\bar l)$. More precisely, each appearance of a d.e. in one of these intervals means that there is an additional interactive effect. Sometimes it is between links engaged at a previous level (see below) and sometimes it involves new links. (The total number of links associated with a d.e. is always at most $2\bar l$.) Pictorially, this is caused by the fact that a d.e. is either (1) a $(2,1)$-quasimessage, in which case other packets of the same parent message join the queues for other links at the

same time, or (2) a (2,2)-packet, in which case there is the additional factor of
information from the previous link to be considered. D.e.s that appeared in the
corresponding queues during busy periods have to be taken into account. This
procedure has to be repeated several times to guarantee that we account for all
the various possibilities.

It is convenient to classify the d.e.s by 'levels': a d.e. of the first level is one
that appears in the busy periods associated with $\mathbf{x}_s$. A d.e. of the second level is
one that appeared initially in the busy periods associated with a d.e. of the first
level, and so on. We are interested in proving that the process of creating new
d.e.s eventually stops, and the stopping-time distribution is controlled uniformly
in N. This is the ultimate goal of the majorisation procedure.

To this end we can define a random process on a (rooted) *Galton–Watson
tree* (i.e., the graphical representation of a stochastic Galton–Watson branching
process) by

(1) assigning to each vertex v of the tree a pair of random variables $(\mathbf{L}_v, \mathbf{D}_v)$
constructed from an independent copy of the collection $\{\psi^+(1),\dots,\psi^+(2\bar{l})\}$;
here, $\mathbf{L}_v$ is distributed as $\mathbf{L}$, and $\mathbf{D}_v$, as the number of points in the corresponding copy of flow $\beta^+ \vee \gamma^+$ on the time interval $\mathbf{B}_v$ of length $\mathbf{L}_v$,
and

(2) generating precisely $\mathbf{D}_v$ new branches from vertex v.

We denote the process $\{(\mathbf{L}_v, \mathbf{D}_v)\}$ on the Galton–Watson tree by Ξ. The
vertices of the tree are classified in a standard way by their 'levels' indicating
the number of generations needed to create a given vertex. Ξ provides a majorant
for the process of d.e.s constructed for message $\mathbf{x}_s$ from flow $\alpha_{0,1}$, as stated in
Theorem 3.2.

Theorem 3.2 *Under conditions* (1.3)–(1.5) *and* (3.9), *for each* (i,j) *and* s *there
exists a joint distribution of flow* α *and the process on the Galton–Watson tree
Ξ such that, for N large enough, the following holds. With probability one, for
each $K \geq 1$ and each d.e. of K-th level descending from message $\mathbf{x}_s$ from flow
$\alpha_{i,j}$, there exists a vertex v of the Galton–Watson tree of level K with the three
properties given below.*

(1) *The lengths of the busy periods in which any of the 1- and (2,1)-quasi-
messages, and the (2,2)-packets, created by the parent message of this
packet sequence fall, are at most $\mathbf{L}_v$.*

(2) *The number of d.e.s of the $(K+1)$th generation that are adjacent to this
d.e. is no more than $\mathbf{D}_v$.*

(3) *Different d.e.s correspond with different vertices of the Galton–Watson tree.
Moreover, the above majorisation procedure may be performed jointly for
any finite number of messages $\mathbf{x}_s$ from $\alpha_{i,j}$.*

We omit the proof of Theorem 3.2 which is based on the same principles
as before, in conjunction with an argument used in Karpelevich *et al.* (1994)
(Theorem 1 and its proof), to avoid lengthy technical constructions. Theorem 3.2

allows us to control the influence of the past and future on the waiting-time vector of any particular message. The related analysis of truncated waiting times is the subject of the next section.

4   The proof of Theorem 2.1 and Theorem 2.2: a truncation procedure

The proofs of Theorems 2.1 and 2.2 are based on properties of the majorising flows analysed in the previous section. We shall discuss the proof of Theorem 2.1 only; the proof of Theorem 2.2 follows a similar line. The proof of assertion (1) is essentially reduced to proving that the workloads $S(\,\cdot\,,\,\cdot\,)$ in (2.2) derive from a rate $2\nu^*$ Poisson flow $(\beta^*_{i,j} \vee \gamma^*_{i,j})$ which is independent of $\{[t_s; x_s]\}$, a sample of flow $\alpha_{i,j}$. In other words, firstly we have to check that the joint distribution of flows $\alpha^0_{i,j}$, $\beta_{i,j}$ and $\gamma_{i,j}$ converges to the product of the marginal distributions of $\alpha^0_{i,j}$, $\beta^*_{i,j}$ and $\gamma^*_{i,j}$. Secondly, we need to verify that the convergence is strong enough to guarantee that the waiting times w^0_s assigned, in flow $\bar{\alpha}_{i,j}$, to the messages $[t_s; x_s]$ from $\alpha_{i,j}$ (or, equivalently, to the 1-quasimessages $[t_s; l^0_s]$ from flow $\alpha^0_{i,j}$) have the limiting distribution described in Theorem 2.1.

It is again convenient to set $i = 0$, $j = 1$ and omit the subscript $_{0,1}$ whenever possible. We will use the majorisation procedures discussed in the previous section in conjunction with a *truncation* procedure. The latter means that the VWT process $V(\tau; \bar{\alpha})$ is replaced by

$$V^{(Q)}(\tau; \bar{\alpha}) = \left[\sup_{\tau - Q \leq \tilde{\tau} \leq \tau} [R(\tilde{\tau}, \tau; \bar{\alpha}) - (\tau - \tilde{\tau})] \right]^+ . \qquad (4.1)$$

(In fact, all VWT processes $V(\tau; \bar{\alpha}._{,\cdot})$ in the network are to be replaced by their truncated counterparts, $V^{(Q)}(\tau; \bar{\alpha}._{,\cdot})$, defined by similar formulae.)

Clearly, $V^{(Q)}(\tau; \bar{\alpha}) \leq V(\tau; \bar{\alpha})$ for any $\tau \in \mathbb{R}$ and $Q > 0$. The values $V^{(Q)}(t_s; \bar{\alpha})$ calculated at the times of generation of the messages from flow $\alpha_{0,1}$ give 'truncated' waiting times, $w^{0(Q)}_s$.

Our argument is based on the following two theorems. (p_0 is determined in part by the condition for the extinction of a Galton–Watson process: see Harris (1989).)

Theorem 4.1 *Suppose the conditions* (1.3)–(1.5), *with p_0 close enough to 1, hold. Then there exists N_0 (depending on λ, μ and p_0) for any $s \in \mathbb{Z}$ and $s' = 1, 2, \ldots$ such that the probability of the following event tends to 0 as $Q \to \infty$, uniformly in $N \geq N_0$: in flow $\bar{\alpha}$, the waiting time w^0_s and the truncated waiting time $w^{0(Q)}_{\tilde{s}}$ are different for at least one $\tilde{s}$ from $\{s, s+1, \ldots, s+s'\}$.*

Theorem 4.2 *Assume again that the conditions* (1.3)–(1.5) *are satisfied, and p_0 is close enough to 1. Then for any $Q > 0$, $s \in \mathbb{Z}$ and $s' = 1, 2, \ldots$ the joint distribution of the truncated waiting times $w^{0(Q)}_{\tilde{s}}$, $\tilde{s} = s, s+1, \ldots, s+s'$, converges as $N \to \infty$. The limiting random variables are described in the same fashion as in statement* (1) *of Theorem 2.1, with formula* (2.2) *replaced by*

$$w_s^{0(Q)} = \max\left[\left(w_{s-1}^{0(Q)} + l_{s-1}^0 + S(t_{s-1} \wedge (t_s - Q), t_s) - Q\right)\mathbf{1}_{(t_s - t_{s-1} \le Q)},\right.$$

$$\left.\sup_{\tilde{t}:\, t_{s-1} \wedge (t_s - Q) < \tilde{t} < t_s} (S(\tilde{t}, t_s) - (t_s - \tilde{t}))\right]^+.$$

$$(4.2)$$

Here $r \wedge r'$ stands for the minimum of r and r'. Assertion (1) of Theorem 2.1 follows immediately from Theorems 4.1 and 4.2.

The proof of Theorem 4.1 is relatively simple and requires the step one majorisation construction only. In fact, for N large enough the processes $V(\tau; \bar{\alpha})$ and $V^{(Q)}(\tau; \bar{\alpha})$ are majorised by $V(\tau; \psi^>)$. The probability that in flow $\bar{\alpha}$ $w_{\tilde{s}}^0 > w_{\tilde{s}}^{0(Q)}$ for at least one $\tilde{s} \in \{s, s+1, \ldots, s+s'\}$ is less than or equal to the probability of the following event: in flow $\psi^>$ (coupled with α, as described in the proof of Theorem 3.1) process $V(\tau; \psi^>)$ does not assume the value 0 in the time interval $(t_{\tilde{s}} - Q, t_{\tilde{s}})$ for at least one $\tilde{s}$. The last probability tends to 0 as $Q \to \infty$.

The proof of Theorem 4.2 is more elaborate and requires both steps of the majorisation construction. The first observation is that we need to control flows β and γ only on the interval $(t_s - Q, t_{s+s'})$. By using step one, we can assess how many quasimessages and packets appear in flows β and γ respectively on this time interval. More particularly, the probability that the number of $(2, 1)$-quasimessages that appear in β plus the number of $(2, 2)$-packets that appear in γ during interval $(t_s - Q, t_{s+s'})$ exceeds M tends to 0 as $M \to \infty$ uniformly in $N \ge N_0$.

This allows us to focus on a fixed number of packet sequences which appear in flows β and γ after time $t_s - Q$. As was noted, such a packet sequence may be a $(2, 1)$-quasimessage $\bar{x} = [\bar{t}; \bar{l}]$ from β, in which case it comes from a Poisson flow $\alpha_{0,k}$, $k \ne 0, 1$. In view of Theorem 3.2, the number of flows $\bar{\alpha}_{i,j}$ which are engaged by the 1- and $(2, 1)$-quasimessages and $(2, 2)$-packets that belong to the same parent message is bounded in probability uniformly in $N \ge N_0$. Furthermore, the length of the time interval during which these flows need to be analysed is also bounded in probability, uniformly in $N \ge N_0$. If, on the other hand, such a packet sequence is a $(2, 2)$-packet, $[\bar{t}'; 1]$, then it was generated in a flow $\alpha_{k,1}$, $k \ne 0, 1$. Again by Theorem 3.2, the number of flows $\bar{\alpha}_{i,j}$ engaged by the 1- and $(2, 1)$-quasimessages and $(2, 2)$-packets belonging to the same parent message is bounded in probability uniformly in $N \ge N_0$, as is the duration of the time interval where these flows are to be examined. We conclude from all of this that the total number of exogenous flows $\alpha_{i,j}$ engaged by all packet sequences under investigation (appearing in flows β and γ after $t_s - Q$) is also bounded in probability uniformly in $N \ge N_0$. More precisely, the probability of the following event tends to 0 as $n_1, Q_1 \to \infty$, uniformly in $N \ge N_0$: the times, $\bar{t}$ and $\bar{t}'$, of arrival of the packet sequences under investigation in their respective flows β and γ, are determined by a number of flows $\bar{\alpha}_{i,j}$ greater than n_1 and on a time interval containing $(t_s - Q - Q_1, t_{s+s'} + Q_1)$.

The next (and final) observation is that, as $N \to \infty$, the probability that the exogenous flows engaged in the process of determining the above times $\bar{t}$ and $\bar{t}'$ will interact with one other on time interval $(t_s - Q - Q_1, t_{s+s'} + Q_1)$ tends to 0 as $N \to \infty$. This is a direct consequence of our assumptions about the way the packets of the messages choose their paths. This completes the proof of Theorem 4.2.

To prove assertion (2) (and thereby complete the proof of Theorem 2.1), we use an argument which is a modification of the preceding one. Namely, given $s \in \mathbb{Z}$ and $s' = 1, 2, \ldots$ we want to trace the progress of all $(2, 1)$-quasimessages and $(2, 2)$-packets that belong to the messages $[t_{\tilde{s}}; x_{\tilde{s}}]$, $\tilde{s} = s, s + 1, \ldots, s + s'$. It is not difficult to formulate appropriate analogues of Theorems 4.1 and 4.2, which are then proved along similar lines. Note that the condition that p_0 is close to 1 plays a crucial role throughout the whole argument, beginning with the majorisation construction: it is this condition that ensures that dimensions of interest (the number of exogenous flows, and the duration of the time interval on which they have to be analysed) are bounded in probability uniformly in N for $N \geq N_0$. A relaxation of this condition remains a challenging task for future publications, but the simulation studies already undertaken by one of the coauthors suggest it is possible.

Acknowledgements The authors thank Dr N.D. Vvedenskaya for stimulating discussions that inspired this work, and the referee for the very helpful remarks made.

Bibliography

1. Baccelli, F. and Brémaud, P. (1987). *Palm Probabilities and Stationary Queues*. Springer-Verlag, Berlin.

2. Brandt, A., Franken, P., and Lisek, B. (1990). *Stationary Stochastic Models*. Wiley, Chichester.

3. Dobrushin, R. L., Kelbert, M. Ya., Rybko, A. N., and Suhov, Yu. M. (1990). Qualitative methods of queueing network theory. In R. L. Dobrushin, V. I. Kryukov, and A. L. Toom, editors, *Stochastic cellular systems: ergodicity, memory, morphogenesis*, pages 183–224. Manchester University Press, Manchester.

4. Franken, P., König, D., Arndt, U., and Schmidt, V. (1982). *Queues and Point Processes*. Wiley, Chichester.

5. Graham, C. and Méléard, S. (1993). Propagation of chaos for a fully connected loss network with alternate routing. *Stochastic Processes and their Applications*, **44**(1), 159–180.

6. Graham, C. and Méléard, S. (1994a). Fluctuations for a fully connected loss network with alternate routing. *Stochastic Processes and their Applications*, **53**(1), 97–115.

7. Graham, C. and Méléard, S. (1994b). Chaos hypothesis for a system interacting through shared resources. *Probability Theory and Related Fields*,

100(2), 157–174.

8. Harris, T.E. (1989). *The Theory of Branching Processes*. Dover, New York.

9. Karpelevich, F. I., Kelbert, M. Ya., and Suhov, Yu. M. (1994). Higher-order Lindley equations. *Stochastic Processes and their Applications*, **53**(1), 65–96.

10. Kel'bert, M. Ya. and Sukhov, Yu. M. (1990). Mathematical theory of queuing networks. *Journal of Soviet Mathematics*, **50**(3), 1527–1600.

11. Kelly, F. P. (1986). Blocking probabilities in large circuit-switched networks. *Advances in Applied Probability*, **18**(2), 473–505.

12. Kleinrock, L. (1964). *Communication Nets: Stochastic Message Flow and Delay*. McGraw-Hill, New York.

13. Kleinrock, L. (1975). *Queueing Systems*, volume I: *Theory*. Wiley, New York.

14. Lindley, D. V. (1952). The theory of queues with a single server. *Proceedings of the Cambridge Philosophical Society*, **48**(2), 277–289.

15. Loynes, R. M. (1962). The stability of a queue with non-independent inter-arrival and service times. *Proceedings of the Cambridge Philosophical Society*, **58**(3), 497–520.

16. Prabhu, N. U. (1965). *Queues and Inventories*. Wiley, New York.

20

A bibliographical guide to self-similar traffic and performance modeling for modern high-speed networks

Walter Willinger

Bellcore

Murad S. Taqqu

Boston University

Ashok Erramilli

Bellcore

Abstract

This paper provides a bibliographical guide to researchers and traffic engineers who are interested in self-similar traffic modeling and analysis. It lists some of the most recent network traffic studies and includes surveys and research papers in the areas of data analysis, statistical inference, mathematical modeling, queueing and performance analysis. It also contains references to other areas of applications (e.g., hydrology, economics, geophysics, biology and biophysics) where similar developments have taken place and where numerous results have been obtained that can often be directly applied in the network traffic context. Heavy-tailed distributions, their relation to self-similar modeling, and corresponding estimation techniques are also covered in this guide.

A fundamental feature of *self-similar* or *fractal* phenomena is that they encompass a wide range of time scales. In the teletraffic literature, the notion of *burstiness* is often used in this context. Mathematical models that attempt to capture and describe self-similar, fractal or bursty phenomena in a parsimonious manner include certain self-similar stochastic processes and appropriately chosen dynamical systems. A common characteristic of these models is that their space-time dynamics is governed *parsimoniously* by *power-law* distribution functions (the "Noah Effect") and *hyperbolically* decaying autocorrelations (the "Joseph Effect"). In sharp contrast, traditional approaches to modeling fractal phenom-

This work was partially supported by the NSF Grants DMS-9404093 and NCR-9404931 at Boston University.

ena typically rely on *highly parameterized* multilevel hierarchies of conventional models which, in turn, are characterized by distribution and autocorrelation functions that decay *exponentially fast.*

Although bursty or fractal phenomena have been observed in virtually all branches of science and engineering, and fractal models have been applied with some success in areas such as hydrology, financial economics and biophysics, they are new to teletraffic theory and represent a recent addition to the already large class of alternative models for describing traffic in packet switched networks. While most applications of fractal models in science and engineering have been based on empirical findings, they have almost exclusively focused on the models' powerful descriptive capabilities; their engineering implications and analyses have been largely ignored, mainly because fractal models are generally viewed to be very difficult to analyze. In contrast, the success of fractal models in teletraffic theory will only partly depend on how well they describe actual network traffic, but will also depend to a large degree on the ability to use these models in network analysis and control. To this end, this bibliographical guide brings together many of the references that (i) report on real-time traffic measurements from working networks and emphasize the importance of data analytic, statistical inference and mathematical modeling methods for modern traffic and performance modeling, (ii) demonstrate how traditional modeling approaches, applied with great success in conventional telephony (i.e., circuit switching), have dealt with the increasingly bursty behavior of traffic resulting from modern packet communications, and (iii) illustrate the emergence and impact of fractal processes in modern high-speed network traffic modeling and performance analysis. Thus, this guide lists some of the most recent network traffic studies, combines the most relevant survey and research papers in the areas of data analysis, statistical inference and mathematical modeling, and contains references to other areas of applications (i.e., hydrology, economics and biophysics) where similar developments have taken place. Special emphasis is given to recent works that take fractal models beyond their descriptive phase and pursue genuine engineering applications in the area of modern high-speed network design, management, and control; e.g., queueing models, performance analysis, and control theory. While we do not claim that this is a comprehensive bibliography (e.g., note that we did not attempt to adequately cover the traditional teletraffic literature, and including the very latest developments in this rapidly growing field is nearly impossible), we hope that it will serve as a useful, up-to-date reference for researchers and traffic engineers who are interested in self-similar traffic modeling and analysis.

Historically, traffic modeling has its origins in conventional telephony, and has been based almost exclusively on *Poisson* (or, more generally, *Markovian*) assumptions about traffic arrival patterns and on *exponential* assumptions about resource holding requirements (e.g., [144, 209]). However, the emergence of modern high-speed packet networks combines drastically new and different transmission and switching technologies with dramatically heterogeneous mixtures of services and applications. As a result, packet traffic is generally expected to be more complex or bursty than voice traffic, simply because it is spanning vastly different time scales, from microseconds to seconds and minutes. Traditional traffic modeling has responded to these developments with a steady supply of novel and increasingly sophisticated stochastic models: fluid flow models [9], Markov-modulated Poisson processes [180], different variations of packet train models [96, 210, 349, 350], the versatile Markovian arrival processes (MAPs) [316, 317], batched Markovian arrival process models (BMAPs) [277], TES (Transform–Expand–Sample) models [305]. While the development of these and other models has been mainly driven by the desire of maintaining analytic tractability of related queueing and performance problems, the resulting models are almost never judged by how well they fit actual traffic data in a statistical sense [12] (for a critical discussion, see [327, 348, 413]).

While the availability of actual traffic measurements from working packet networks has been a serious problem in the past (see [329, 330], where it is noted that between 1966 and 1987, several thousand papers on queueing problems have been published, but only about 50 on traffic measurement results), more recently, enormous volumes of traffic data from working networks have been collected and made available to researchers: CCSN/SS7 [106, 107], ISDN [122, 304], Ethernet LANs [173–175, 255–260, 293, 378, 409–412], WANs and NSFNet [41, 68, 69, 88, 181, 220, 314, 331–333, 335], and VBR traffic [26, 145, 146, 186, 199]. Other traffic measurement studies we are aware of include [346] (Ethernet traffic to a file server), [5] (FASTPAC, an Australian high-speed data network), [59] (DQDB MAN environment), and [78] (World-Wide-Web traffic). Some of these data consist of high-resolution traffic measurements over hours and days/weeks (e.g., [26, 41, 106, 146, 174, 186, 220, 255, 256, 260, 304, 331, 332, 346]), others provide information on coarser time scales over time periods ranging from weeks to months/years (e.g., [69, 333]). The former are typically used for traffic characterization purposes, and the latter yield insight into long-term growth trends and network utilizations. Extensive recent statistical analyses of high time-resolution traffic measurements reported in [26, 107, 122, 146, 150, 199, 220, 256, 331, 410, 411] have provided convincing evidence that actual traffic data from working packet networks are consistent with statistical *self-similarity* or *fractal* characteristics. Moreover, these empirically observed features often distinguish clearly between traffic generated by traditional models and measured data [409, 410]. The main reason for this clear distinction is a subtle difference in the underlying dependence structure; while traditional packet traffic models are *short-range dependent* (i.e., have exponentially decaying autocorrelations), measured packet traffic data are consistent with *long-range dependence* (i.e., hyperbolically decaying autocorrelations).

The probability theory of self-similarity and long-range dependence is discussed in [22, 24, 77, 171, 177, 287, 389, 390, 402]. The books [130, 178, 337, 338] also contain large sections on self-similar processes, and extensive bibliographies can be found in [14, 22, 24, 287, 374, 389]. Self-similar stochastic processes were introduced by Kolmogorov [239] in a theoretical context and brought to the attention of probabilists and statisticians by Mandelbrot and his co-workers

[287–292]. They have been used in hydrology [200–202, 214, 254, 302, 310–312], geophysics [40, 322], biophysics [143, 267–269] and biology [339, 340]. An area of application where self-similarity and long-range dependence continue to play a significant role and where many results of practical relevance for traffic engineering have been discovered is economics or, more precisely, financial economics [15, 17, 31, 50, 51, 54, 75, 83, 167, 168, 273, 284, 297, 342, 383]. The paper [14] provides an overview. For an early application of the self-similarity concept and related topics to communications systems, see the seminal paper by Mandelbrot [283]. Enlightening philosophical discussions centering around the issues of traditional mathematical modeling (based on Markov processes) versus unconventional fractal modeling (based on concepts, such as self-similarity or long-range dependence), as well as technical issues related to the problem of stationarity and long-range dependence versus non-stationarity, can be found in [103, 104, 205, 219, 246, 266, 303]. Methods for dealing with non-stationarity are developed in [18, 343, 399].

From a modeling viewpoint, the two major families of self-similar time series models are *fractional Gaussian noises* (i.e., the increment processes of *fractional Brownian motion*) [24, 288, 289, 374] and *fractional ARIMA processes* [166, 195, 196], a generalization of the popular ARIMA time series models [33, 37]. Techniques for identifying fractional ARIMA models (also called FARIMA or ARFIMA) are discussed in [23, 310]. Other stochastic approaches to modeling self-similar features are considered in [28, 156, 160, 274, 275, 369, 371] (based on shot-noise processes), [373] (linear models with long-range dependence), [30, 261, 262, 276, 284, 411, 412] (renewal reward processes and their superposition), [116, 283, 401] (renewal processes or "zero-rate" processes), [165] (aggregation of simple short-range-dependent models), and [135, 294, 414, 416] (wavelet analysis). Further models are considered in [16, 19, 176, 313, 379, 386, 403, 417]. A radically different approach to modeling self-similar phenomena relies on ideas from the theories of chaos and fractals [73, 97, 118–120, 124, 125, 171, 250, 287, 337, 344, 345, 377]; for a general discussion on chaos, probability and statistics, see [29, 46, 47].

An overview of statistical inference methods for self-similar models and random processes with long-range dependence can be found in [22, 24], the papers [392–394] listing additional techniques. More specifically, R/S analysis is discussed in [18, 24, 26, 28, 130, 200, 258, 272, 273, 286, 288, 290–292, 302, 310, 394] (see also [10, 131]), variance–time analysis in [24, 26, 77, 258, 310, 331, 394, 399] and for spectral domain methods using periodograms, see [24, 26, 48, 84, 140, 149, 157, 159, 183, 203, 204, 206, 249, 253, 353, 357–366, 393, 407, 418].

Examples of new statistical techniques in this area include [3, 7, 20, 21, 25, 27, 52, 53, 57, 58, 80–82, 86, 98, 99, 154, 155, 189, 190, 197, 247, 258, 381, 382, 410, 415]. For a practical evaluation of the different techniques see [392–394]. The paper [76] provides a general overview on the statistical analysis of time series, and references [123, 301, 355] comment on some of the shortcomings of traditional time series analysis in the presence of large sets of traffic measurements. The problem of estimating a linear or polynomial regression when the errors have long-range dependence is considered in [85, 154, 241, 370, 418, 419]. Prediction problems in the context of long-range dependence are addressed in [24, 169, 336, 352].

The theoretical background behind many of these statistical tools is based on central and non-central limit theorems for random sequences with long-range dependence [34, 100, 139, 141, 142, 152, 153, 158, 159, 161, 162, 164, 182, 188, 191–193, 280, 387, 388, 395–397]. The proofs require an understanding of the structure of moments of non-linear functions of Gaussian random variables and linear processes [34, 94, 158, 182, 280, 385, 398]. Some of the results have been extended to random fields, that is, to processes where the "time" parameter is viewed as a "space" parameter and is multidimensional [11, 148, 179, 185, 208, 278, 279, 341, 375].

Besides the statistical and practical aspects of self-similar or fractal models, there is the ever-present desire for a physical or phenomenological "explanation" for the fractal nature of empirically observed data. For recent work on this topic in the context of high-speed network traffic modeling and how it relates to the *infinite variance syndrome* or *heavy-tailed behavior* (the "Noah Effect") of individual mechanisms that are responsible for the self-similarity property at the aggregation level, see [107, 220, 248, 258, 326, 331, 409–412] and related earlier results reported in [77, 261, 284]. Explaining and validating (with actual data) self-similarity on physical grounds in a network context result in (i) less resistance to non-traditional modeling approaches [77, 103], (ii) new insight into the essential characteristics of modern high-speed network traffic, and (iii) novel approaches for dealing with problems related to network traffic management and control. In this context, statistical methods for dealing with heavy-tailed phenomena and appropriate modeling techniques are of crucial importance. Moreover, in terms of modeling, the insight gained into the relationship between the "Noah Effect" exhibited by the individual mechanisms and the "Joseph Effect" observed at the aggregation level provides convincing evidence in favor of applying *the principle of parsimony* or *Ockham's Razor* [117, 147, 212, 411–413]; see [32, 107] for a particular example involving call holding time modeling for ordinary telephony.

The relevance of heavy-tailed modeling for teletraffic data is also the subject of the recent survey paper [355]. Empirically, the heavy-tailed counterpart of the Gaussian distribution is the stable distribution [126, 128, 132, 420]. While a Gaussian distribution is always symmetric around the mean or median, a stable distribution can be either symmetric or skewed. It has three parameters, α, β and μ; α characterizes the heaviness of the tail, β the skewness and μ the drift. The stable distribution is symmetric around μ when $\beta = 0$ and is maximally skewed to the right when $\beta = 1$. Numerical tables of stable distributions can be found in [38, 108, 194, 296, 300, 324, 325, 374]. The books [211, 374] provide a systematic treatment of stable time series and processes. There are infinite variance counterparts to fractional Gaussian noise [372–374], FARIMA [232, 233, 236, 374] but also pulse-based models [60–62]. Infinite variance models are also of interest in economics and finance [126, 127, 281, 282, 297, 307–309, 347, 371]. For random fields with infinite variance, see [225–227, 231].

The covariance, which is used to describe the dependence, is not defined when the variance is infinite. Alternatives include the covariation which enters in formulas for regression [43, 45, 63–67, 136, 137, 374] and the codifference which also characterizes the "ergodic" properties of the time series [13, 170, 211, 230, 231, 237, 321, 367, 368, 374].

Estimation of the exponent α characterizing the heaviness of the distribution is of central importance and is one of the main themes in [355]. One way of estimating α is to use regression in a log–log plot [87, 91, 112, 187, 244, 298, 354, 380]. Other methods are considered by [39, 109–111, 129, 242, 243, 295, 299, 315, 355]. Note that even though the variance is infinite one can still use spectral methods to estimate the unknown parameters in a model [92, 221–223, 228, 238]. In particular, the Whittle-type estimator is still applicable [224, 234, 235, 238, 306]. One can also use M-estimators [90], in particular estimators based on least absolute deviation [89], and also the bootstrap [93]. For prediction, see [42, 44, 70, 71, 229].

While from a statistical viewpoint the distinction between traditional traffic models and measured network traffic is significant and intriguing, there is also mounting evidence that the empirically observed fractal features of actual traffic (in particular, long-range dependence and heavy-tailed distributions) have practical implications for a wide range of network design and engineering problems. Traditional Markovian (or more general, short-range-dependent) input streams to queues are known to impact queueing performance (see, for example, [6, 104, 113, 133, 134, 207, 240, 263, 264, 271, 351, 384, 400]), and a range of techniques (e.g., [101, 163, 251, 265, 317, 405]) are by now available to quantify these impacts and their implications for network management and control. For example, considerable attention has been paid in the recent past to the problem of call admission control in high-speed networks based on the notion of *effective bandwidth*, e.g., [49, 55, 95, 114, 150, 151, 172, 216–218, 406].

In contrast to the well-developed field of Markovian queueing models, only few theoretical results exist to date for queueing systems with long-range-dependent inputs. For recent work in this area and on the general problem of the relevance of fractal traffic in practice, see [4, 5, 102, 105, 115–117, 121, 122, 138, 146, 184, 199, 258, 318–320, 328, 335, 356]. In this context, see also the discussions in [117, 318, 413] related to some practical experience with the first-generation ATM buffers reported in [79]. While there is considerable scope for future research in the area of queueing models with long-range-dependent inputs, queueing in the presence of heavy-tailed service time distributions (and, in general, independent arrivals) is relatively well understood; e.g., see [1, 2, 56, 72, 213, 323, 408]. For specific results relating the behavior of simple queues fed by a single or by many *ON/OFF* sources exhibiting heavy-tailed *ON-* or *OFF*-periods to that of queues with fractional Brownian input streams (as considered in [318]), see, for example, [8, 35, 36, 270, 344, 345].

Given the shortage of theoretical results for long-range-dependent queueing models, the ability to generate synthetic traces is of particular importance in the context of teletraffic theory and practice. There exist numerous methods to date for generating self-similar traffic traces. Exact methods, which are based on the Durbin–Levinson algorithm [37, 394] are discussed in [24, 146, 196, 198, 394]. They are generally impractical for long time series. Approximate methods are described in [30, 74, 120, 215, 245, 252, 258, 285, 290, 331, 334, 337, 356, 374, 376, 404, 412]; some of these methods rely on earlier results reported in [77, 165, 391], derived for a different purpose and re-interpreted here in the context of synthetic traffic generation. These methods are generally very fast and feasible for even very long time series. However, the statistical quality of the generated sequences is, in general, not well understood [252].

Bibliography

1. J. Abate, G. L. Choudhury, and W. Whitt. Calculation of the GI/G/1 waiting time distribution and its cumulants from Pollaczek's formulas. *Archiv für Elektronik und Übertragungstechnik*, 47:311–321, 1993.

2. J. Abate, G. L. Choudhury, and W. Whitt. Waiting-time tail probabilities in queues with long-tail service-time distributions. *Queueing Systems*, 16:311–338, 1994.

3. P. Abry. *Transformées en ondelettes – Analyses multirésolution et signaux de pression en turbulence*. PhD thesis, Université Claude Bernard, Lyon 1, France, 1994.

4. A. Adas and A. Mukherjee. On resource management and QoS guarantees for long-range dependent traffic. Preprint, 1994.

5. R. Addie, M. Zuckerman, and T. Neame. Fractal traffic: measurements modelling and performance. Preprint, 1994.

6. R. G. Addie and M. Zuckerman. A Gaussian traffic model for a B-ISDN statistical multiplexer. In *Proceedings of the IEEE Globecom '92*, 1992.

7. C. Agiakloglou and P. Newbold. Lagrange multiplier tests for fractional difference. *Journal of Time Series Analysis*, 15:253–262, 1994.

8. V. Anantharam. On the sojourn time of sessions at an ATM buffer with long-range dependent input traffic. Preprint, University of California, Berkeley, 1995.

9. D. Anick, D. Mitra, and M. M. Sondhi. Stochastic theory of a data-handling system with multiple sources. *Bell System Technical Journal*, 61:1871–1894, 1982.

10. A. A. Anis and E. H. Lloyd. The expected value of the adjusted rescaled Hurst range of independent normal summands. *Biometrika*, 63:111–116, 1976.

11. M. A. Arcones. Limit theorems for non-linear functionals of a stationary Gaussian sequence of vectors. *The Annals of Probability*, 22:2242–2274, 1994.

12. A. Arvidsson and R. Harris. On the performance of some commonly employed models of bursty traffic. Preprint, 1994.

13. A. Astrauskas, J. B. Levy, and M. S. Taqqu. The asymptotic dependence structure of the linear fractional Lévy motion. *Lietuvos Matematikos Rinkinys (Lithuanian Mathematical Journal)*, 31(1):1–28, 1991.

14. R. T. Baillie. Long memory processes and fractional integration in econometrics. *Journal of Econometrics*, 1995. To appear.

15. R. T. Baillie and T. Bollerslev. Cointegration, fractional cointegration, and exchange rate dynamics. *The Journal of Finance*, 49:737–745, 1994.

16. R. T. Baillie, T. Bollerslev, and H. O. Æ. Mikkelsen. Fractionally integrated generalized autoregressive conditional heteroskedasticity. *Journal of Econometrics*, 1995. To appear.

17. R. T. Baillie, C. F. Chung, and M. A. Tieslau. Analyzing inflation by the fractionally integrated ARFIMA-GARCH model. *Journal of Applied Econometrics*, 1995. To appear.

18. R. Ballerini and D. C. Boes. Hurst behavior of shifting level processes. *Water Resources Research*, 21:1642–1648, 1985.

19. L. Bel, G. Oppenheim, L. Robbiano, and M. C. Viano. Densité spectrale, covariance, régularité et propriétés de mélange de processes distributions de type ARMA fractionnaire. Preprint, Journée Longue Portée, Groupe d'Automatique d'Orsay, Paris, May 11 1995.

20. J. Beran. A test of location for data with slowly decaying serial correlations. *Biometrika*, 76:261–269, 1989.

21. J. Beran. A goodness of fit test for time series with long-range dependence. *Journal of the Royal Statistical Society, Series B*, 54:749–760, 1992.

22. J. Beran. Statistical methods for data with long-range dependence. *Statistical Science*, 7(4):404–416, 1992. With discussions and rejoinder, pages 404-427.

23. J. Beran. Maximum likelihood estimation of the differencing parameter for invertible short- and long-memory ARIMA models. *Journal of the Royal Statistical Society, Series B*, 57:659–672, 1995.

24. J. Beran. *Statistics for Long-Memory Processes*. Chapman & Hall, New York, 1994.

25. J. Beran and H. Künsch. Location estimators for processes with long-range dependence. Preprint, 1985.

26. J. Beran, R. Sherman, M. S. Taqqu, and W. Willinger. Long-range dependence in variable-bit-rate video traffic. *IEEE Transactions on Communications*, 43:1566–1579, 1995.

27. J. Beran and N. Terrin. Estimation of the long memory parameter, based on a multivariate central limit theorem. *Journal of Time Series Analysis*, 15:269–278, 1994.

28. L. I. Berge, N. Rakotomalala, J. Feder, and T. Jøssamg. Cross-over in R/S analysis and power spectrum: measurements and simulations. Preprint, 1993.

29. L. M. Berliner. Statistics, probability and chaos. *Statistical Science*, 7(1):69–90, 1992.

30. D. C. Boes. Schemes exhibiting Hurst behavior. In J. N. Srivastava, editor, *Probability and Statistics: Essays in honor of Franklin A. Graybill*, pages 21–36, The Netherlands, Elsevier Science Publishers B.V. (North-Holland), 1988.

31. T. Bollerslev and H. O. Æ. Mikkelsen. Modeling and pricing long-memory in stock market volatility. Technical Report 164, Kellogg Graduate School of Management, Northwestern University, 1994. Working paper.

32. V. A. Bolotin. Modeling call holding time distributions for CCS network design and performance analysis. *IEEE Journal of Selected Areas in Communications*, 12(3):433–438, 1994.

33. G. E. P. Box, G. M. Jenkins, and C. Reinsel. *Time Series Analysis: Forecasting and Control*. Prentice Hall, Englewood Cliffs, N.J., third edition, 1994.

34. P. Breuer and P. Major. Central limit theorems for non-linear functionals of Gaussian fields. *Journal of Multivariate Analysis*, 13:425–441, 1983.

35. F. Brichet, J. W. Roberts, A. Simonian, and D. Veitch. Heavy traffic analysis of a fluid queue fed by On/Off with long-range dependence. Technical Report TD(95)03v1, COST 242, 1995.

36. F. Brichet, J. W. Roberts, A. Simonian, and D. Veitch. Heavy traffic analysis of a storage model with long-range dependent On/Off sources. Preprint, 1995.

37. P. J. Brockwell and R. A. Davis. *Time Series: Theory and Methods*. Springer-Verlag, New York, 2nd edition, 1991.

38. K. M. Brothers, W. H. DuMouchel, and A. S. Paulson. Fractiles of the stable laws. Technical Report, Rensselaer Polytechnic Institute, Troy, 1983.

39. D. J. Buckle. Bayesian inference for stable distributions. *Journal of the American Statistical Society*, 90(430):605–613, 1995.

40. L. F. Burlaga and L. W. Klein. Fractal structure of the interplanetary magnetic field. *Journal of Geophysical Research*, 91(A1):347–350, 1986.

41. R. Caceres, P. B. Danzig, S. Jamin, and D. J. Mitzel. Characteristics of wide-area TCP/IP conversations. In *Proceedings of the ACM Sigcomm'91*, pages 101–112, 1991.

42. S. Cambanis and I. Fakhre-Zakeri. On prediction of heavy-tailed autoregressive sequences: forward versus reversed time. Technical Report 383, Center for Stochastic Processes at the University of North Carolina, Chapel Hill, 1993.

43. S. Cambanis and S. B. Fotopoulos. On the conditional variance for scale mixtures of normal distributions. Preprint, 1995.

44. S. Cambanis and A. R. Soltani. Prediction of stable processes: spectral and moving average representation. *Zeitschrift für Wahrscheinlichkeitstheorie und verwandte Gebiete*, 66:593–612, 1984.

45. S. Cambanis and W. Wu. Multiple regression on stable vectors. *Journal of Multivariate Analysis*, 41:243–272, 1992.

46. M. Casdagli. Chaos and deterministic versus stochastic non-linear modeling. *Journal of the Royal Statistical Society, Series B*, 54:303–328, 1991.

47. S. Chatterjee and M. R. Yilmaz. Chaos, fractal and statistics. *Statistical Science*, 7(1):49–68, 1992.

48. G. Chen, B. Abraham, and S. Peiris. Lag window estimation of the degree of differencing in fractionally integrated time series models. *Journal of Time Series Analysis*, 15:473–487, 1994.

49. C. S. Cheng and J. A. Thomas. Effective bandwidth in high-speed digital networks. *IEEE Journal on Selected Areas in Communications*, 13:1091–1100, 1995.

50. Y.-W. Cheung. *Long memory in foreign exchange rates and sampling properties of some statistical procedures related to long memory series*. PhD thesis, University of Pennsylvania, Philadelphia, 1990.

51. Y.-W. Cheung. Long memory in foreign exchange rates. *Journal of Business and Economic Statistics*, 11:93–101, 1993.

52. Y.-W. Cheung. Tests for fractional integration: a Monte Carlo investigation. *Journal of Time Series Analysis*, 14:331–345, 1993.

53. Y.-W. Cheung and F. X. Diebold. On maximum likelihood estimation of the differencing parameter of fractionally-integrated noise with unknown mean. *Journal of Econometrics*, 62:301–316, 1994.

54. Y.-W. Cheung and K. S. Lai. Do gold market returns have long memory? *The Financial Review*, 28:181–202, 1993.

55. G. L. Choudhury, D. M. Lucantoni, and W. Whitt. On the effectiveness of effective bandwidths for admission control in ATM networks. In J. Labetoulle and J.W. Roberts, editors, *The Fundamental Role of Teletraffic in The Evolution of Telecommunications Networks (Proceedings of ITC-14, Antibes Juan-les-Pins, France, June 1994)*, pages 411–420. Elsevier, Amsterdam, 1994.

56. G. L. Choudhury and W. Whitt. Long-tail buffer-content distributions in broadband networks. Preprint, 1995.

57. C. F. Chung and R. T. Baillie. Small sample bias in conditional sum-of-squares estimators of fractionally integrated ARMA models. *Empirical Economics*, 18:791–806, 1993.

58. K. L. Chung and P. Schmidt. The minimum distance estimator for fractionally integrated ARMA models. Preprint, 1995.

59. M. Cinotti, E. Dalle Mese, S. Giordano, and F. Russo. Long-range dependence in Ethernet traffic offered to interconnect DQDB MANs. Preprint, University of Pisa, Italy, 1994.

60. R. Cioczek-Georges and B. B. Mandelbrot. Alternative micropulses and FBM. Preprint, 1994.

61. R. Cioczek-Georges and B. B. Mandelbrot. A class of micropulses and antipersistent fractional Brownian motion. Preprint, 1994.

62. R. Cioczek-Georges, B. B. Mandelbrot, G. Samorodnitsky, and M. S. Taqqu. Stable fractal sums of pulses: the cylindrical case. *Bernoulli*, 1:201–216, 1995.

63. R. Cioczek-Georges and M. S. Taqqu. Does asymptotic linearity of the regression extend to stable domains of attraction? *Journal of Multivariate Analysis*, 48:70–86, 1994.

64. R. Cioczek-Georges and M. S. Taqqu. How do conditional moments of stable vectors depend on the spectral measure? *Stochastic Processes and their Applications*, 54:95–111, 1994.

65. R. Cioczek-Georges and M. S. Taqqu. Sufficient conditions for the existence of conditional moments of stable random variables. Preprint, 1994.

66. R. Cioczek-Georges and M. S. Taqqu. Form of the conditional variance for symmetric stable random variables. *Statistica Sinica*, 5:351–361, 1995.

67. R. Cioczek-Georges and M. S. Taqqu. Necessary conditions for the existence of conditional moments of stable random variables. *Stochastic Processes and their Applications*, 56:233–246, 1995.

68. K. Claffy, H.-W. Braun, and G. C. Polyzos. Application of sampling methodologies to network traffic characterization. In *Proceedings of the ACM Sigcomm '93*, pages 194–203, San Francisco, 1993.

69. K. Claffy, H.-W. Braun, and G. C. Polyzos. Tracking long-term growth of the NSFNET backbone. In *Proceedings Infocom '93*, San Francisco, 1993.

70. D. B. H. Cline. Linear prediction of ARMA processes with infinite variance. *Stochastic Processes and their Applications*, 19:281–296, 1985.

71. D. B. H. Cline and P. J. Brockwell. Linear prediction of ARMA processes with infinite variance. *Stochastic Processes and their Applications*, 19:281–296, 1985.

72. J. W. Cohen. Some results on regular variation for distributions in queueing and fluctuation theory. *Journal of Applied Probability*, 10:343–353, 1973.

73. P. Collet and J.-P. Eckmann. Iterated maps on the interval as dynamical systems. In *Progress in Physics I*. Birkhauser, Boston, 1980.

74. F. Comte. Simulation and estimation of long memory continuous time models. Technical Report 9410, Centre de Recherche en Economie et Statistique, Institut National de la Statistique et des Etudes Economiques, 1994. Working paper.

75. F. Comte and E. Renault. Long memory continuous time models. Preprint, Journée Longue Portée, Groupe d'Automatique d'Orsay, Paris, May 11 1995.

76. D. R. Cox. Statistical analysis of time series: some recent developments. *Scandinavian Journal of Statistics*, 8:93–115, 1981.

77. D. R. Cox. Long-range dependence: a review. In H. A. David and H. T. David, editors, *Statistics: An Appraisal*, pages 55–74. Iowa State University Press, 1984.

78. M. E. Crovella and A. Bestavros. Self-similarity in world wide web traffic: evidence and possible causes. In *Proceedings of the 1996 ACM SIGMETRICS. International Conference on Measurement and Modeling of Computer Systems*,

May 1996. To appear.

79. M. Csenger. Early ATM users lose data. *Communications Week*, May 16 1994.

80. S. Csörgő and J. Mielniczuk. Density estimation under long-range dependence. *The Annals of Statistics*, 23:990–999, 1995.

81. S. Csörgő and J. Mielniczuk. Distant long-range dependent sums and regression estimation. *Stochastic Processes and their Applications*, 59:143–155, 1995.

82. S. Csörgő and J. Mielniczuk. Nonparametric regression under long-range dependent normal errors. *The Annals of Statistics*, 23:1000–1014, 1995.

83. N. J. Cutland, P. E. Kopp, and W. Willinger. Stock price returns and the Joseph effect: a fractional version of the Black-Scholes model. In E. Bolthausen, M. Dozzi, and F. Russo, editors, *Seminar on Stochastic Analysis, Random Fields and Applications*, pages 327–351. Birkhauser, Boston, 1995.

84. R. Dahlhaus. Efficient parameter estimation for self similar processes. *The Annals of Statistics*, 17(4):1749–1766, 1989.

85. R. Dahlhaus. Efficient location and regression estimation for long range dependent regression models. *The Annals of Statistics*, 23:1029–1047, 1995.

86. R. Dahlhaus and L. Giraitis. The bias and the mean squared error in semi-parametric models for locally stationary time-series. Preprint, 1995.

87. J. Danielsson and C. G. de Vries. Robust tail index and quantile estimation. Preprint, Tinbergen Institute, Rotterdam, The Netherlands, 1995.

88. P. Danzig, S. Jamin, R. Caceres, D. Mitzel, and D. Estrin. An empirical workload model for driving wide-area tcp/ip network simulations. *Internetworking: Research and Experience*, 3:1–26, 1992.

89. R. A. Davis. Gauss-Newton and m-estimation for ARMA processes with infinite variance. Preprint, 1995.

90. R. A. Davis, K. Knight, and J. Liu. M-estimation for autoregressions with infinite variance. *Stochastic Processes and their Applications*, 40(1):145–180, 1992.

91. R. A. Davis and S. I. Resnick. Tail estimates motivated by extreme value theory. *The Annals of Statistics*, 12:1467–1487, 1984.

92. R. A. Davis and S. I. Resnick. More limit theory for the sample correlation function of moving averages. *Stochastic Processes and their Applications*, 20:257–279, 1985.

93. R. A. Davis and W. Wu. Bootstrapping m-estimates in regression and autoregression with infinite variance. Preprint, 1994.

94. A. C Davison and D. R. Cox. Some simple properties of sums of random variables having long-range dependence. *Proceedings of the Royal Society London*, A424:255–262, 1989.

95. G. de Veciana, G. Kesidis, and J. Walrand. Resource management in wide-area ATM networks using effective bandwidths. *IEEE Journal on Selected Areas in Communications*, 13:1081–1090, 1995.

96. A. Descloux. Contention probabilities in packet switching networks with strung input processes. In *Proceedings of the 12th ITC*, Torino, Italy, 1988.

97. R. L. Devaney. *An Introduction to Chaotic Dynamical Systems*. Addison-Wesley, New York, 1989.

98. F. Diebold and G. Rudebusch. Long memory and persistence in aggregate output. *Journal of Monetary Economics*, 24:189–209, 1989.

99. F. Diebold and G. Rudebusch. On the power of Diskery-Fuller tests against fractional alternatives. *Economics Letters*, 35:155–160, 1991.

100. R. L. Dobrushin and P. Major. Non-central limit theorems for non-linear func-

tions of Gaussian fields. *Zeitschrift für Wahrscheinlichkeitstheorie und verwandte Gebiete*, 50:27–52, 1979.

101. N. G. Duffield. Exponential bounds for queues with Markovian arrivals. Preprint, 1993.

102. N. G. Duffield. Economies of scale in queues with sources having power-law large deviation scalings. Preprint, Dublin Institute of Advanced Studies, Dublin, Ireland, DIAS-APG-94-27, 1994.

103. N. G. Duffield, J. T. Lewis, N. O'Connell, R. Russell, and F. Toomey. Statistical issues raised by the Bellcore data. Preprint, 1994.

104. N. G. Duffield, J. T. Lewis, N. O'Connell, R. Russell, and F. Toomey. Entropy of ATM traffic streams: a tool for estimating QoS parameters. *IEEE Journal on Selected Areas in Communications*, 13:981–990, 1995.

105. N. G. Duffield and N. O'Connell. Large deviation and overflow probabilities for the general single-server queue, with applications. *Mathematical Proceedings of the Cambridge Philosophical Society*, 118:363–375, 1995.

106. D. E. Duffy, A. A. McIntosh, M. Rosenstein, and W. Willinger. Analyzing telecommunications traffic data from working common channel signaling subnetworks. In M. E. Tarter and M. D. Lock, editors, *Statistical Applications of Expanding Computer Facilities*, volume 25, pages 156–165. Interface Foundation of North America, 1993. *Computing Science and Statistics*.

107. D. E. Duffy, A. A. McIntosh, M. Rosenstein, and W. Willinger. Statistical analysis of CCSN/SS7 traffic data from working CCS subnetworks. *IEEE Journal on Selected Areas in Communications*, 12:544–551, 1994.

108. W. H. DuMouchel. *Stable distributions in statistical inference*. PhD thesis, University of Ann Arbor, Univ. Microfilms, Ann Arbor, MI, 1971.

109. W. H. DuMouchel. On the asymptotic normality of the maximum-likelihood estimate when sampling from a stable distribution. *The Annals of Statistics*, 1:948–957, 1973.

110. W. H. DuMouchel. Stable distributions in statistical inference: 1. Symmetric stable distributions compared to other long-tailed distributions. *Journal of the American Statistical Association*, 68:469–477, 1973.

111. W. H. DuMouchel. Stable distributions in statistical inference: 2. Information from stably distributed samples. *Journal of the American Statistical Association*, 70:386–393, 1975.

112. W. H. DuMouchel. Estimating the stable index in order to measure tail thickness: a critique. *The Annals of Statistics*, 11:1019–1031, 1983.

113. A. Elwalid, D. P. Heyman, T. V. Lakshman, D. Mitra, and A. Weiss. Fundamental bounds and approximations for ATM multiplexers with applications to video teleconferencing. *IEEE Journal on Selected Areas in Communications*, 13:1004–1016, 1995.

114. A. I. Elwalid and D. Mitra. Effective bandwidth of general Markovian traffic sources and admission control of high-speed networks. *IEEE/ACM Transactions on Networking*, 1(3):329–343, 1993.

115. A. Erramilli, J. Gordon, and W. Willinger. Applications of fractals in engineering for realistic traffic processes. In J. Labetoulle and J. W. Roberts, editors, *The Fundamental Role of Teletraffic in The Evolution of Telecommunications Networks (Proceedings of ITC-14, Antibes Juan-les-Pins, France, June 1994)*, pages 35–44. Elsevier, Amsterdam, 1994.

116. A. Erramilli, D. D. Gosby, and W. Willinger. Engineering for realistic traffic: a fractal analysis of burstiness. In *Proceedings of the Bangalore Regional ITC*

Seminar, Bangalore, India, 1993.

117. A. Erramilli, O. Narayan, and W. Willinger. Experimental queueing analysis with long-range dependent packet traffic. *IEEE/ACM Transactions on Networking,* 4:209–223, 1996.

118. A. Erramilli and R. P. Singh. Application of deterministic chaotic maps to characterize broadband traffic. In *Proceedings of the ITC 7th Specialist Seminar, Livingston, NJ,* 1990.

119. A. Erramilli, R. P. Singh, and P. Pruthi. Chaotic maps as models of packet traffic. In J. Labetoulle and J. W. Roberts, editors, *The Fundamental Role of Teletraffic in the Evolution of Telecommunications Networks (Proceedings of ITC-14, Antibes, Juan-les-Pins, France, June 1994),* pages 329–338. Elsevier, Amsterdam, 1994.

120. A. Erramilli, R. P. Singh, and P. Pruthi. An application of deterministic chaotic maps to model packet traffic. *Queueing Systems,* 20:171–206, 1995.

121. A. Erramilli and J. Wang. Monitoring packet traffic levels. In *Proceedings of the IEEE Globecom '94,* pages 274–280, San Francisco, CA, 1994.

122. A. Erramilli and W. Willinger. Fractal properties in packet traffic measurements. In *Proceedings of the St. Petersburg Regional ITC Seminar, St. Petersburg, Russia,* pages 144–158, 1993.

123. A. Erramilli and W. Willinger. A case for fractal traffic modeling. In *Proceedings of the Australian Telecommunication Networks & Applications Conference 1995,* pages XV–XX, Sydney, Australia, 1995.

124. A. Erramilli, W. Willinger, and P. Pruthi. Fractal traffic flows in high-speed communications networks. *Fractals,* 2(3):409–412, 1994.

125. K. Falconer. *Fractal Geometry: Mathematical Foundations and Applications.* Wiley, New York, 1993.

126. E. Fama. Mandelbrot and the stable Paretian hypothesis. *Journal of Business,* 36:420–429, 1963. Reprinted in *The Random Character of Stock Market Prices,* P. Cootner, editor, MIT Press, 1964, pages 297–306.

127. E. Fama. The behavior of stock market prices. *Journal of Business,* 38:34–105, 1965.

128. E. Fama and R. Roll. Some properties of symmetric stable distributions. *Journal of the American Statistical Association,* 63:817–836, 1968.

129. E. Fama and R. Roll. Parameter estimates for symmetric stable distributions. *Journal of the American Statistical Association,* 66:331–338, 1971.

130. J. Feder. *Fractals.* Plenum Press, New York, 1988.

131. W. Feller. The asymptotic distributions of the range of sums of independent random variables. *Annals of Mathematical Statistics,* 22:427–432, 1951.

132. W. Feller. *An Introduction to Probability Theory and its Applications,* volume 2. Wiley, New York, 2nd edition, 1971.

133. K. W. Fendick, V. R. Saksena, and W. Whitt. Dependence in packet queues. *IEEE Transactions on Communications,* 37:1173–1183, 1989.

134. K. W. Fendick, V. R. Saksena, and W. Whitt. Investigating dependence in packet queues with the index of dispersion for work. *IEEE Transactions on Communications,* 39:1231–1244, 1991.

135. P. Flandrin. Wavelet analysis and synthesis of fractional Brownian motion. *IEEE Transactions on Information Theory,* 38:910–917, 1992.

136. S. B. Fotopoulos and S. Cambanis. Conditional variance for stable random vectors. Technical Report 426, Center for Stochastic Processes at the University of

North Carolina, Chapel Hill, 1994.

137. S. B. Fotopoulos and L. He. Form of the conditional variance-covariance matrix for α-stable scale mixtures of normal distributions. Preprint, 1995.

138. H. J. Fowler and W. E. Leland. Local area network traffic characteristics with implications for broadband network congestion management. *IEEE Journal on Selected Areas in Communications*, 9:1139–1149, 1991.

139. R. Fox and M. S. Taqqu. Non-central limit theorems for quadratic forms in random variables having long-range dependence. *The Annals of Probability*, 13:428–446, 1985.

140. R. Fox and M. S. Taqqu. Large-sample properties of parameter estimates for strongly dependent stationary Gaussian time series. *The Annals of Statistics*, 14:517–532, 1986.

141. R. Fox and M. S. Taqqu. Central limit theorems for quadratic forms in random variables having long-range dependence. *Probability Theory and Related Fields*, 24:213–240, 1987.

142. R. Fox and M. S. Taqqu. Multiple stochastic integrals with dependent integrators. *Journal of Multivariate Analysis*, 21:105–127, 1987.

143. A. S. French and L. L. Stockbridge. Fractal and Markov behavior in ion channel kinetics. *Canadian Journal of Physiology and Pharmacology*, 66:967–970, 1988.

144. V. Frost and B. Melamed. Traffic modeling for telecommunications networks. *IEEE Communications Magazine*, 32:70–80, 1994.

145. M. W. Garrett. *Contributions Toward Real-Time Services on Packet Switched Networks*. PhD thesis, Columbia University, New York, 1993.

146. M. W. Garrett and W. Willinger. Analysis, modeling and generation of self-similar VBR video traffic. In *Proceedings of the ACM Sigcomm '94, London, UK*, pages 269–280, 1994.

147. H. G. Gauch. Prediction, parsimony and noise. *American Scientist*, 81:468–478, 1993.

148. R. Gay and C. C. Heyde. On a class of random field models which allows long range dependence. *Biometrika*, 77:401–403, 1990.

149. J. Geweke and S. Porter-Hudak. The estimation and application of long memory time series models. *Journal of Time Series Analysis*, 4:221–238, 1983.

150. R. J. Gibbens. Traffic characterisation and effective bandwidths for broadband network traces. This volume, 1996.

151. R. J. Gibbens and P. J. Hunt. Effective bandwidths for the multi-type UAS channel. *Queueing Systems*, 9:17–28, 1991.

152. L. Giraitis. Central limit theorem for functionals of a linear process. *Lithuanian Mathematical Journal*, 25:25–35, 1985.

153. L. Giraitis. Central limit theorem for polynomial forms I. *Lithuanian Mathematical Journal*, 29:109–128, 1989.

154. L. Giraitis, H. Koul, and D. Surgailis. Asymptotic normality of regression estimators with long memory errors. *Statistics and Probability Letters*, 1995. To appear.

155. L. Giraitis and R. Leipus. A generalized fractionally differencing approach in long-memory modelling. *Lithuanian Mathematical Journal*, 35:53–65, 1995.

156. L. Giraitis, S. A. Molchanov, and D. Surgailis. Long memory shot noises and limit theorems with application to Burgers' equation. In D. Brillinger, P. Caines, J. Geweke, E. Parzen, M. Rosenblatt, and M. S. Taqqu, editors, *New Directions in Time Series Analysis, Part II*, pages 153–176. IMA Volumes in Mathematics

and its Applications, Volume 46, Springer-Verlag, New York, 1992.

157. L. Giraitis, A. Samarov, and P. M. Robinson. Rate optimal semiparametric estimation of the memory parameter of the Gaussian time series with long range dependence. Technical Report, Beiträge für Statistik, Universität Heidelberg, 1995.

158. L. Giraitis and D. Surgailis. Multivariate Appell polynomials aúd the central limit theorem. In E. Eberlein and M. S. Taqqu, editors, *Dependence in Probability and Statistics*. Birkhauser, New York, 1986.

159. L. Giraitis and D. Surgailis. A central limit theorem for quadratic forms in strongly dependent linear variables and application to asymptotical normality of Whittle's estimate. *Probability Theory and Related Fields*, 86:87–104, 1990.

160. L. Giraitis and D. Surgailis. On shot noise processes with long range dependence. In B. Grigelionis, Y. U. Prohorov, V. V. Sazonov, and V. Statulevičius, editors, *Probability Theory and Mathematical Statistics. Proceedings of the Fifth Vilnius Conference June 25–July 1, 1989*, pages 401–408. VSP BV Press, Vol. 1, Utrecht, The Netherlands, 1990.

161. L. Giraitis and M. S. Taqqu. Limit theorem for bivariate Appell polynomials: Part I. Central limit theorems. Preprint, 1995.

162. L. Giraitis and M. S. Taqqu. Central limit theorems for quadratic forms with time-domain conditions. Preprint, 1996.

163. P. W. Glynn and W. Whitt. Logarithmic asymptotics for steady-state tail probabilities in a single-server queue. *Journal of Applied Probability*, 31:131–156, 1994.

164. V. V. Gorodetskii. On convergence to semi-stable Gaussian processes. *Theory of Probability and its Applications*, 22:498–508, 1977.

165. C. W. J. Granger. Long memory relationships and aggregation of dynamic models. *Journal of Econometrics*, 14:227–238, 1980.

166. C. W. J. Granger and R. Joyeux. An introduction to long-memory time series and fractional differencing. *Journal of Time Series Analysis*, 1:15–30, 1980.

167. M. T. Greene and B. D. Fielitz. Long-term dependence in common stock returns. *Journal of Financial Economics*, 4:339–349, 1977.

168. M. T. Greene and B. D. Fielitz. The effect of long-term dependence on risk return models of common stocks. *Operations Research*, 27:944–951, 1979.

169. G. Gripenberg and I. Norros. On the prediction of fractional Brownian motion. *Journal of Applied Probability*, 1996. To appear.

170. A. Gross. Some mixing conditions for stationary symmetric stable stochastic processes. *Stochastic Processes and their Applications*, 51:277–295, 1994.

171. D. Guégan. *Séries Chronologiques Non Linéaires à Temps Discret*. Statistique Mathématique et Probabilité. Economica, 49, rue Héricart, 75015 Paris, 1994.

172. R. Guerin, H. Admadi, and M. Naghshineh. Equivalent capacity and its application to bandwidth allocation in high-speed networks. *IEEE Journal on Selected Areas in Communications*, 9:968–981, 1991.

173. R. Gusella. *A Characterization of the Variability of Packet Arrival Processes in Workstation Networks*. PhD thesis, University of California at Berkeley, 1990.

174. R. Gusella. A measurement study of diskless workstation traffic on an ethernet. *IEEE Transactions on Communications*, 38:1557–1568, 1990.

175. R. Gusella. Characterizing the variability of arrival processes with indices of dispersion. *IEEE Journal on Selected Areas in Communications*, 9(2):968–981, 1991.

176. D. Halford. A general mechanical model for $|f|^\alpha$ spectral density random noise

with special reference to flicker noise $1/|f|$. *Proceedings of the IEEE*, 56(3):251–258, 1968.

177. F. R. Hampel. Data analysis and self-similar processes. In *Proceedings of the 46th Session of the International Statistical Institute*, Tokyo, Japan, September 1987. International Statistical Institute.

178. F. R. Hampel, E. M. Ronchetti, P. J. Rousseeuv, and W. A. Stahel. *Robust Statistics*. Wiley, New York, 1986.

179. J. Haslett and A. E. Raftery. Space-time modelling with long-memory dependence: assessing Ireland's wind power resource. *Applied Statistics*, 38:1–50, 1989. Includes discussion.

180. H. Heffes and D.M. Lucantoni. A Markov modulated characterization of packetized voice and data traffic and related statistical multiplexer performance. *IEEE Journal on Selected Areas in Communications*, 4:856–868, 1986.

181. S. A. Heimlich. Traffic characterization of the NSFNET national backbone. In *Proceedings of the 1990 Usenix Conference*, pages 207–227, 1990.

182. P. Heinrich. Zero-one laws for polynomials in Gaussian random variables. Preprint, 1995.

183. M. Henry and P. M. Robinson. Bandwidth choice in Gaussian semiparametric estimation of long range dependence. In *Proceedings of the Athens Conference on Applied Probability and Time Series Analysis*. Springer-Verlag, New York, 1996. Time series volume in honor of E. J. Hannan. To appear.

184. C. C. Heyde. Some results on inference for stationary processes and queueing systems. In U. N. Bhatt and I. V. Basawa, editors, *Queueing and Related Models*, pages 337–345. Oxford University Press (Clarendon Press), Oxford, UK, 1992.

185. C. C. Heyde and R. Gay. Smoothed periodogram asymptotics and estimation for processes and fields with possible long-range dependence. *Stochastic Processes and their Applications*, 45:169–182, 1993.

186. D. P. Heyman, A. Tabatabai, and T. V. Lakshman. Statistical analysis and simulation study of video teleconference traffic in ATM networks. *IEEE Transactions on Circuits and Systems for Video Technology*, 2:49–59, 1992.

187. B. M. Hill. A simple general approach to inference about the tail of a distribution. *The Annals of Statistics*, 3:1163–1174, 1975.

188. H. C. Ho. On limiting distributions of nonlinear functions of noisy Gaussian sequences. *Stochastic Analysis and Applications*, 10:417–430, 1992.

189. H. C. Ho. On central and non-central limit theorems in density estimation for sequences of long-range dependence. Preprint, 1995.

190. H. C. Ho. On the strong uniform consistency of density estimation for strongly dependent sequences. *Statistics & Probability Letters*, 22:149–156, 1995.

191. H. C. Ho and T. Hsing. On the asymptotic expansion of the empirical process of long memory moving averages. Preprint, 1995.

192. H. C. Ho and T. C. Sun. A central limit theorem for non-instantaneous filters of a stationary Gaussian process. *Journal of Multivariate Analysis*, 22:144–155, 1987.

193. H. C. Ho and T. C. Sun. A mixture-type limit theorem for nonlinear functions of Gaussian sequences. *Journal of Theoretical Probability*, 4:407–415, 1991.

194. D. R. Holt and E. L. Crow. Tables and graphs of the stable probability density functions. *Journal of Research of the National Bureau of Standards*, 77B:143–198, 1973.

195. J. R. M. Hosking. Fractional differencing. *Biometrika*, 68(1):165–176, 1981.

196. J. R. M. Hosking. Modeling persistence in hydrological time series using fractional differencing. *Water Resources Research*, 20:1898–1908, 1984.

197. J. R. M. Hosking. Asymptotic distributions of the sample mean, autocovariances and autocorrelations of long-memory time series. *Journal of Econometrics*, 1995. To appear.

198. C. Huang, M. Devetsikiotis, I. Lambadaris, and A. R. Kaye. Fast simulation for self-similar traffic in ATM networks. In *Proceedings of the ICC '95*, pages 438–444, Seattle, WA, 1995.

199. C. Huang, M. Devetsikiotis, I. Lambadaris, and A. R. Kaye. Modeling and simulation of self-similar variable bit rate compressed video: A unified approach. *Computer Communications Review*, 25:114–125, 1995. Proceedings of the ACM/SIGCOMM'95, Cambridge, MA, August 1995.

200. H. E. Hurst. Long-term storage capacity of reservoirs. *Transactions of the American Society of Civil Engineers*, 116:770–808, 1951.

201. H. E. Hurst. Methods of using long-term storage in reservoirs. *Proceedings of the Institution of Civil Engineers, Part I*, pages 519–577, 1955.

202. H. E. Hurst, R. P. Black, and Y. M. Simaika. *Long-Term Storage: An Experimental Study*. Constable, London, 1965.

203. C. M. Hurvich and K. I. Beltrao. Asymptotics for the low-frequency ordinates of the periodogram of a long-memory time series. *Journal of Time Series Analysis*, 14:455–472, 1993.

204. C. M. Hurvich and K. I. Beltrao. Automatic semiparametric estimation of the memory parameter of a long memory time series. *Journal of Time Series Analysis*, 15:285–302, 1994.

205. C. M. Hurvich and B. K. Ray. Estimation of the long-memory parameter for nonstationary or noninvertible fractionally integrated processes. *Journal of Time Series Analysis*, 16:17–41, 1995.

206. C.M. Hurvich, R. Deo, and J. Brodsky. The mean squared error of Geweke and Porter-Hudak's estimator of the memory parameter of a long memory time series. Preprint, 1995.

207. C. L. Hwang and S. Q. Li. On input state space reduction and buffer noneffective region. In *Proceedings of IEEE Infocom '94*, pages 1018–1028, 1994.

208. A. V. Ivanov and N. N. Leonenko. *Statistical Analysis of Random Fields*. Kluwer Academic Publishers, Dordrecht/Boston/London, 1989. Translated from the Russian, 1986 edition.

209. D. L. Jagerman, B. Melamed, and W. Willinger. Stochastic modeling of traffic processes. In J. Dshalalow, editor, *Frontiers in Queueing: Models, Methods and Problems*. CRC Press, 1996. To appear.

210. R. Jain and S. A. Routhier. Packet trains: measurements and a new model for computer network traffic. *IEEE Journal on Selected Areas in Communications*, 4:986–995, 1986.

211. A. Janicki and A. Weron. *Simulation and Chaotic Behavior of α-stable Stochastic Processes*. Marcel Dekker, New York, 1994.

212. W. H. Jefferys and J. O. Berger. Ockham's razor and Bayesian analysis. *American Scientist*, 80:64–72, 1992.

213. P. R. Jelenkovic and A. A. Lazar. Subexponential asymptotics of a network multiplexer. Preprint, CTR, Columbia University, New York, 1995.

214. C. Jimenez, K. Hipel, and A. I. McLeod. Developments in modelling long term persistence. Preprint, 1988.

215. N. J. Kasdin. Discrete simulation of colored noise and stochastic processes and

$1/f^\alpha$ power law noise generation. *Proceedings of the IEEE*, 83(5):802–827, 1995.

216. F. P. Kelly. Effective bandwidths at multi-class queues. *Queueing Systems*, 9:5–15, 1991.

217. F. P. Kelly. Notes on effective bandwidths. This volume, 1996.

218. G. Kesidis, J. Walrand, and C. S. Chang. Effective bandwidths for multiclass Markov fluids and other ATM sources. *IEEE/ACM Transactions on Networking*, 1:424–428, 1993.

219. V. Klemeš. The Hurst phenomenon: a puzzle? *Water Resources Research*, 10:675–688, 1974.

220. S. Klivansky, A. Mukherjee, and C. Song. Factors contributing to self-similarity over nsfnet. Preprint, Georgia Institute of Technology, 1994.

221. C. Klüppelberg and T. Mikosch. Spectral estimates and stable processes. *Stochastic Processes and their Applications*, 47:323–344, 1993.

222. C. Klüppelberg and T. Mikosch. Some limit theory for the self-normalized periodogram of stable processes. *Scandinavian Journal of Statistics*, 21:485–492, 1994.

223. C. Klüppelberg and T. Mikosch. The integrated periodogram for stable processes. *The Annals of Statistics*, 1995. To appear.

224. C. Klüppelberg and T. Mikosch. Self-normalized and randomly centered spectral estimates. In *Proceedings of the Athens Conference on Applied Probability and Time Series Analysis*. Springer-Verlag, New York, 1996. Time series volume in honor of E. J. Hannan. To appear.

225. S. M. Kogon and D. G. Manolakis. Infrared scene modeling and interpolation using fractional Lévy stable motion. *Fractals*, 2(6):303–306, 1994.

226. S. M. Kogon and D. G. Manolakis. Fractal-based modeling interpolation of non-Gaussian images. Preprint. Presented at SPIE: Visual Communications and Image Processing '94, Chicago, IL, 1995.

227. S. M. Kogon and D. G. Manolakis. Linear parametric models for signals with long-range dependence and infinite variance. Preprint. Submitted to ICASSP95, 1995.

228. P. Kokoszka and T. Mikosch. The integrated periodogram for long-memory processes with finite or infinite variance. Preprint, 1995.

229. P. S. Kokoszka. Prediction of infinite variance fractional ARIMA. *Probability and Mathematical Statistics*, 16, 1995. To appear.

230. P. S. Kokoszka and M. S. Taqqu. Asymptotic dependence of stable self-similar processes of Chentsov type. In R. M. Dudley, M. G. Hahn, and J. Kuelbs, editors, *Probability in Banach Spaces, 8: Proceedings of the Eighth International Conference*, pages 152–165. Birkhauser, Boston, 1992.

231. P. S. Kokoszka and M. S. Taqqu. Asymptotic dependence of moving average type self-similar stable random fields. *Nagoya Mathematical Journal*, 130:85–100, 1993.

232. P. S. Kokoszka and M. S. Taqqu. Infinite variance stable ARMA processes. *Journal of Time Series Analysis*, 15:203–220, 1994.

233. P. S. Kokoszka and M. S. Taqqu. New classes of self-similar symmetric stable random fields. *Journal of Theoretical Probability*, 7:527–549, 1994.

234. P. S. Kokoszka and M. S. Taqqu. The asymptotic behavior of quadratic forms in heavy-tailed strongly dependent random variables. *Stochastic Processes and their Applications*, 1996. To appear.

235. P. S. Kokoszka and M. S. Taqqu. Discrete time parametric models with long

memory and infinite variance. Preprint, 1995.

236. P. S. Kokoszka and M. S. Taqqu. Fractional ARIMA with stable innovations. *Stochastic Processes and their Applications*, 60:19–47, 1995.

237. P. S. Kokoszka and M. S. Taqqu. Infinite variance stable moving averages with long memory. *Journal of Econometrics*, 73, 1996. To appear.

238. P. S. Kokoszka and M. S. Taqqu. Parameter estimation for infinite variance fractional ARIMA. To appear in *The Annals of Statistics*, 1996.

239. A. N. Kolmogorov. Local structure of turbulence in an incompressible liquid for very large Reynolds numbers. *Comptes Rendus (Doklady) de l'Académie des Sciences de l' URSS (N.S.)*, 30:299–303, 1941. Reprinted in S. K. Friedlander and L. Topper *Turbulence: classic papers on statistical theory.* Interscience, New York, 1961.

240. T. Konstantopoulos and V. Anantharam. Optimal flow control schemes that regulate the burstiness of traffic. *IEEE/ACM Transactions on Networking*, 3:423–432, 1995.

241. H. L. Koul and K. Mukherjee. Asymptotics of R-, MD- and LAD-estimators in linear regression models with long range dependent errors. *Probability Theory and Related Fields*, 95:535–553, 1993.

242. I. A. Koutrouvelis. Regression-type estimation of the parameters of stable laws. *Journal of the American Statistical Association*, 75:918–928, 1980.

243. I. A. Koutrouvelis. An iterative procedure for the estimation of the parameters of the stable law. *Communication in Statistics-Simulation and Computation*, 10:17–28, 1981.

244. M. Kratz and S. I. Resnick. The qq-estimator and heavy tails. Preprint, School of ORIE, Cornell University, Ithaca, NY, 1995.

245. W. M. Kruger, S. D. Jost, and U. Axen. On synthesizing discrete fractional Brownian motion. Preprint, 1992.

246. H. Künsch. Statistical aspects of self-similar processes. *Proceedings of the First World Congress of the Bernoulli Society*, 1:67–74, 1987.

247. H. Künsch, J. Beran, and F. Hampel. Contrasts under long-range correlations. *The Annals of Statistics*, 21:943–964, 1993.

248. T. G. Kurtz. Limit theorems for workload input models. This volume, 1996.

249. G. Lang and J.-M. Azaïs. Nonparametric estimation of the strong dependence exponent for Gaussian processes. Preprint, Journée Longue Portée, Groupe d'Automatique d'Orsay, Paris, May 11 1995.

250. A. Lasota and M. C. Mackey. *Chaos, Fractals, and Noise – Stochastic Aspects of Dynamics.* Springer-Verlag, New York, 1994.

251. G. Latouche and V. Ramaswami. A logarithmic reduction algorithm for quasi birth and death processes. *Journal of Applied Probability*, 30:650–674, 1993.

252. W.-C. Lau, A. Erramilli, J. L. Wang, and W. Willinger. Self-similar traffic generation: the random midpoint displacement algorithm and its properties. In *Proceedings of the ICC '95*, pages 466–472, Seattle, WA, 1995.

253. W.-C. Lau, A. Erramilli, J. L. Wang, and W. Willinger. Self-similar traffic parameter estimation: a semi-parametric periodogram-based algorithm. In *Proceedings of the IEEE Globecom '95*, pages 2225–2231, Singapore, 1995.

254. A. J. Lawrence and N. T. Kottegoda. Stochastic modelling of riverflow time series. *Journal of the Royal Statistical Society*, A 140(1):1–47, 1977.

255. W. E. Leland. LAN traffic behavior from milliseconds to days. In *Proceedings of the ITC 7th Specialist Seminar*, Morristown, NJ, 1990.

256. W. E. Leland, M. S. Taqqu, W. Willinger, and D. V. Wilson. On the self-similar nature of Ethernet traffic. *Computer Communications Review*, 23:183–193, 1993. Proceedings of the ACM/SIGCOMM'93, San Francisco, September 1993. Reprinted in *Trends in Networking – Internet*, the conference book of the Spring 1995 Conference of the National Unix User Group of the Netherlands (NLUUG). Also reprinted in *Computer Communications Review*, **25**, Nb. 1 (1995), 202–212, a special anniversary issue devoted to "Highlights from 25 years of the Computer Communications Review".

257. W. E. Leland, M. S. Taqqu, W. Willinger, and D. V. Wilson. Statistical analysis of high time-resolution Ethernet LAN traffic measurements. In M. E. Tarter and M. D. Lock, editors, *Statistical Applications of Expanding Computer Facilities*, Volume 25, pages 146–155. Interface Foundation of North America, 1993. *Computing Science and Statistics*.

258. W. E. Leland, M. S. Taqqu, W. Willinger, and D. V. Wilson. On the self-similar nature of Ethernet traffic (Extended version). *IEEE/ACM Transactions on Networking*, 2:1–15, 1994.

259. W. E. Leland, W. Willinger, M. S. Taqqu, and D. V. Wilson. Statistical analysis and stochastic modeling of self-similar data traffic. In J. Labetoulle and J. W. Roberts, editors, *The Fundamental Role of Teletraffic in the Evolution of Telecommunications Networks*, pages 319–328, Amsterdam, 1994. Proceedings of the 14th International Teletraffic Congress (ITC '94), Elsevier Science B.V.

260. W. E. Leland and D. V. Wilson. High time-resolution measurement and analysis of LAN traffic: implications for LAN interconnection. In *Proceedings of the Infocom '91*, pages 1360–1366, Bal Harbour, FL, 1991.

261. J. Levy and M. S. Taqqu. On renewal processes having stable inter-renewal intervals and stable rewards. *Les Annales des Sciences Mathematiques du Quebec*, 11:95–110, 1987.

262. J. B. Levy and M. S. Taqqu. A characterization of the asymptotic behavior of stationary stable processes. In S. Cambanis, G. Samorodnitsky, and M. S. Taqqu, editors, *Stable Processes and Related Topics*, Volume 25 of *Progress in Probability*, pages 181–198. Birkhauser, Boston, 1991.

263. S. Q. Li, S. Chong, and C. L. Hwang. Link capacity allocation and network control by filtered input rate in high-speed networks. *IEEE/ACM Transactions on Networking*, 3:10–25, 1995.

264. S. Q. Li and C. L. Hwang. Queue response to input correlation functions continuous spectral analysis. *IEEE/ACM Transactions on Networking*, 1:678–692, 1993.

265. S. Q. Li and C. L. Hwang. Queue response to input correlation functions: discrete spectral analysis. *IEEE/ACM Transactions on Networking*, 1:522–533, 1993.

266. L. S. Liebovitch. Testing fractal and Markov models of ion channel kinetics. *Biophysics Journal*, 55:373–377, 1989.

267. L. S. Liebovitch, J. Fischbarg, and J. P. Koniarek. Ion channel kinetics: a model based on fractal scaling rather than Markov processes. *Mathematical Biosciences*, 84:37–68, 1987.

268. L. S. Liebovitch, J. Fischbarg, J. P. Koniarek, I. Todorova, and M. Wang. Fractal model of ion-channel kinetics. *Biochimica Biophysica Acta*, 896:173–180, 1987.

269. L. S. Liebovitch and J. M. Sullivan. Fractal analysis of a voltage-dependent potassium channel from cultured mouse hippocampal neurons. *Biophysics Journal*, 52:979–988, 1987.

270. N. Likhanov, B. Tsybakov, and N. D. Georganas. Analysis of an ATM buffer with self-similar ("fractal") input traffic. In *Proceedings of the IEEE Infocom '95*,

pages 985–992, Boston, MA, 1995.

271. M. Livny, B. Melamed, and A. K. Tsiolis. The impact of autocorrelation on queueing systems. *Management Science*, 39:322–339, 1993.

272. E. H. Lloyd and D. Warren. The discrete Hurst range for skew independent two-valued inflows. *Stochastic Hydrology and Hydraulics*, 1:53–66, 1987.

273. A. W. Lo. Long-term memory in stock market prices. *Econometrica*, 59:1279–1313, 1991.

274. S. B. Lowen and M. C. Teich. Power-law shot noise. *IEEE Transactions on Information Theory*, IT-36(6):1302–1318, 1990.

275. S. B. Lowen and M. C. Teich. Doubly stochastic Poisson point process driven by fractal shot noise. *Physical Review A*, 43:4192–4215, 1991.

276. S. B. Lowen and M. C. Teich. Fractal renewal processes generate $1/f$ noise. *Physical Review E*, 47:992–1001, 1993.

277. D. M. Lucantoni. The BMAP/G/1 queue. In L. Donatiello and R. Nelson, editors, *Models and Techniques for Performance Evaluation of Computer and Communication Systems*, pages 330–358, Lecture Notes in Computer Science. Springer-Verlag, New York, 1993.

278. C. Ludena. Estimation of integrals with respect to the logarithm of the spectral density of stationary Gaussian processes with long range dependence. Preprint, Journée Longue Portée, Groupe d'Automatique d'Orsay, Paris, May 11 1995.

279. T. Lundahl, W. J. Ohley, S. M. Kay, and R. Siffert. Fractional Brownian motion: a maximum likelihood estimator and its application to image texture. *IEEE Transactions on Pattern Analysis and Machine Intelligence*, MI-5(3):152–161, 1986.

280. P. Major. *Multiple Wiener-Itô Integrals*, Volume 849. Springer Lecture Notes in Mathematics. Springer-Verlag, New York, 1981.

281. B. B. Mandelbrot. The Pareto-Lévy law and the distribution of income. *International Economic Review*, 1:79–106, 1960.

282. B. B. Mandelbrot. The variation of certain speculative prices. *Journal of Business*, 36:394–419, 1963. Reprinted in *The Random Character of Stock Market Prices*, P. Cootner, editor, pages 307–332. MIT Press, 1964.

283. B. B. Mandelbrot. Self-similar error clusters in communications systems and the concept of conditional systems and the concept of conditional stationarity. *IEEE Transactions on Communications Technology*, COM-13:71–90, 1965.

284. B. B. Mandelbrot. Long-run linearity, locally Gaussian processes, H-spectra and infinite variances. *International Economic Review*, 10:82–113, 1969.

285. B. B. Mandelbrot. A fast fractional Gaussian noise generator. *Water Resources Research*, 7:543–553, 1971.

286. B. B. Mandelbrot. Limit theorems on the self-normalized range for weakly and strongly dependent processes. *Zeitschrift für Wahrscheinlichkeitstheorie und verwandte Gebiete*, 31:271–285, 1975.

287. B. B. Mandelbrot. *The Fractal Geometry of Nature*. W.H. Freeman, San Francisco, 1982.

288. B. B. Mandelbrot and M. S. Taqqu. Robust R/S analysis of long-run serial correlation. In *Proceedings of the 42nd Session of the International Statistical Institute*, Manila, 1979. *Bulletin of the International Statistical Institute*, Vol 48, Book 2, pages 69–104.

289. B. B. Mandelbrot and J. W. Van Ness. Fractional Brownian motions, fractional noises and applications. *SIAM Review*, 10:422–437, 1968.

290. B .B. Mandelbrot and J. R. Wallis. Computer experiments with fractional Gaussian noises, Parts 1,2,3. *Water Resources Research*, 5:228–267, 1969.

291. B. B. Mandelbrot and J. R. Wallis. Robustness of the rescaled range R/S in the measurement of noncyclic long-run statistical dependence. *Water Resources Research*, 5:967–988, 1969.

292. B. B. Mandelbrot and J. R. Wallis. Some long-run properties of geophysical records. *Water Resources Research*, 5:321–340, 1969.

293. W. T. Marshall and S. P. Morgan. Statistics of mixed data traffic on a local area network. *Computer Networks and ISDN Systems*, 10:185–194, 1985.

294. E. Masry. The wavelet transform of stochastic processes with stationary increments and its application to fractional Brownian motion. *IEEE Transactions on Information Theory*, 39(1):260–264, 1993.

295. J. H. McCulloch. Simple consistent estimators of stable distribution parameters. *Communications in Statistics-Computation and Simulation*, 15:1109–1136, 1986.

296. J. H. McCulloch. Numerical approximation of the symmetric stable distribution and density. Preprint, 1994.

297. J. H. McCulloch. Financial applications of stable distributions. In G. S. Maddala and C. R. Rao, editors, *Statistical Methods in Finance*, Volume 14 of *Handbook of Statistics*. Elsevier Science, 1996. To appear.

298. J. H. McCulloch. Measuring tail thickness in order to estimate the stable index α: a critique. *Journal of Business and Economic Statistics*, 1996. To appear.

299. J. H. McCulloch, J. P. Nolan, and A. K. Panorska. Estimation of stable spectral measures. Preprint, 1996.

300. J. H. McCulloch and D. B. Panton. Precise tabulation of the maximally-skewed stable distributions and densities. *Computational Statistics and Data Analysis*, 1996. To appear.

301. A. A. Mcintosh. Analyzing telephone network data. Preprint, 1995.

302. A. I. McLeod and K. W. Hipel. Preservation of the rescaled adjusted range, Parts 1,2,3. *Water Resources Research*, 14:491–518, 1978.

303. O. B. McManus, D. S. Weiss, C. E. Spivak, A. L. Blatz, and K. L. Magleby. Fractal models are inadequate for the kinetics of four different ion channels. *Biophysics Journal*, 54:859–870, 1988.

304. K. Meier-Hellstern, P. E. Wirth, Y.-L. Yan, and D. A. Hoeflin. Traffic models for ISDN data users: office automation application. In A. Jensen and V. B. Iversen, editors, *Teletraffic and Datatraffic in a Period of Change (Proceedings of ITC-13, Copenhagen, Denmark)*, pages 167–172. North-Holland, Amsterdam, 1991.

305. B. Melamed. An overview of TES processes and modeling methodology. In L. Donatiello and R. Nelson, editors, *Models and techniques for Performance Evaluation of Computer and Communications Systems*, pages 359–393, Lecture Notes in Computer Science. Springer-Verlag, New York, 1993.

306. T. Mikosch, T. Gadrich, C. Klüppelberg, and R. J. Adler. Parameter estimation for ARMA models with infinite variance innovations. *The Annals of Statistics*, 23:305–326, 1995.

307. S. Mittnik and S. T. Rachev. Stable distributions for asset returns. *Applied Mathematics Letters*, 2:301–304, 1989.

308. S. Mittnik and S. T. Rachev. Alternative multivariate stable distributions and their applications to financial modeling. In S. Cambanis, G. Samorodnitsky, and M. S. Taqqu, editors, *Stable Processes and Related Topics*, Volume 25 of *Progress in Probability*, pages 107–119. Birkhauser, Boston, 1991.

309. S. Mittnik and S. T. Rachev. Modeling asset returns with alternative stable

distributions. *Econometric Review*, 12:261–330, 1993.

310. A. Montanari, R. Rosso, and M. S. Taqqu. Fractionally differenced ARIMA models applied to hydrologic time series: identification, estimation and simulation. Preprint, 1995.

311. A. Montanari, R. Rosso, and M. S. Taqqu. A seasonal fractionally different ARIMA model: an application to the Nile River monthly flows at Aswan. Preprint, 1995.

312. A. Montanari, R. Rosso, and M. S. Taqqu. Some long-run properties of rainfall records in Italy. Preprint, 1995.

313. E. W. Montroll and M. F. Shlesinger. On 1/f noise and other distributions with long tails. *Proceedings of the National Academy of Sciences of the USA*, 79:3380–3383, 1982.

314. A. Mukherjee. On the dynamics and significance of low frequency components of internet load. Preprint, 1992.

315. D. K. Nassiuma. Symmetric stable sequences with missing observations. *Journal of Time Series Analysis*, 15:313–323, 1994.

316. M. F. Neuts. A versatile Markovian point process. *Journal of Applied Probability*, 18:764–779, 1979.

317. M. F. Neuts. *Structured Stochastic Matrices of M/G/1 Type and Their Applications*. Marcel Dekker, New York, 1989.

318. I. Norros. A storage model with self-similar input. *Queueing Systems and their Applications*, 16:387–396, 1994.

319. I. Norros. On the use of fractional Brownian motion in the theory of connectionless networks. *IEEE Journal on Selected Areas in Communications*, 13:953–962, 1995.

320. I. Norros, A. Simonian, D. Veitch, and J. Virtamo. A Beneš formula for the fractional Brownian storage. Technical Report TD(95)004v2, COST 242, 1995.

321. J. Nowicka and A. Weron. Numerical approximation of dependence structure for symmetric stable AR(2) processes. Preprint, 1995.

322. S. Painter and L. Paterson. Fractional Lévy motion as a model for spatial variability in sedimentary rock. *Geophysical Research Letters*, 21(25):2857–2860, 1994.

323. A. G. Pakes. On the tails of waiting-time distributions. *Journal of Applied Probability*, 12:555–564, 1975.

324. D. B. Panton. Cumulative distribution function values for symmetric standardized stable distributions. *Communication in Statistics-Simulation and Computation*, 21:485–492, 1992.

325. D. B. Panton. Distribution function values for logstable distributions. *Computers and Mathematics with Applications*, 9:17–24, 1993.

326. K. Park, G. Kim, and M. Crovella. On the cause and effect of self-similar network traffic. Preprint, Boston University, 1996.

327. C. Partridge. The end of simple traffic models. *IEEE Network*, September 1993. Editor's Note.

328. M. Parulekar and A. M. Makowski. Buffer overflow probabilities for a multiplexer with self-similar traffic. Preprint, University of Maryland, College Park, MD, 1995.

329. P. F. Pawlita. Traffic measurements in data networks, recent measurement results, and some implications. *IEEE Transactions of Communication COM-29*, pages 525–535, 1981.

330. P. F. Pawlita. Two decades of data traffic measurements: a survey of published

results, experiences and applicability. In *Proceedings of the 12th ITC*, Torino, Italy, 1988.

331. V. Paxon and S. Floyd. Wide-area traffic: the failure of Poisson modeling. In *Proceedings of the ACM Sigcomm '94*, pages 257–268, London, UK, 1994.

332. V. Paxson. Empirically derived analytic models of wide-area TCP connections. *IEEE/ACM Transactions on Networking*, 2(4):316–336, 1994.

333. V. Paxson. Growth trends in wide-area TCP connections. *IEEE Network*, pages 8–17, July/August 1994.

334. V. Paxson. Fast approximation of self-similar network traffic. Preprint, 1995.

335. V. Paxson and S. Floyd. Wide area traffic: the failure of Poisson modeling. *IEEE/ACM Transactions on Networking*, 3:226–244, 1995.

336. M. S. Peiris and B. J. C. Perera. On prediction with fractionally differenced ARIMA models. *Journal of Time Series Analysis*, 9:215–220, 1988.

337. H.-O. Peitgen, H. Juergens, and D. Saupe. *Chaos and Fractals: New Frontiers of Science*. Springer-Verlag, New York, 1992.

338. H.-O. Peitgen and D. Saupe, editors. *The Science of Fractal Images*. Springer-Verlag, New York, 1988.

339. C.-K. Peng, S. V. Buldyrev, A. L. Goldberger, S. Havlin, F. Sciortino, M. Simons, and H. E. Stanley. Long-range correlations in nucleotide sequences. *Nature*, 356:168–170, 1992.

340. C.-K. Peng, S. Havlin, H. E. Stanley, and A. L. Goldberger. Quantification of scaling exponents and crossover phenomena in nonstationary heartbeat time series. *Chaos*, 5:82–87, 1995.

341. A. P. Pentland. Fractal-based description of natural scenes. *IEEE Transactions on Pattern Analysis and Machine Intelligence*, PAMI-6(4):661–674, 1984.

342. E. E. Peters. *Chaos and Order in the Capital Market*. Wiley, New York, 1991.

343. M. B. Priestley and T. S. Rao. A test for non-stationarity of time-series. *Journal of the Royal Statistical Society, Series B*, 31:140–149, 1969.

344. P. Pruthi. *An application of chaotic maps to packet traffic modeling*. PhD thesis, Royal Institute of Technology, Stockholm, Sweden, 1995.

345. P. Pruthi and A. Erramilli. Heavy-tailed on/off source behavior and self-similar traffic. In *Proceedings of the ICC '95*, pages 445–450, Seattle, WA, 1995.

346. K. E. E. Raatikainen. Symptoms of self-similarity in measured arrival process of ethernet packets to a file server. Technical Report Series of Publications C-1994-4, University of Helsinki, Dept. of Computer Science, 1994.

347. S. T. Rachev and G. Samorodnitsky. Option pricing formulae for speculative prices modelled by subordinated stochastic processes. *Serdica*, 19:175–190, 1993.

348. V. Ramaswami. Traffic performance modeling for packet communications: whence where and whither? In *Proceedings of the 3rd Australian Teletraffic Seminar*, Adelaide, Australia, 1988.

349. V. Ramaswami and G. Latouche. Modeling packet arrivals from asynchronous input lines. In *Proceedings of the 12th ITC*, Torino, Italy, 1988.

350. V. Ramaswami, M. Rumsewicz, W. Willinger, and T. Eliazov. Comparison of some traffic models for ATM performance studies. In A. Jensen and V. B. Iversen, editors, *Teletraffic and Datatraffic in a Period of Change (Proceedings of ITC-13, Copenhagen, Denmark)*, pages 7–12. North-Holland, Amsterdam, 1991.

351. V. Ramaswami and W. Willinger. Efficient traffic performance strategies for packet multiplexers. *Computer Networks and ISDN Systems*, 20:401–407, 1990.

352. B. K. Ray. Modeling long-memory processes for optimal long-range prediction.

Journal of Time Series Analysis, 14(5):511–525, 1993.

353. V. A. Reisen. Estimation of the fractional difference parameter in the ARIMA (p, d, q) model using the smoothed periodogram. *Journal of Time Series Analysis*, 15:335–350, 1994.

354. S. Resnick and C. Starica. Consistency of Hill's estimator for dependent data. *Journal of Applied Probability*, 32:139–167, 1995.

355. S. I. Resnick. Heavy tail modeling and teletraffic data. Preprint, School of ORIE, Cornell University, Ithaca, NY, 1995.

356. S. I. Resnick and G. Samorodnitsky. Performance decay in a single server exponential queueing model with long range dependence. *Operations Research*, 1995. to appear.

357. P. M. Robinson. Automatic frequency domain inference on semiparametric and nonparametric models. *Econometrica*, 59:1329–1363, 1991.

358. P. M. Robinson. Nonparametric function estimation for long memory time series. In W. A. Barnett, J. Powell, and G. E. Tauchen, editors, *Nonparametric and Semiparametric Methods in Econometrics and Statistics: Proceedings of the Fifth International Symposium in Economic Theory and Econometrics*, pages 437–457. Cambridge University Press, 1991.

359. P. M. Robinson. Efficient tests of nonstationarity hypotheses. *Journal of the American Statistical Association*, 89:1420–1437, 1994.

360. P. M. Robinson. Rates of convergence and optimal bandwidth in spectral analysis of processes with long range dependence. *Probability Theory and Related Fields*, 99:443–473, 1994.

361. P. M. Robinson. Semiparametric analysis of long-memory time series. *The Annals of Statistics*, 22:515–539, 1994.

362. P. M. Robinson. *Time series with strong dependence*, Volume 1 of *Advances in Econometrics. Sixth World Congress*, Chapter 2, pages 47–95. Cambridge University Press, 1994.

363. P. M. Robinson. Gaussian semiparametric estimation of long range dependence. *The Annals of Statistics*, 23:1630–1661, 1995.

364. P. M. Robinson. Log-periodogram regression of time series with long range dependence. *The Annals of Statistics*, 23:1048–1072, 1995.

365. P. M. Robinson and F. J. Hidalgo. Time series regression with long range dependence. Preprint, 1995.

366. P. M. Robinson and C. Velasco. Autocorrelation – robust inference. To appear In *Handbook of Statistics. Volume on Robust Inference*, 1995.

367. J. Rosiński and T. Żak. The equivalence of ergodicity and weak mixing for infinitely divisible processes. Preprint, 1995.

368. J. Rosiński and T. Żak. Simple conditions for mixing of infinitely divisible processes. Preprint, 1995.

369. B. K. Ryu and S. B. Lowen. Modeling self-similar traffic using the fractal-shot-noise-driven-Poisson-process. Preprint, 1994.

370. A. Samarov and M. S. Taqqu. On the efficiency of the sample mean in long memory noise. *Journal of Time Series Analysis*, 9:191–200, 1988.

371. G. Samorodnitsky. A class of shot noise models for financial applications. Preprint, 1995.

372. G. Samorodnitsky and M. S. Taqqu. The various linear fractional Lévy motions. In T. W. Anderson, K. B. Athreya, and D. L. Iglehart, editors, *Probability, Statistics and Mathematics: Papers in Honor of Samuel Karlin*, pages 261–270.

Academic Press, Boston, 1989.

373. G. Samorodnitsky and M. S. Taqqu. Linear models with long-range dependence and finite or infinite variance. In D. Brillinger, P. Caines, J. Geweke, E. Parzen, M. Rosenblatt, and M. S. Taqqu, editors, *New Directions in Time Series Analysis, Part II*, pages 325–340, IMA Volumes in Mathematics and its Applications, Volume 46. Springer-Verlag, New York, 1992.

374. G. Samorodnitsky and M. S. Taqqu. *Stable Non-Gaussian Processes: Stochastic Models with Infinite Variance*. Chapman and Hall, New York, London, 1994.

375. M. V. Sanchez de Naranjo. Central limit theorem for non-linear functionals of stationary vector Gaussian process. Preprint, 1994.

376. D. Saupe. Algorithm for random fractals. In H.-O. Peitgen and D. Saupe, editors, *The Science of Fractal Images*, Chapter 2, pages 71–113. Springer-Verlag, New York, 1988.

377. H. G. Schuster. *Deterministic Chaos: An Introduction*. VCH, New York, 1988. 2nd Edition.

378. J. F. Shoch and J. A. Hupp. Measured performance of an ethernet local network. *Communications of the ACM*, 23(12):711–721, 1980.

379. E. Slud. Some applications of counting process models with partially observed covariates. Preprint, 1995.

380. R. L. Smith. Estimating tails of probability distributions. *The Annals of Statistics*, 15:1174–1207, 1987.

381. F. B. Sowell. The fractional unit-root distribution. *Econometrica*, 58:495–505, 1990.

382. F. B. Sowell. Maximum likelihood estimation of stationary univariate fractionally integrated time series models. *Journal of Econometrics*, 53:165–188, 1992.

383. F. B. Sowell. Modeling long run behavior with the fractional ARIMA model. *Journal of Monetary Economics*, 29:277–302, 1992.

384. K. Sriram and W. Whitt. Characterizing superposition arrival processes in packet multiplexers for voice and data. *IEEE Journal on Selected Areas in Communications*, 4:833–846, 1989.

385. D. Surgailis. On Poisson multiple stochastic integral and associated equilibrium Markov process. In *Theory and Applications of Random Fields*, pages 233–238, Lecture Notes in Control and Information Science, Volume 49. Springer-Verlag, Berlin, 1983.

386. H. Takayasu. $f^{-\beta}$ power spectrum and stable distribution. *Journal of the Physical Society of Japan*, 56(4):1257–1260, 1987.

387. M. S. Taqqu. Weak convergence to fractional Brownian motion and to the Rosenblatt process. *Zeitschrift für Wahrscheinlichkeitstheorie und verwandte Gebiete*, 31:287–302, 1975.

388. M. S. Taqqu. Convergence of integrated processes of arbitrary Hermite rank. *Zeitschrift für Wahrscheinlichkeitstheorie und verwandte Gebiete*, 50:53–83, 1979.

389. M. S. Taqqu. A bibliographical guide to self-similar processes and long-range dependence. In E. Eberlein and M. S. Taqqu, editors, *Dependence in Probability and Statistics*, pages 137–162. Birkhauser, Boston, 1986.

390. M. S. Taqqu. Self-similar processes. In S. Kotz and N. Johnson, editors, *Encyclopedia of Statistical Sciences*, pages 352–357, Volume 8. Wiley, New York, 1988.

391. M. S. Taqqu and J. Levy. Using renewal processes to generate long-range dependence and high variability. In E. Eberlein and M. S. Taqqu, editors, *Dependence in Probability and Statistics*, pages 73–89. Birkhauser, Boston, 1986.

392. M. S. Taqqu and V. Teverovsky. Robustness of Whittle-type estimates for time series with long-range dependence. Preprint, 1995.

393. M. S. Taqqu and V. Teverovsky. Semi-parametric graphical estimation techniques for long-memory data. In *Proceedings of the Athens Conference on Applied Probability and Time Series Analysis.* Springer-Verlag, New York, 1996. Time series volume in honor of E. J. Hannan. To appear.

394. M. S. Taqqu, V. Teverovsky, and W. Willinger. Estimators for long-range dependence: an empirical study. *Fractals*, 3(4):785–798, 1995.

395. N. Terrin and M. S. Taqqu. A noncentral limit theorem for quadratic forms of Gaussian stationary sequences. *Journal of Theoretical Probability*, 3:449–475, 1990.

396. N. Terrin and M. S. Taqqu. Convergence in distribution of sums of bivariate Appell polynomials with long-range dependence. *Probability Theory and Related Fields*, 90:57–81, 1991.

397. N. Terrin and M. S. Taqqu. Convergence to a gaussian limit as the normalization exponent tends to 1/2. *Statistics and Probability Letters*, 11:419–427, 1991.

398. N. Terrin and M. S. Taqqu. Power counting theorem on R^n. In R. Durrett and H. Kesten, editors, *Spitzer Festschrift*, pages 425–440. Birkhauser, Boston, 1991.

399. V. Teverovsky and M. S. Taqqu. Testing for long-range dependence in the presence of shifting means or a slowly declining trend using a variance-type estimator. Preprint, 1995.

400. D. N. C. Tse, R. G. Gallager, and J. N. Tsitsiklis. Statistical multiplexing of multiple time-scale Markov streams. *IEEE Journal on Selected Areas in Communications*, 13:1028–1038, 1995.

401. D. Veitch. Novel methods of broadband traffic. In *Proceedings of Globecom '93*, pages 1057–1061, Houston, TX, 1993.

402. W. Vervaat. Properties of general self-similar processes. *Bulletin of the International Statistical Institute*, 52(Book 4):199–216, 1987.

403. M. C. Viano, Cl. Deniau, and G. Oppenheim. Long-range dependence and mixing for discrete time fractional processes. *Journal of Time Series Analysis*, 16:323–338, 1995.

404. R. F. Voss. Fractals in nature: from characterization to simulation. In H.-O. Peitgen and D. Saupe, editors, *The Science of Fractal Images*, Chapter 1, pages 21–70. Springer-Verlag, New York, 1988.

405. A. Weiss. An introduction to large deviations for communication networks. *IEEE Journal on Selected Areas in Communications*, 13:938–952, 1995.

406. W. Whitt. Tail probabilities with statistical multiplexing and effective bandwidths in multiclass queues. *Telecommunication Systems*, 2:71–107, 1993.

407. P. Whittle. *Hypothesis Testing in Time Series Analysis.* Hafner, New York, 1951.

408. E. Willekens and J. L. Teugels. Asymptotic expansion for waiting time probabilities in an M/G/1 queue with long-tailed service time. *Queueing Systems*, 10:295–312, 1992.

409. W. Willinger. Traffic modeling for high-speed networks: theory versus practice. In F. P. Kelly and R. J. Williams, editors, *Stochastic Networks*, Volume 71, pages 395–409, IMA Volumes in Mathematics and its Applications. Springer-Verlag, New York, 1995.

410. W. Willinger, M. S. Taqqu, W. E. Leland, and V. Wilson. Self-similarity in high-speed packet traffic: analysis and modeling of Ethernet traffic measurements. *Statistical Science*, 10:67–85, 1995.

411. W. Willinger, M. S. Taqqu, R. Sherman, and D. V. Wilson. Self-similarity

through high-variability: statistical analysis of Ethernet LAN traffic at the source level. *Computer Communications Review*, 25:100–113, 1995. Proceedings of the ACM/SIGCOMM'95, Boston, August 1995.

412. W. Willinger, M. S. Taqqu, R. Sherman, and D. V. Wilson. Self-similarity through high-variability: statistical analysis of Ethernet LAN traffic at the source level (Extended Version). Preprint. The paper also contains the mathematical proof that the superposition of strictly alternating ON/OFF sources converges to fractional Brownian motion, 1995.

413. W. Willinger, D. V. Wilson, W. E. Leland, and M. S. Taqqu. On traffic measurements that defy traffic models (and vice versa): self-similar traffic modeling for high-speed networks. *ConneXions*, 8(11):14–24, 1994.

414. G. W. Wornell. Wavelet-based representations for the $1/f$ family of fractal processes. *Procceedings of the IEEE*, 81:1428–1450, 1992.

415. G. W. Wornell and A. V. Oppenheim. Estimation of fractal signals from noisy measurements using wavelets. *IEEE Transactions on Information Theory*, 40(3):611–623, 1992.

416. G. W. Wornell and A. V. Oppenheim. Wavelet-based representations for a class of self-similar signals with application to fractal modulation. *IEEE Transactions on Information Theory*, 38(2):785–800, 1992.

417. W. Wyss. Fractional noise. *Foundations of Physics Letters*, 4(3):235–246, 1991.

418. Y. Yajima. On estimation of a regression model with long-memory stationary errors. *The Annals of Statistics*, 16:791–807, 1988.

419. Y. Yajima. Asymptotic properties of LSE in a regression model with long-memory stationary errors. *The Annals of Statistics*, 19:158–177, 1991.

420. V. M. Zolotarev. *Odnomernye ustoichivye raspredeleniya*. Nauka, Moscow, 1983. Subsequently translated as *One-dimensional Stable Distributions*, American Mathematical Society, 1986.